Advances in Electrical and Electronic Engineering

Advances in Electrical and Electronic Engineering

Edited by Norman Schultz

CLANRYE
INTERNATIONAL
www.clanryeinternational.com

Clanrye International,
750 Third Avenue, 9th Floor,
New York, NY 10017, USA

ISBN: 978-1-63240-797-9

Cataloging-in-Publication Data

Advances in electrical and electronic engineering / edited by Norman Schultz.
    p. cm.
Includes bibliographical references and index.
ISBN 978-1-63240-797-9
1. Electrical engineering. 2. Electronics. 3. Engineering. I. Schultz, Norman.
TK145 .A38 2019
621.3--dc23

For information on all Clanrye International publications
visit our website at www.clanryeinternational.com

ℒLANRYE
ℐNTERNATIONAL

Contents

Permissions

List of Contributors

Index

Preface

This book aims to highlight the current researches and provides a platform to further the scope of innovations in this area. This book is a product of the combined efforts of many researchers and scientists, after going through thorough studies and analysis from different parts of the world. The objective of this book is to provide the readers with the latest information of the field.

Electrical engineering studies and applies the principles of electromagnetism, electricity and electronics. One of the sub-disciplines of electrical engineering is electronic engineering. It is concerned with the design of electronic circuits, devices and systems by integrating different electrical components like diodes, transistors and integrated circuits. This is achieved by integrating the concepts and principles of several related fields like solid-state physics, radio engineering, telecommunications, control systems, signal processing, etc. This book elucidates the concepts and innovative models around prospective developments with respect to electrical and electronic engineering. It strives to provide a fair idea of these disciplines and to help develop a better understanding of these fields. It will be advantageous to electronic and electrical engineers, researchers and students working in these domains.

I would like to express my sincere thanks to the authors for their dedicated efforts in the completion of this book. I acknowledge the efforts of the publisher for providing constant support. Lastly, I would like to thank my family for their support in all academic endeavors.

Editor

COMPARISON OF RESERVATION PROTOCOLS FOR SOA AND MEMS TECHNOLOGY

Michaela SOLANSKA, Miroslav MARKOVIC, Milan DADO

Department of Telecommunications and Multimedia, Faculty of Electrical Engineering,
University of Zilina, Univerzitna 8215/1, 010 26 Zilina, Slovakia

michaela.solanska@fel.uniza.sk, miroslav.markovic@fel.uniza.sk, milan.dado@fel.uniza.sk

Abstract. *The data transmission by high-speed optical networks has an upward trend. The most effective data transmission can be achieved by powerful reservation protocols, which are together with optical switches very important parts of high–speed optical networks. This paper deals with new reservation protocol called Search & Compare, which is designed according to the well–known Segment-based Robust Fast Optical Reservation Protocol. In this paper, we present time analyses of both reservation protocols. We focused mainly on intra–segment time analyses. During the reservation of network resources, we used the optical cross-connect as the core node, which is based on microelectromechanical system and semiconductor optical amplifier. These two technologies are becoming the dominant technologies in optical switching.*

Keywords

Microelectromechanical systems, optical switches, reservation protocols, semiconductor optical amplifiers.

1. Introduction

Network service providers have to design their core networks to satisfy increasing user claims in the future. To make this possible, they have to use the multiplex methods. Time-division multiplexing (TDM) and wavelength-division multiplexing (WDM) belong among the most popular multiplex methods [1]. WDM allows more efficient data transmission via multiple wavelengths transmitting in single optical fiber. The advantage of WDM is the ability to transmit the data with different transmission speed and modulation format in every single wavelength [2].

With growing demands of IP services for transmission capacity and speed, the optical burst switching (OBS) presents the solution for future high-speed WDM optical networks. OBS networks need high-performance nodes, which can handle the growing flexibility and efficiency. Important parts of high-performance nodes are the reservation protocols and optical switches.

The major role of the reservation protocols is a discovery of the most suitable path (or the wavelength) followed by the reservation of the node resources. The best-known reservation protocols are Segment-based Robust Fast Optical Reservation Protocol (S-RFORP), Robust Fast Optical Reservation Protocol (RFORP), Resource Reservation Protocol-Traffic Engineering (RSVP-TE) and Intermediate-node Initiated Reservation (IIR). The S-RFORP protocol has the best features in comparison to the mentioned reservation protocols [3].

The optical switching plays an major role in resources reservation. Optical switches provide the optical path and improve the optical network reliability. Currently, several switching technologies are available, e.g. optomechanical switches, microelectromechanical system (MEMS) based switches, electrooptical switches, thermooptical switches, liquid-crystal switches, bubble switches, acoustooptical switches, switches based on semiconductor optical amplifier (SOA), switches based on fiber Bragg grating (FBG). From abovementioned switching technologies the MEMS and SOA are the most widely used. These technologies allow us to build the cost-effective and high-capacity optical cross-connects [4] and [5].

The paper is organized as follows. Section 2. describes the usage of MEMS technology in optical networks such as optical switches. Optical switches based on SOA technology are mentioned in Section 3. Section 4. describes time analysis of S-RFORP protocol and suggested reservation protocol. The performance comparison of the reservation protocols is reported in Section 5. The conclusion is drawn in Section 6.

2. MEMS Technology

Optical MEMS switches can be categorized into three groups: MEMS switches using micromirror, MEMS switches using membranes, MEMS switches using plane moving waveguides. The first two groups represent free space switches because they use space as the transmission medium. The last group represents waveguide switches that require moving certain parts of the switch once functioning. Most of the optical MEMS switches use micromirrors which can be divided into two groups, namely, two-dimensional MEMS (2D MEMS) and three-dimensional MEMS (3D MEMS) [6] and [7].

2.1. 2D MEMS Optical Switches

In 2D MEMS optical switches the micromirrors are arranged in a crossbar configuration. Micromirrors work in a digital mode, it means that each micromirror has only two positions, so their position is bistable (ON/OFF). The bistable position of micromirrors greatly simplifies the control mechanism. Typically, the control mechanism consists of simple transistor-transistor-logic (TTL) gates and appropriate amplifiers to apply an adequate voltage to actuate micromirrors [8], [9] and [10].

Micromirrors are placed on an electrostatic actuator that is suspended on a torsion spring. When the switch voltage (only a few microwatts) is applied, the actuator rotates around the axis of the torsion spring so that the micromirror moves downward into the optical beam. The 2D MEMS switch consists of two or more collimator arrays that are actively aligned with the micromirrors (a collimator is an optical element that transforms the optical mode of a single mode fiber into a light beam) [8], [9] and [10].

2D MEMS technology can deliver a range of applications including medium-sized and large optical cross-connects, wavelength selective optical cross- connects, wavelength add-drop multiplexing, optical service monitoring, and optical protection switching. MEMS technology is an important key to ensuring reliability and flexibility of a network [11] and [12].

3. SOA Technology

An SOA gate array is an array of devices monolithically integrated on the same substrate. By changing the electric current, the SOA array can act as "ON/OFF" switch. If electric current falls near zero, the input signals are absorbed ("OFF" position). In another case, if the current grows, SOA will amplify the input signals ("ON" position). The combination of amplification in "ON" position and absorption in "OFF" position make SOA capable of achieving very high extinction ratio [13], [14] and [15]. Due to the nonlinear characteristics, the SOAs are versatile devices used in optical networks. SOA technology is used not only to optical switching but also for all-optical wavelength conversion, regeneration, wavelength selection, booster and in-line amplification, in-node optical preamplification and mid-span spectral inversion in optical networks [13], [14] and [15].

4. Reservation Protocols

The reservation protocols are the important part of high-speed optical networks. The main role of reservation protocols in nodes is to reserve its resources for some time period. Currently, several reservation protocols are designed, which are trying to use the resources of nodes most efficiently with the lowest blocking probability of wavelength assignment. Good reservation protocols can save a big amount of data losses.

4.1. Reservation Protocol S-RFORP

The reservation protocol S-RFORP consists of two phases. During the first phase, the phase of wavelength discovery, all available wavelengths for each segment are discovered and then one of them is chosen for the reservation. During the second phase, the phase of wavelength reservation, the chosen wavelength is reserved. S-RFORP uses the parallel inter-segment discovery and reservation to minimize the wavelength assignment delay and the serial intra-segment discovery and reservation [16] and [17].

The time of the intra-segment discovery of S-RFORP is given by the sum of time that is necessary for wavelength discovery in a given node and time that is necessary for wavelengths comparison, as seen from equation Eq. (1).

$$D_t = D_{t_{S1}} + (D_{t_{S2}} + D_{t_{C2}}) + (D_{t_{S3}} + D_{t_{C3}}) + \ldots \\ + (D_{t_{Sn}} + D_{t_{Cn}}), \tag{1}$$

where D_t is the discovery time, $D_{t_{S1}}$ is the discovery time necessary for the discovery of wavelength in the segment, $D_{t_{cC}}$ is the discovery time necessary for comparison of two wavelengths [16] and [17].

From Eq. (1) it is seen that the first node in a segment only needs time for available wavelengths discovery since it has nothing to compare with the discovered wavelengths. If we separate from equation Eq. (1) the time that is needed for wavelength discovery and time that is necessary for wavelengths comparison we can write equations Eq. (2) and Eq. (3):

$$D_{t_{S1}} + D_{t_{S2}} + D_{t_{S3}} + ... + D_{t_{Sn}} = \sum_{i=1}^{n} D_{t_{Si}}, \qquad (2)$$

$$D_{t_{C2}} + D_{t_{C3}} + D_{t_{C4}} + ... + D_{t_{Cn}} = \sum_{j=2}^{n} D_{t_{Cj}}. \qquad (3)$$

If we substitute Eq. (2) and Eq. (3) to Eq. (1), we will have Eq. (4), which describes the total time needed for intra-segment discovery.

$$D_{t_{S-RFORP}} = \sum_{i=1}^{n} D_{t_{Si}} + \sum_{j=2}^{n} D_{t_{Cj}}, \qquad (4)$$

where $D_{t_{S-RFORP}}$ is the total time of the intra-segment discovery of S-RFORP, n is the number of active nodes in the segment [16] and [17].

When equally powerful nodes are presented in the given segment, we can write Eq. (5) and Eq. (6). Equation (7) describes the total time, which is necessary for intra-segment discovery in the reservation protocol S-RFORP, but only on condition that all the nodes in a segment are equally powerful.

$$D_{t_{S1}} = D_{t_{S2}} = D_{t_{S3}} = D_{t_{S4}} \rightarrow \sum_{i=1}^{n} D_{t_{Si}} \rightarrow nD_{t_{Si}}, \quad (5)$$

$$D_{t_{C2}} = D_{t_{C3}} = D_{t_{Cn}} \rightarrow \sum_{j=2}^{n} D_{t_{Cj}} \rightarrow (n-1)D_{t_{Cj}}, \quad (6)$$

$$\begin{aligned} D_{t_{S-RFORP}} &= nD_{t_{S1}} + (n-1)D_{t_{Cj}} = \\ &= nD_{t_{S1}} + nD_{t_{Cj}} - D_{t_{Cj}} = \qquad (7) \\ &= n(D_{t_{S1}} + D_{t_{Cj}}) - D_{t_{Cj}}. \end{aligned}$$

The time of the intra-segment reservation of S-RFORP is given as a sum of time intervals which are necessary for resource reservation in each node that is involved in the transfer, as seen from Eq. (8). If all nodes in the segment are equally powerful, the total time needed for intra-segment reservation is given by Eq. (9):

$$R_{t_{S-RFORP}} = R_{t_1} + R_{t_2} + R_{t_3} + ... + R_{t_n}, \qquad (8)$$

$$R_{t_1} = R_{t_2} = R_{t_3} = R_{t_i} \rightarrow R_{t_{S-RFORP}} = nR_{t_n}, \quad (9)$$

where $R_{t_{S-RFORP}}$ is the reservation time of S-RFORP protocol, R_{t_1} is the reservation time in the first node in the segment, R_{t_n} is the reservation time in the last node in the segment, n is the number of active nodes in the segment [16] and [17].

4.2. Reservation Protocol S&C

The suggested reservation protocol S&C is based on S-RFORP. Inter-segment discovery and reservation of S&C protocol are identical with S-RFORP protocol. S&C uses parallel segment-based discovery and parallel link-based reservation within the segment. It is possible to achieve a shorter time of intra-segment discovery and reservation [16] and [17].

The total time of intra-segment discovery of S&C is given as the sum of time for wavelengths discovery in the table of the main reservation node and the time for comparison of discovered wavelengths, as seen from the Eq. (10):

$$D_{t_{S\&C}} = t_s + t_c, \qquad (10)$$

where $D_{t_{S\&C}}$ is the discovery time of S&C protocol, t_s is the discovery time, t_c is the comparison time [16] and [17].

The total time necessary for intra-segment reservation of S&C is given by the sum of time, which is necessary for verification (if the discovered wavelengths are still available) and the reservation time of the slowest node in the segment, which is participated in the transfer, as seen from Eq. (11). In case that all the nodes in a segment are equally powerful, the time $R_{t_{S\&C}}$ is populated from any node in the given segment.

$$R_{t_{S\&C}} = t_v + R_{t_S}, \qquad (11)$$

where $R_{t_{S\&C}}$ is the reservation time of S&C protocol, t_v is the verification time, R_{t_S} is the time for reservation of the slowest node [16] and [17].

5. Performance Results

The performance evaluation of S&C reservation protocol is based on the intra-segment discovery time and on the following intra-segment wavelength reservation. The performance S&C is compared with S-RFORP. The numerical computer network model was executed in MATLAB development environment.

The network model was based on the topology of the Pan-European network [18], which was divided into the three segments. Each segment is different in the number of active nodes Fig. 1.

The optical cross-connects are used as OBS core nodes with different switching technologies: MEMS-based switching nodes and SOA-based switching nodes. The switching time of each 2D MEMS switch was 10 ms and SOA switch was 3 ns for following calculations [6] and [19].

In the network model it was required to set the main reservation node in each segment the distances between

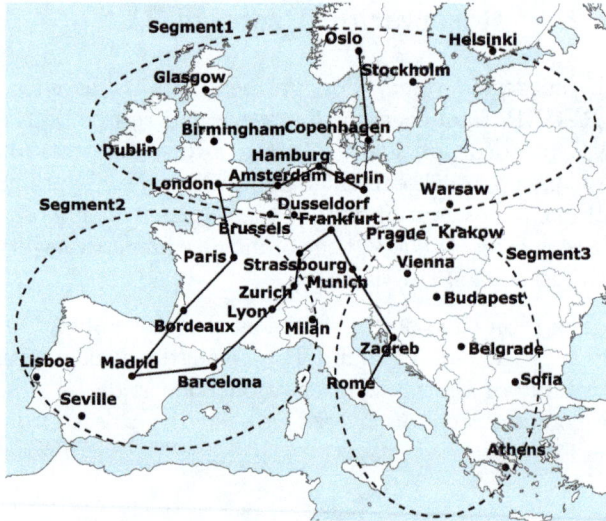

Fig. 1: The proposed node topology.

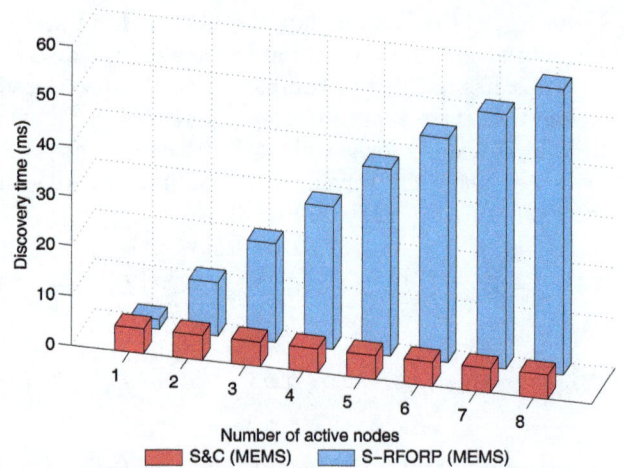

Fig. 2: Discovery time in Segment 2 with MEMS-based optical switch.

the network nodes (the distances are listed in [18]), the type of optical fiber (G.652), bitrate (40 Gbit·s^{-1}), and wavelength (1550 nm). It was crucial to set the time for discovery of wavelength (2 ms), the time for comparison of wavelengths (3 ms), the reservation time of the node (2D MEMS switch 10 ms, SOA switch 3 ns). It was also needed to set the time for discovery of wavelengths in the table of the main reservation node (2 ms), the time for comparison of discovered wavelengths (2.9 ms), the verification time (0.1 ns), the time for reservation of the slowest node (2D MEMS switch 10 ms, SOA switch 3 ns).

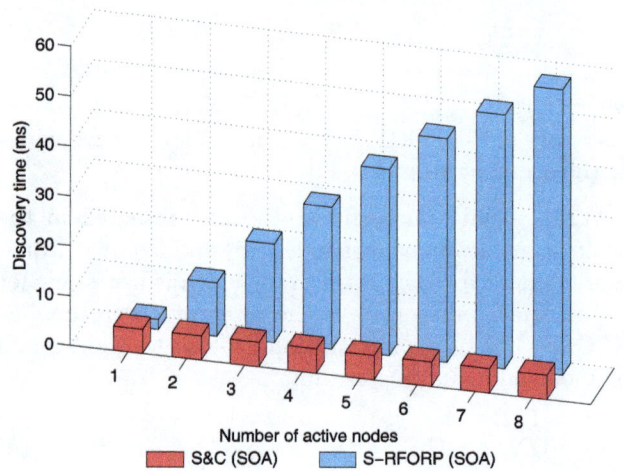

Fig. 3: Discovery time in Segment 2 with SOA-based optical switch.

5.1.   Intra-Segment Discovery Time

As can be seen in Fig. 2 and Fig. 3, the time necessary to intra-segment discovery in S-RFORP protocol is longer than the time necessary for intra-segment discovery in S&C protocol. It is caused by the fact that the time which is necessary for wavelength discovery in the given segment, is markedly dependent on the number of active nodes in the given segment. The greater amount of nodes in the given segment, the longer discovery time.

The time needed for wavelength discovery in both compared reservation protocols is independent of the chosen switching technology Fig. 4.

5.2.   Intra-Segment Reservation Time

As can be seen in Fig. 5 and Fig. 6, the time necessary for intra-segment reservation in S-RFORP protocol is again longer than the time necessary for intra-segment reservation in S&C protocol. The reservation time of

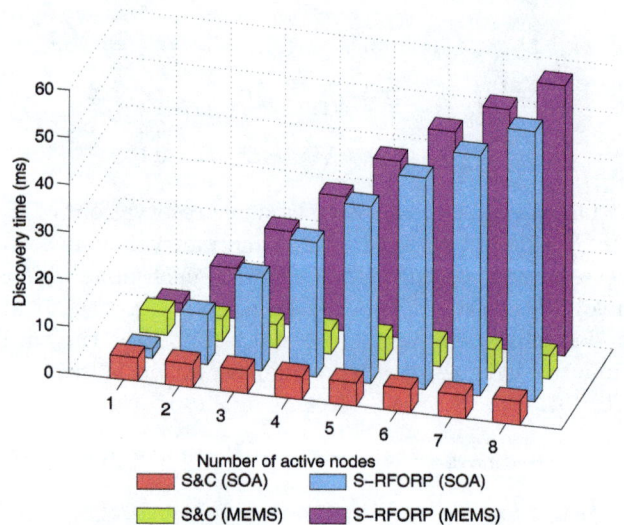

Fig. 4: Total time of discovery for Segment 2 with MEMS and SOA-based optical switch.

S-RFORP protocol is dependent on the number of active nodes in the given segment.

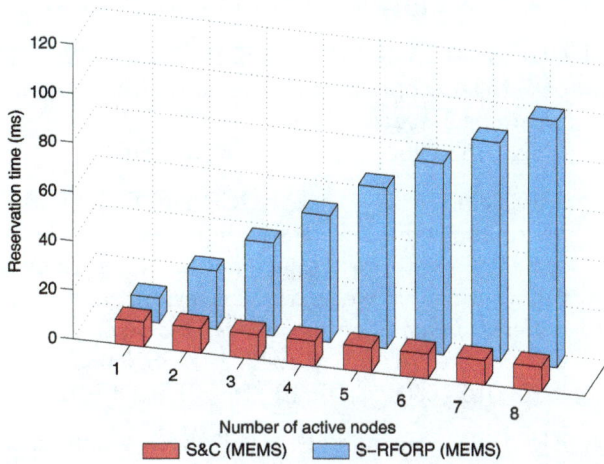

Fig. 5: Reservation time in Segment 2 with MEMS-based optical switch.

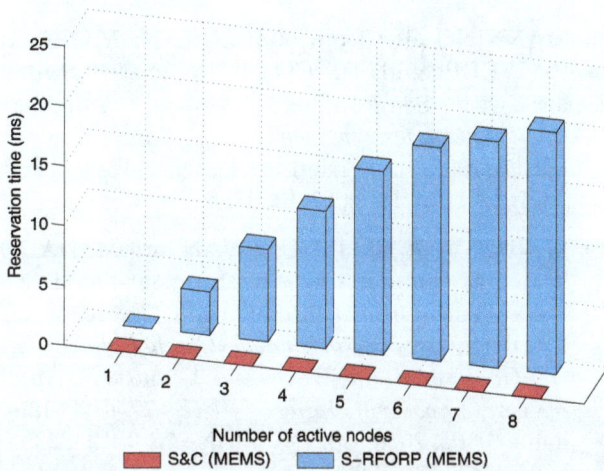

Fig. 6: Reservation time in Segment 2 with SOA-based optical switch.

On the switching technology is dependent only the time necessary for wavelength reservation Fig. 7. From the results could be seen, that the optical bursts can be handled with the SOA-based optical switches in the range of a few ns. That means that the optical switches based on SOA technology are well suited for optical switching due to their switching time.

The results show that the intra-segment discovery and reservation time of S-RFORP is dependent on the number of active nodes in the given segment. The time necessary for discovery and reservation in protocol S&C is independent of the number of active nodes in the segment, but is dependent on the speed of the main reservation node and from the slowest node in the given segment. The reservation time of both protocols depends on the chosen switching technology.

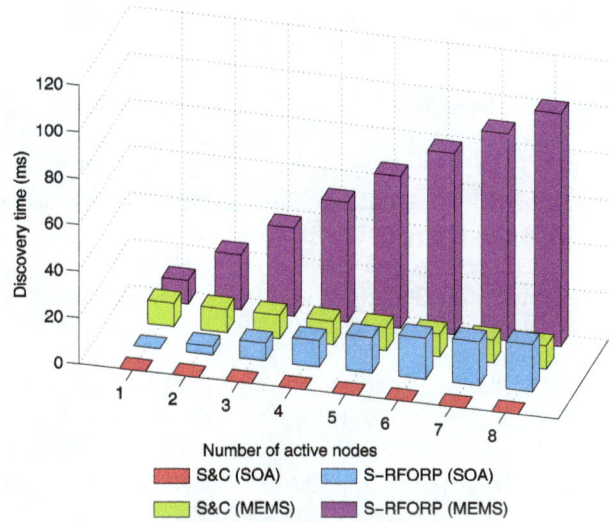

Fig. 7: Total time of reservation for Segment 2 with MEMS and SOA-based optical switch.

6. Conclusion

In this paper, we present the performance comparison of the currently known and proposed reservation protocol. From the analytical and numerical models could be seen, that proposed reservation protocol needs much shorter time, which is necessary to wavelength discovery and reservation in the given segment. Therefore proposed reservation protocol S&C is more powerful than the reservation protocol S-RFORP. From the results could be also seen that the optical switching plays an important role in the resource reservation. SOA belongs to the most attractive candidates to realize high-speed optical switching.

Acknowledgment

This work is supported by the Slovak Research and Development Agency under the project APVV-0025-12 ("Mitigation of stochastic effects in high-bitrate all-optical networks") and the European Regional Development Fund and the Slovak state budget for the project "Research Centre of University of Zilina", ITMS 26220220183.

References

[1] POBORIL, R., J. LATAL, P. KOUDELKA, J. VITASEK, P. SISKA, J. SKAPA and V. VASINEK. A Concept of a Hybrid WDM/TDM Topology Using the Fabry-Perot Laser in the Optiwave Simulation Environment. *Advances in Electrical and Electronic Engineering.* 2011,

vol. 9, no. 4, pp. 167–178. ISSN 1336-1376. DOI: 10.15598/aeee.v9i4.537.

[2] BENEDIKOVIC, D., J. LITVIK, M. KUBA, M. DADO and J. DUBOVAN. Influence of nonlinear effects in WDM system with non-equidistant channel spacing using different types of high-order PSK and QAM modulation formats. In: *Optical Modelling and Design II: SPIE 8429*. Bellingham: SPIE, 2012, pp. 1–6. ISBN 978-081949121-3. DOI: 10.1117/12.921836.

[3] XU, H., O. YU, L. YIN and M. LIAO. Segment-Based Robust Fast Optical Reservation Protocol. In: *High-Speed Networks Workshop*. Anchorage: IEEE, 2007, pp. 36–40. ISBN 1-4244-1580-2. DOI: 10.1109/HSNW.2007.4290542.

[4] PAPADIMITRIOU, G. I., C. PAPAZOGLOU and A. POMPORTSIS. *Optical switching*. Hoboken: Wiley-Interscience, 2007. ISBN 978-0-471-68596-8.

[5] XIAOHUA, M. and G. S. KUO. Optical switching technology comparison: Optical MEMS vs. other technologies. *IEEE Communications Magazine*. 2003, vol. 41, iss. 11, pp. 16–23. ISSN 0163-6804. DOI: 10.1109/MCOM.2003.1244924.

[6] THAKULSUKANANT, K. MEMS Technology for Optical Switching. *Walailak Journal of Science and Technology*. 2013, vol. 10, iss. 1, pp. 9–18. ISSN 1686-3933.

[7] SOLGAARD, O. Optical MEMS: From Micromirrors to Complex Systems. *Journal of Microelectromechanical Systems*. 2014, vol. 23, iss. 3, pp. 517–538. ISSN 1057-7157. DOI: 10.1109/JMEMS.2014.2319266.

[8] YANG, Y. J. and B. T. LIA. A novel 4 × 4 optical switching using an anisotropically etched micromirror array and a bistable mini–actuator array. *IEEE Photonics Technology Letters*. 2009, vol. 21, iss. 2, pp. 115–117. ISSN 1041-1135. DOI: 10.1109/LPT.2008.2009127.

[9] FAN, K. C., W. L. LIN, L. H. CHAING and S. H. CHEN. A 2 × 2 mechanical optical switch with a thin MEMS mirror. *Journal of Lightwave Technology*. 2009, vol. 27, iss. 9, pp. 1155–1161. ISSN 0733-8724. DOI: 10.1109/JLT.2008.928955.

[10] DE DOBBELAERE, P., K. FALTA and S. GLOECKNER. Advances in integrated 2D MEMS-based solutions for optical network applications. *IEEE Communications Magazine*. 2003, vol. 41, iss. 5, pp. 16–23. ISSN 0163-6804. DOI: 10.1109/MCOM.2003.1200101.

[11] YEOW, T. W., K. L. E. LAW and A. GOLDENBERG. MEMS optical switches. *IEEE Communications Magazine*. 2001, vol. 39, iss. 11, pp. 158–163. ISSN 0163-6804. DOI: 10.1109/35.965375.

[12] CHU, P. B., S. S. LEE and S. PARK. MEMS: The path to large optical crossconnects. *IEEE Communications Magazine*. 2002, vol. 40, iss. 3, pp. 80–87. ISSN 0163-6804. DOI: 10.1109/35.989762.

[13] BUCHTA, H., C. M. GAUGER and E. PATZAK. Maximum size and throughput of SOA-based optical burst switching nodes with limited tuning-range wavelength converters and FDL buffers. *Journal of Lightwave Technology*. 2008, vol. 26, iss. 16, pp. 2919–2927. ISSN 0733-8724. ISSN 0163-6804. DOI: 10.1109/JLT.2008.923329.

[14] BUCHTA, H. and E. PATZAK. Analysis of the physical impairments on maximum size and throughput of SOA-based optical burst switching nodes. *Journal of Lightwave Technology*. 2008, vol. 26, iss. 16, pp. 2821–2830. ISSN 0733-8724. DOI: 10.1109/JLT.2008.923277.

[15] MANNING, R. J., R. GILLER, X. YANG, R. P. WEBB and D. COTTER. Faster Switching with Semiconductor Optical Amplifiers. In: *Photonics in Switching*. San Francisco: IEEE, 2007, pp. 145–146. ISBN 1-4244-1121-1. DOI: 10.1109/PS.2007.4300786.

[16] MARKOVIC, M., J. DUBOVAN and M. DADO. Segment time analysis of Segment-based Robust Fast Optical Reservation Protocol (S-RFORP). In: *17th Slovak-Czech-Polish Optical Conference on Wave and Quantum Aspects of Contemporary Optics: SPIE 7746*. Bellingham: SPIE, 2010, pp. 1–5. ISBN 978-081948236-5. DOI: 10.1117/12.882400.

[17] MARKOVIC, M., J. DUBOVAN and M. DADO. Search&Compare (S&C)-Reservation protocol in High-Speed Optical Networks. *Elektronika ir Elektrotechnika*. 2011, vol. 114, iss. 8, pp. 39–42. ISSN 2029-5731. DOI: 10.5755/j01.eee.114.8.693.

[18] BETKER, A., C. GERLACH, R. HULSERMANN and M. JAGER. Reference transport network scenarios: Technical report German Ministry of Education and Research within the MultiteraNet. In: *Institut fur Kommunikationsnetze und Rechnersysteme* [online]. Stuttgart: IKR, 2004. Available at: http://www.ikr.uni-stuttgart.de/IKRSimLib/Usage/Referenz_Netze_v14_full.pdf.

[19] YADAV, R. and R. R. AGGARWAL. Survey and Comparison of Optical Switch Fabrication Techniques and Architectures. *Journal of Computing*. 2010, vol. 2, iss. 4, pp. 133–137. ISSN 2151-9617.

About Authors

Michaela SOLANSKA was born in Ruzomberok, Slovakia in 1987. In 2012, she finished M.Sc. at University of Zilina, Faculty of Electrical Engineering, Department of Telecommunications. Currently, she studies doctor degree; her research interests include reservation protocols and optical switching in high-speed optical networks.

Miroslav MARKOVIC was born in Cadca, Slovakia in 1984. In 2007, he finished M.Sc. at University of Zilina, Faculty of Electrical Engineering, Department of Telecommunications. In 2010, he reached Ph.D. in Telecommunications. Currently, he is an employee of Department of Telecommunications. His professional interests involve reservation protocols, optical switching in high-speed optical networks and optical access networks and their design.

Milan DADO was born in Krupina, Slovakia in 1951. He is Full Professor with the Department of Telecommunication and Multimedia. He has been actively involved in European research and education programs TEMPUS, COST, LEONARDO, Socrates, 5th, 6th and 7th Framework Program, European University Association projects and has managed national projects related to the information and communication technologies, advanced optical fibers, the high-speed optical networks, telecommunication and information technologies and services, intelligent transportation systems, regional innovation strategies and e-learning. He is project coordinator of ERA CHAIR project in Intelligent Transport Systems at the University of Zilina at the present time. He has been a member of a number of international committees and is COST program national coordinator and member of COST Committee of Senior Officials at the present time.

2

Application of $Cl_2/BCl_3/Ar$ Plasma Treatment in the Improvement of Ti/Al/Mo/Au Ohmic Contacts

Jacek GRYGLEWICZ, Wojciech MACHERZYNSKI, Andrzej STAFINIAK,
Bogdan PASZKIEWICZ, Regina PASZKIEWICZ

Department of Microelectronics and Nanotechnology, Faculty of Microsystem Electronics and Photonics,
Wroclaw University of Technology, Janiszewskiego 11/17, 50-372 Wroclaw, Poland

jacek.gryglewicz@pwr.edu.pl, wojciech.macherzynski@pwr.edu.pl, andrzej.stafiniak@pwr.edu.pl,
regina.paszkiewicz@pwr.edu.pl

Abstract. *Significant improvement of Ti/Al/Mo/Au ohmic contacts deposited on previously $Cl_2/BCl_3/Ar$ plasma treated surface was observed. The standard deviation of contact resistance was crucially reduced due to the incorporation of $Cl_2/BCl_3/Ar$ plasma treatment. The Cl_2:BCl_3:Ar gas mixture was used in order to thin the top of AlGaN layer prior to deposition of Ti/Al/Mo/Au ohmic contacts. The surface morphology of AlGaN was investigated using scanning electron microscopy and atomic force microscopy. TLM measurements revealed a consequential decrease of contact resistivity.*

Keywords

AlGaN, GaN, ohmic metallization, recess, Ti/Al/Mo/Au.

1. Introduction

Gallium nitride and aluminium gallium nitride are the materials used for high frequency power devices including high electron mobility transistors (AlGaN/GaN HEMTs). The fabrication of advanced AlGaN/GaN HEMTs requires elaborating of low-resistance ohmic contacts to AlGaN/GaN heterostructures [1]. In spite of technological advance achieved in recent years [2] there are still some challenges regarding the improvement of ohmic contacts parameters, especially in case of Ti/Al based contacts. It is a common practice to introduce thin AlN layer to suppress Al alloy scattering in HEMTs. However, by incorporation of wide band gap material it is even more difficult to create high quality ohmic metallization. One of the available

technological approaches is BCl_3-based plasma treatment [1], [2], [3], [4] due to deoxidizing of heterostructure surface. Without sputter desorption it is possible to deposit B_x-Cl_y which contributes to the increase of contact resistance [5]. The addition of Cl_2/Ar enhances the process of AlGaN etching due to sputtering effect. In result, the distance between the metallization and two dimensional electron gas (2DEG) is decreased which affects contact resistance.

2. Experiment

The $Al_{0.2}Ga_{0.8}N$/GaN heterostructures were deposited on 2" sapphire substrates using low pressure MOVPE process (3×2"). The heterostructures consisted of about 50 nm thick $Al_xGa_{1-x}N$, AlN spacer (1.6 nm) and 2.35 µm thick unintentionally doped GaN layer. The surface was etched in H_2SO_4 ($t = 3$ min), then exposed to N_2O ($t = 3$ min) and N_2 ($t = 3$ min) plasma in order to get rid of contamination.

After surface pre-treatment the heterostructures were exposed to plasma in RIE system using the following conditions: $P = 150$ W, $p = 20$ mTorr (2.66 Pa), $T = 7$ °C, Cl_2:BCl_3:Ar (7:3:5) in parallel plate reactor. The etch rate evaluation was based on measuring etch depth using atomic force microscope (AFM). For mentioned conditions the etch rate of $Al_{0.2}Ga_{0.8}N$ was 5 ± 1 nm·min^{-1} [3]. By modifying processing time, the thickness of the top AlGaN layer was varied for $Al_{0.2}Ga_{0.8}N$/GaN heterostructures.

Three samples (A, B, C) were etched in such conditions in order to decrease AlGaN thickness and to strip the native oxide of the surface. For reference, sample O (unetched) was examined. The remaining thicknesses of plasma treated AlGaN layers were presented

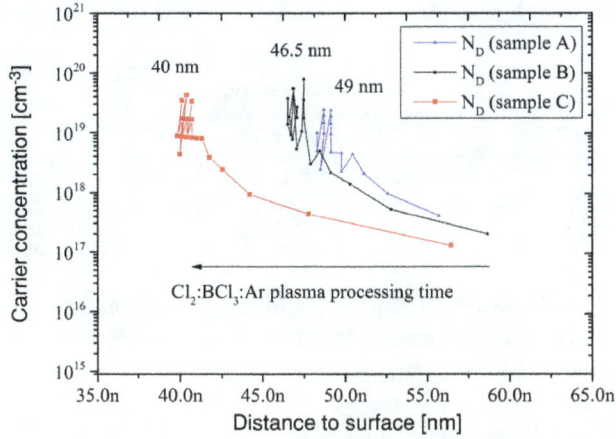

(a) Carrier concentration (N_D).

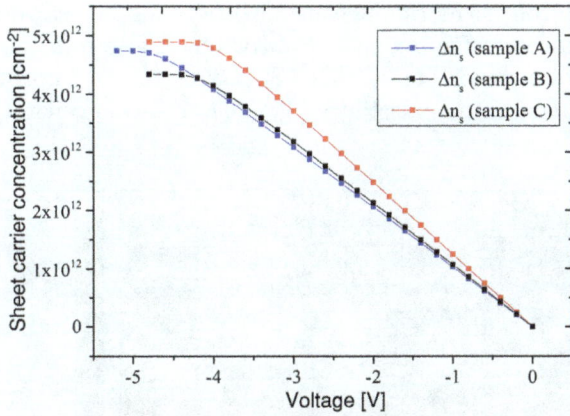

(b) Sheet carrier concentration.

Fig. 1: Carrier concentration (N_D) in function of distance to surface and sheet carrier concentration of 2DEG in function of applied voltage. Evaluation was based on C-V Hg-probe measurement.

in Fig. 1. The C-V measurement of carrier concentration and sheet charge concentration using Hg probe gave an information about remaining thicknesses for investigated heterostructures. The heterostructures were annealed in a nitrogen ambient at 825 °C ($t = 60$ s) in order to improve heterostructure properties.

After the definition of an active region (mesa etching), the TLM (Transfer Length Method) [6] structures were deposited on previously etched AlGaN surface. The metallization consisted of Ti/Al/Mo/Au (230/1000/ 450/1700) [7]. After that, the heterostructures were annealed once again in a nitrogen ambient at 825 °C in order to form ohmic contacts.

3. Results and Discussion

The evaluation of etch depth was based on AFM measurements and performed C-V measurements. From C-V curve it was possible to derive carrier concentration

profile (Fig. 1(a)). The width of depletion region under mercury probe was evaluated under assuming it was a parallel plate capacitor. The sheet carrier concentration (n_s) was evaluated using the integration of carrier concentration profile. From the slope of the variation of 2DEG sheet carrier concentration (Fig. 1(b)) it was also possible to evaluate thickness of AlGaN layer after etching.

Significant improvement of Ti/Al/Mo/Au contact resistance was observed for contacts deposited on previously plasma treated and pre-annealed $Al_{0.2}Ga_{0.8}N$/GaN heterostructures. Contact resistance (R_c), contact resistivity (ρ_c) and transfer length (L_T) were calculated using TLM method which relies on calculation of total resistance (R_T) in function of distance (L) between adjacent metallization pads (Fig. 2(a)) from I-V characteristics (Fig. 2(b)). Values of contact resistance (R_c) and corresponding standard error calculated from linear fitting of curves (Fig. 2(a)) for investigated heterostructures were presented in Tab. 1.

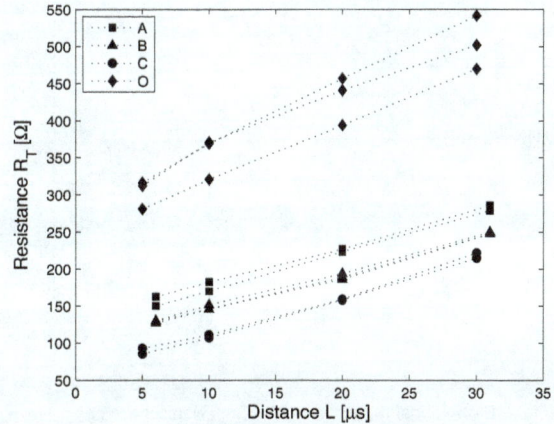

(a) Total resistance R_T.

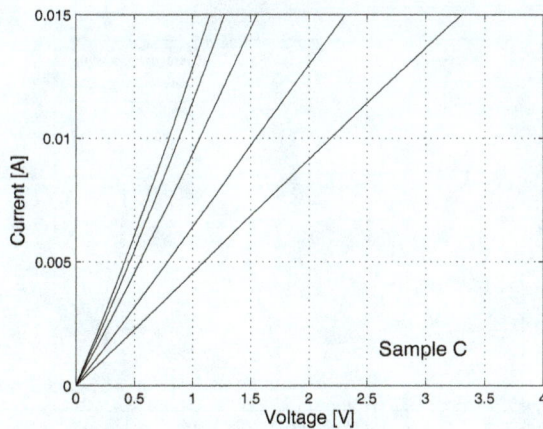

(b) I-V charasteristic.

Fig. 2: Total resistance (R_T) in function of adjacent Ti/Al/Mo/Au pads and corresponding I-V characteristic of sample C.

Tab. 1: Contact resistance (R_c) and corresponding standard error along with proportional reduction of thickness for investigated samples size.

Sample	Contact resistance R_c (Ω)	Standard Error (Ω)	Proportional reduction of $Al_{0.2}Ga_{0.8}N$ thickness (%)
O	135.78	5.70	0 (unetched)
A	62.85	1.53	8.7
B	49.65	3.32	19.5
C	29.02	1.43	32.6

Even though proportional reduction of AlGaN thickness was significant (32.6 %), the surface roughness of plasma treated and as-grown samples was similar ($R_a < 1.5$ nm) as it was depicted in Fig. 3(a). Surface roughness deterioration of AlGaN caused by ion bombarding did not affect contact resistance (R_c). Similar non-affecting influence of surface roughness was observed for specific contact resistivity (ρ_c) and transfer length (L_T) (Fig. 3(b)). Surface of $Al_{0.2}Ga_{0.8}N$ prior and after etching was depicted in Fig. 4.

It was observed that even insignificant reduction of the AlGaN thickness (8.7 %) gives promising results in achieving lower contact resistivity, contact resistance as well as transfer length improvement. Thinning of AlGaN layer caused by deeper etch depths resulted in further decrease of Ti/Al/Mo/Au contact resistance.

Boron trichloride plasma surface treatment not only removes surface oxide efficiently, but it also introduces surface donor states that contribute to the improvement of ohmic resistance [3]. BCl_x radicals generated

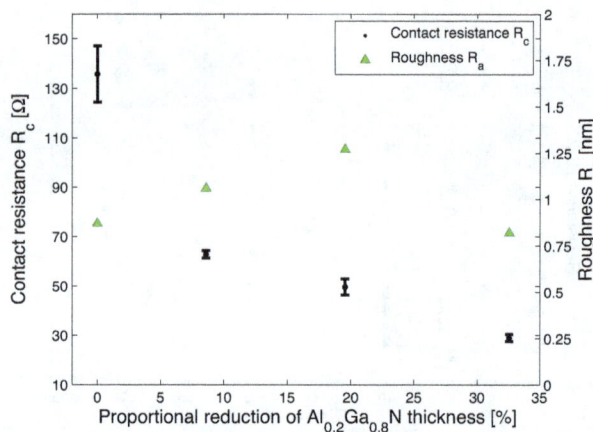

(a) Contact resistance R_c.

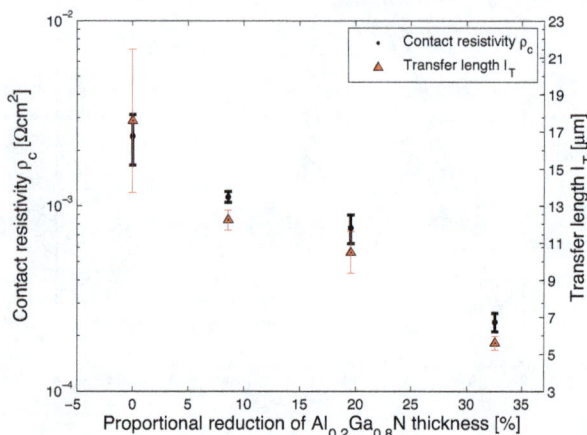

(b) Contact resistivity ρ_c.

Fig. 3: Contact resistance R_c and contact resistivity ρ_c in function of proportional reduction of $Al_{0.2}Ga_{0.8}N$ thickness.

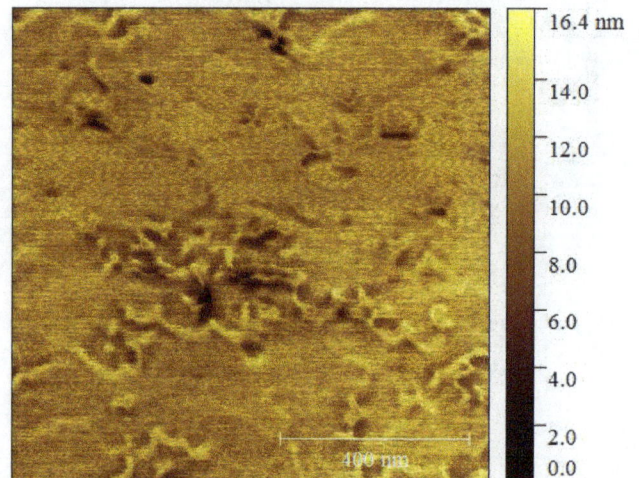

(a) Unetched surface.

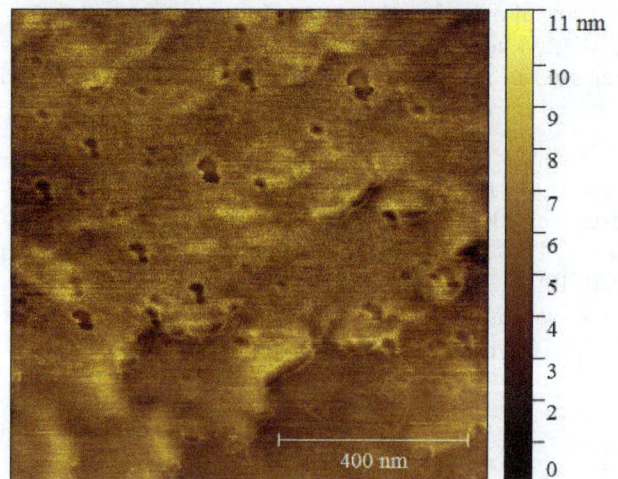

(b) Etched surface.

Fig. 4: AFM pictures of unetched (a) and etched (b) surface of Al0.2Ga0.8N/GaN heterostructure (sample B).

by cascade electron impact ionization enhance oxide layer etching by forming volatile B_xOCl_y and B_xO_y etch products which are removed from surface by accompanying ion bombardment. To increase the ion bombardment contribution, Cl_2/Ar gas mixture was added, which helped in preventing from the deposition of B_x–Cl_y passivation layer reported elsewhere [5]. Results presented in Fig. 3 indicate on dependency that predominant factor in the improvement of contact resistance was the reduction of AlGaN thickness. Further improvement of contact resistance can be obtained by forming Ti/Al/Mo/Au contacts at 850 °C [8].

4.   Conclusion

The influence of AlGaN layer etching in Cl2:BCl3:Ar plasma on the parameters of Ti/Al/Mo/Au ohmic contacts to AlGaN/GaN heterostructure was investigated. By reducing AlGaN thickness and subsequent annealing at 825 °C in nitrogen ambient we observed the significant improvement of Ti/Al/Mo/Au ohmic contact resistance. Although etching caused gentle deterioration of surface roughness, it is believed that surface roughness did not affect contact resistance significantly. Shrinking the distance between Ti/Al/Mo/Au metallization and two.

Acknowledgment

This work was co-financed by the European Union within European Regional Development Fund, through grant Innovative Economy (POIG.01.01.02-00-008/08-05), National Science Centre under the grant no. DEC-2012/07/D/ST7/02583, by National Centre for Research and Development through Applied Research Program grant no. 178782, program LIDER no. 027/533/L-5/13/NCBR/2014, by Wroclaw University of Technology statutory grants and Slovak-Polish International Cooperation Program no. SK-PL-2015-0028.

References

[1] LI, L. K., L. S. TAN and E. F. CHOR. Effects of surface plasma treatment on n-GaN ohmic contact formation. *Journal of Crystal Growth*. 2004, vol. 268, no. 3–4, pp. 499–503. ISSN 0022-0248. DOI: 10.1016/j.jcrysgro.2004.04.080.

[2] GRYGLEWICZ, J., R. PASZKIEWICZ, W. MACHERZYNSKI, A. STAFINIAK and M. WOSKO. Precise etching of AlGaN/GaN HEMT structures with Cl_2/BCl_3/Ar plasma. In: *Proceedings of The Tenth International Conference on Advanced Semiconductor Devices and Microsystems*. Smolenice: IEEE, 2014, pp. 1–4. ISBN 978-1-4799-5474-2. DOI: 10.1109/AS-DAM.2014.6998649.

[3] FUJISHIMA, T., S. JOGLEKAR, D. PIEDRA, H.-S. LEE, Y. ZHANG, A. UEDONO and T. PALACIOS. Formation of low resistance ohmic contacts in GaN-based high electron mobility transistors with BCl_3 surface plasma treatment. *Applied Physics Letters*. 2013, vol. 103, no. 8, pp. 083508-1–083508-4. ISSN 0003-6951. DOI: 10.1063/1.4819334.

[4] WANG, C., S. J. CHO and N. Y. KIM. Optimization of Ohmic Contact Metallization Process for AlGaN/GaN High Electron Mobility Transistor. *Transactions on Electrical and Electronic Materials*. 2013, vol. 14, no. 1, pp. 32–35. ISSN 1229-7607. DOI: 10.4313/TEEM.2013.14.1.32.

[5] KOBELEY, A. A., Y. V. BARSUKOV and N. A. ANDRIANOV. Boron trichloride plasma treatment effect on ohmic contact resistance formed on GaN-based epitaxial structure. *Journal of Physics Conference Series*. 2015, vol. 586, no. 1, pp. 012013-1–012013-4. ISSN 1742-6588. DOI: 10.1088/1742-6596/586/1/012013.

[6] SCHRODER, D. K. *Semiconductor Material and Device Characterization*. 3rd ed. Hoboken: John Wiley & Sons Inc., 2005. ISBN 978-0-471-73906-7.

[7] MACHERZYNSKI, W. and B. PASZKIEWICZ. Development of diffusion barriers for Ti/Al based ohmic contact to AlGaN/GaN heterostructures, In: *Proceedings of Ninth International Conference on Advanced Semiconductor Devices and Microsystems*. Smolenice: IEEE, 2012, pp. 203–206. ISBN 978-1-4673-1197-7. DOI: 10.1109/AS-DAM.2012.6418532.

[8] MACHERZYNSKI, W., J. GRYGLEWICZ, J. PRAZMOWSKA, A. STAFINIAK and R. PASZKIEWICZ. Effect of annealing temperature on the Ti/Al/Mo/Au ohmic contact. In: *Proceedings of Advances In Electronic And Photonic Technologies*. Strbske Pleso: University of Zilina, 2015, pp. 206–209. ISBN 978-80-554-1033-3.

About Authors

Jacek GRYGLEWICZ received his M.Sc. degree in Electrical Engineering from Wroclaw University of Technology (WrUT), Poland in 2009 and Ph.D. degree from the Wroclaw University of Technology in 2015. Now he is assistant professor at WrUT. His research is focused on device processing and parameter evaluation of nitrides-based devices: HEMTs and sensors. He is co-author of 16 scientific publications.

Wojciech MACHERZYNSKI received his M.Sc. degree in Electronic from Wroclaw University of Technology, Poland in 2005 and Ph.D. degree from the Wroclaw University of Technology in 2011. Now he is assistant professor at WrUT. His research is focused on the technology of semiconductors devices in particular on development of the metal-semiconductor junction.

Andrzej STAFINIAK received M.Sc. degree (2008) and Ph.D. degree (2015) in electronics from Wroclaw University of Technology. Since then, he has been assistant professor in Division of Microelectronics and Nanotechnology, WrUT. His current research is focused on development of process technology and measurements of nanostructure based devices.

Bogdan PASZKIEWICZ received his M.Sc. degree in Electrical Engineering from St. Petersburg Electrotechnical University, St. Petersburg, Russia in 1979 and Ph.D. degree from the Wroclaw University of Technology in 1997. Now he is assistant professor at WrUT. His research is focused on the design and parameter evaluation of nitrides-based devices: HEMTs and sensors.

Regina PASZKIEWICZ received her M.Sc. degree in Electrical Engineering from St. Petersburg Electrotechnical University, St. Petersburg, Russia in 1982 and Ph.D. degree from the Wroclaw University of Technology in 1997. Now she is full professor at WrUT. Her research is focused on the technology of (Ga, Al, In)N semiconductors, microwave and optoelectronic devices technological processes development.

3

Enhancement of Color Rendering Index for White Light LED Lamps by Red Y_2O_3:Eu^{3+} Phosphor

Tran Hoang Quang MINH, Nguyen Huu Khanh NHAN,
Thoai Phu VO, Nguyen Doan Quoc ANH

Department of Electronics and Telecommunications, Faculty of Electrical and Electronics Engineering,
Ton Duc Thang University, 19 Nguyen Huu Tho Street, Ho Chi Minh City, Vietnam

tranhoangquangminh@tdt.edu.vn, nguyenhuukhanhnhan@tdt.edu.vn, thoaiphuvo@tdt.edu.vn,
nguyendoanquocanh@tdt.edu.vn

Abstract. *We present an application of the red Y_2O_3:Eu^{3+} dopant phosphor compound for reaching the color rendering index as high as 86. The Multi-Chip White LED lamps (MCW-LEDs) with high Correlated Color Temperatures (CCTs) including 7000 K and 8500 K are employed in this study. Besides, the impacts of the Y_2O_3:Eu^{3+} phosphor on the attenuation of light through phosphor layers of the various packages is also demonstrated based on the Beer-Lambert law. Simulation results provide important conclusion for selecting and developing the phosphor materials in MCW-LEDs manufacturing.*

Keywords

Beer-Lambert law, Color Rendering Index, pc-WLEDs, Y_2O_3:Eu^{3+}.

1. Introduction

There are many benefits of MCW-LEDs to consumers such as high brightness, low power consumption, long lifetime, fast response, climate impact resistance [1]. Correspondingly, MCW-LEDs are considered to be key lighting devices to replace traditional lamps. Luminous output and Correlated Color Temperature (CCT) uniformity are two main factors of white LED lamps. By adding SiO_2 powder having the suitable size and concentration, some researchers have reduced spatial CCT deviation without decreasing lumen output significantly. The lumen output of MCW-LEDs can be enhanced significantly after adding the green $Ce_{0.67}$ $Tb_{0.33}$ $MgAl_{11}$ O_{19}:Ce, Tb phosphor to MCW-LEDs [2].

Furthermore, we cannot but mention the Color Rendering Index (CRI) that is considered as one important characteristic of MCW-LEDs. Several previous studies have applied methods that consist of mixing red-phosphors or doping red LEDs to compensate red-light to MCW-LEDs [3], [4], [5], [6], [7]. Besides, Won et al. presented high CRI multi-chip white LEDs, combining blue LEDs and green $(Ba,Sr)_2SiO_4$:Eu^{2+} and red $CaAlSiN_3$:Eu^{2+} phosphors with the different packages. By doping the missing red component in phosphor-converted MCW-LEDs (pc-WLEDs), the CRI of MCW-LEDs can be enhanced to more than 80, which is an important goal [2].

Red Y_2O_3:Eu^{3+} phosphor is one of cathodoluminescent phosphors, which is employed widely in color displays as a red-light-emitting component. However, Y_2O_3:Eu^{3+} phosphor has not many applications for improving CRI as yet.

In this paper, we introduce the impacts of Y_2O_3:Eu^{3+} phosphor particles in multi-chip white light LEDs with conformal phosphor or in-cup phosphor packages to enhance color rendering ability. It has been found that the participation of Y_2O_3:Eu^{3+} phosphor particles can dominate the red-light emitting event in pc-WLEDs, so that the LED light distribution can be free from the dispersion incurred by LED packages to yield higher Color Rendering Index.

We have divided the work processes into three main stages. The precise MCW-LEDs physical model having average CCTs of 7000 K and 8500 K has been conducted by LightTools software at first. Then, the transmission of light has been decreased after mixing Y_2O_3:Eu^{3+} particles, which has demonstrated the Beer-Lambert law. Finally, we have investigated the effects of Y_2O_3:Eu^{3+} phosphor particles on the color rendering ability of MCW-LEDs according to the simulation results. Based on the results, the proposed method of doping certain amounts of $Y_2O_3:Eu^{3+}$ in the LED packages can improve their CRI significantly.

2. Simulation

The pc-WLEDs is covered by flat silicone layer, which is simulated by using LightTools 8.1.0 program. The key work consists of pc-WLEDs construction and phosphor concentration adjustment. Firstly, the structures of pc-WLEDs such as the Conformal Phosphor Package (CPP) and the In-cup Phosphor Package (IPP) are introduced with five CCTs of 7000 K and 8500 K, see Fig. 1. Secondly, it is necessary to keep the MCW-LEDs work at mean CCTs from 7000 K to 8500 K for achieving the LED product specification. If the weight percentage of the red $Y_2O_3:Eu^{3+}$ phosphor increases, that of yellow YAG:Ce phosphor needs to be decreased to maintain the mean CCT values.

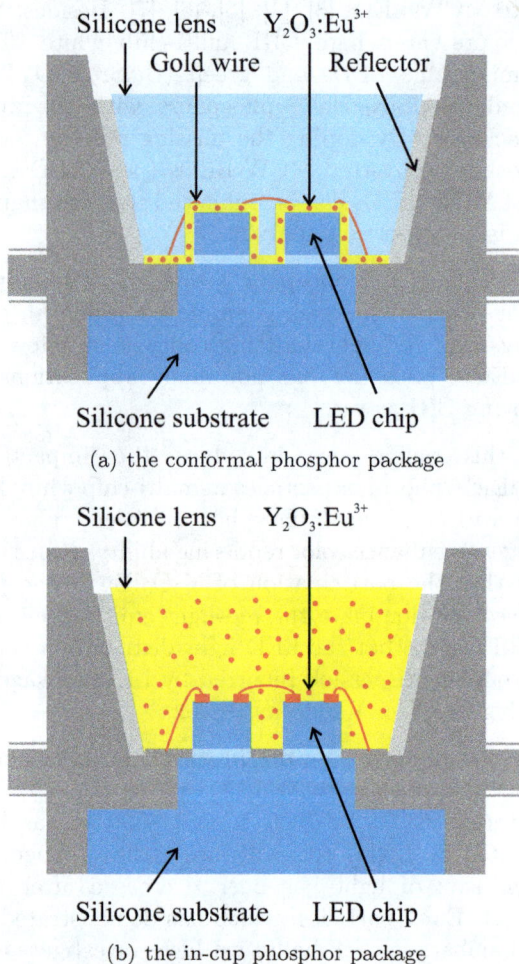

(a) the conformal phosphor package

(b) the in-cup phosphor package

Fig. 1: Illustration of MCW-LEDs.

The optical properties of reflector of CPP and IPP are similar. The depth, inner, and outer radius of the reflector are 2.07 mm, 8 mm and 9.85 mm, respectively. The CPP and the IPP, with the fixed thickness of 0.08 mm and 2.07 mm in turn, cover the nine chips. The blue Led chip has a dimension of 1.14×0.15 mm. The radiant flux of each blue chip is 1.16 W, and the peak wavelength is 453 nm. At the CPP displayed in Fig. 1(a), its phosphor layer is coated conformally on nine LEDs. As for the IPP, its phosphor layer is located in the silicone lens, as displayed in Fig. 1(b).

The absorption, emission and scattering of both YAG:Ce and $Y_2O_3:Eu^{3+}$ phosphor particles, with the peak wavelengths including blue of 453 nm and green-yellow of 555 nm, can be computed by Mie-scattering theory [2]. The phosphor layers consist of YAG:Ce and $Y_2O_3:Eu^{3+}$ powders and the silicone matrix. Their refractive indexes are 1.83, 1.93 and 1.50, respectively. Meanwhile, the mean radii of the phosphor powders are 7.25 μm, which conforms the real particle size.

3. Color Rendering Index

In order to verify the improvement of CRI using $Y_2O_3:Eu^{3+}$ phosphor, we switch the average CCT among the values of 7000 K and 8500 K and change $Y_2O_3:Eu^{3+}$ weight. The corresponding values of CRI are then calculated and displayed on Fig. 2. Figure 2(a) illustrates the impact of $Y_2O_3:Eu^{3+}$ concentration on CRI of CPP structures. It can be observed that the CRI grows with the weight percentage of $Y_2O_3:Eu^{3+}$ phosphor in the continuous range from 0 % to nearly 10 %. The highest color rendering ability is obtained with the $Y_2O_3:Eu^{3+}$ weight range from 8 % to 12 %. In particular, optimal color rendering index that can be achieved exceeds the value of 86 in this case. As for IPP structure, the $Y_2O_3:Eu^{3+}$ concentration ranges continuously from 0 % to approximately 0.4 %. From 0.3 % to 0.4 %, the CRI has a decreasing tendency with the increasing of $Y_2O_3:Eu^{3+}$ weight beyond a point where the red-light starts to be over-dominant, causing color rendering ability to reduce.

The highest color rendering ability with the different CCTs can be obtained when the $Y_2O_3:Eu^{3+}$ percentage ranges from 0.24 % to 0.3 %, as shown in Fig. 2(b). The optimal color rendering index of MCW-LEDs can exceed 84 in this case, which is 25.8 % higher than that of the non $Y_2O_3:Eu^{3+}$ case, i.e. When the $Y_2O_3:Eu^{3+}$ concentration is equal to 0 %. In summary, with the simulated results of CRI, we can demonstrate that the $Y_2O_3:Eu^{3+}$ phosphor having proper concentration can be used for increasing the CRI.

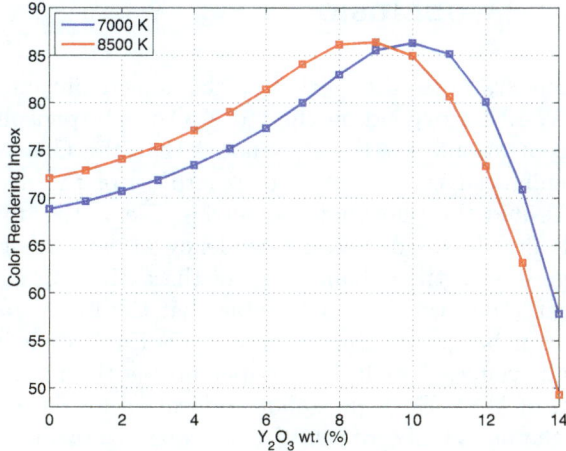

(a) CPP

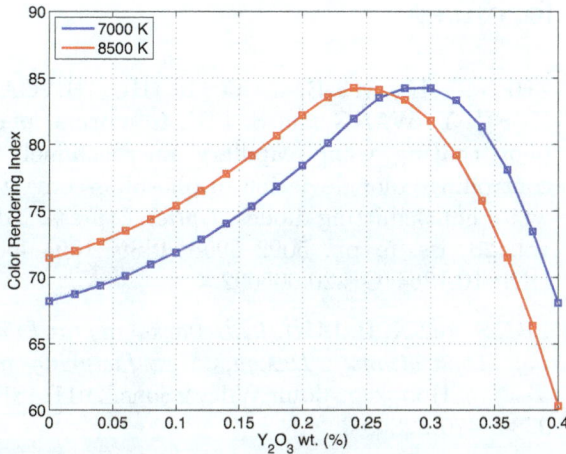

(a) CPP@7000 K

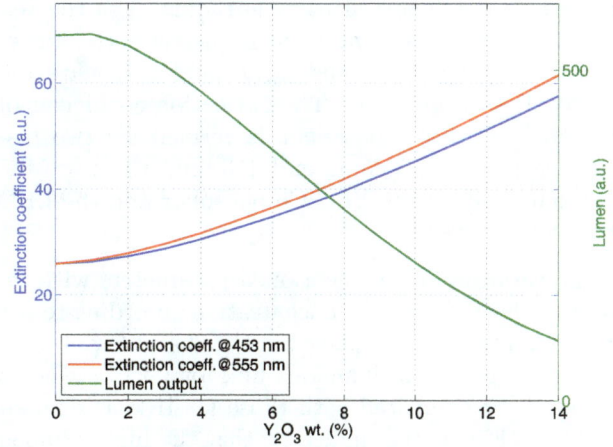

(b) IPP

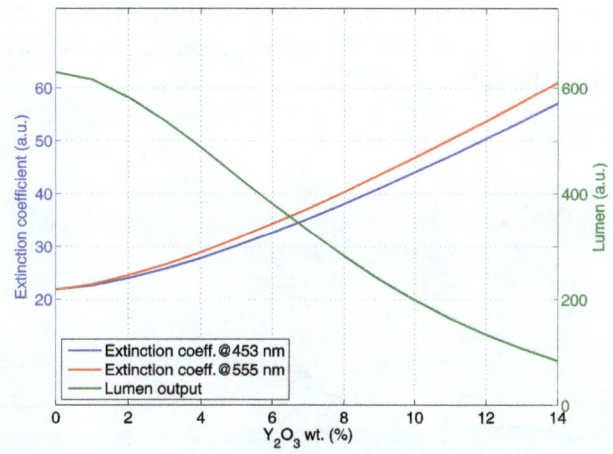

(b) CPP@8500 K

Fig. 2: The color rendering index at average CCTs of 7000 K and 8500 K with various Y_2O_3:Eu^{3+} weight with two phosphor geometries.

Fig. 3: Luminous flux according to computed extinction coefficient of CPPs.

4. Luminous Flux

The effect of Y_2O_3:Eu^{3+} concentration in phosphor compound on the lumen output is also verified together with the CRI, as shown in Fig. 3. The weight percentage of phosphor compound was varied continuously from 0 % to 14 % and to 0.4 % for CPP and IPP, respectively. At low Y_2O_3:Eu^{3+} concentration regime, the extinction coefficient tends to reduce, which results in the enhancement of luminous output. Meanwhile, the luminous flux decreases with the Y_2O_3:Eu^{3+} weight enhancement due to the increase of extinction coefficient.

To verify these results, the relationship of luminous output to the Y_2O_3:Eu^{3+} weight can be formulated according to Mie-scattering theory. The depletion of light is calculated by the Beer-Lambert law:

$$I = I_0 e^{-\mu_{ext} L}, \tag{1}$$

where I is the transmitted light power, I_0 is the incident light power, $\mu_{ext} = N \cdot C_{ext}$ is the extinction coefficient, L is the path length and N is the number of particles per cubic millimeter.

According to Mie-scattering theory, the extinction cross section C_{ext} of phosphor particles can be characterized by the following relationship:

$$C_{ext} = \frac{2\pi a^2}{x^2} \sum_{n=1}^{\infty} (2n+1)\mathrm{Re}(a_n + b_n), \tag{2}$$

where $x = 2\pi a/\lambda$ is the size parameter, a_n and b_n are the expansion coefficients with even symmetry and odd symmetry, respectively. The parameters a_n and b_n are defined as:

$$a_n(x, m) = \frac{\psi'_n(mx)\psi_n(x) - m\psi_n(mx)\psi'_n(x)}{\psi'_n(mx)\xi_n(x) - m\psi_n(mx)\xi'_n(x)}, \tag{3}$$

$$b_n(x, m) = \frac{m\psi'(mx)\psi_n(x) - \psi_n(mx)\psi'_n(x)}{m\psi'_n(mx)\xi_n(x) - \psi_n(mx)\xi'_n(x)}, \tag{4}$$

where a is the spherical particle radius, λ is the relative scattering wavelength, m is the refractive index of scattering particles, and $\psi_n(x)$ and $\xi_n(x)$ are the Riccati–Bessel functions. The extinction coefficient of the red $Y_2O_3{:}Eu^{3+}$ phosphor is verified for two distinct wavelengths, 555 nm and 453 nm, which are the emission peaks of the YAG:Ce phosphor and the LED chips, respectively.

The variation of the mentioned parameters with respect to the $Y_2O_3{:}Eu^{3+}$ concentration according to the above equations are displayed in Fig. 3 and Fig. 4. The simulation results of luminous flux for CPP as shown in Fig. 3 are compared with those for IPP as demonstrated in Fig. 4. It is indicated that the higher lumen output should occur at low $Y_2O_3{:}Eu^{3+}$ concentration, which corresponds to the lower extinction coefficient value. These results can be employed to estimate the influence of $Y_2O_3{:}Eu^{3+}$ concentration on the lumen output from the pc-WLEDs.

5.   Conclusion

Summary, both the CRI and the lumen output of MCW-LEDs depend on the red $Y_2O_3{:}Eu^{3+}$ phosphor concentration. Firstly, it is noted that the CRI can be enhanced to 86 and more regardless of the mean CCTs and the phosphor geometries. Next, the luminous flux has a decreasing tendency at large weight range due to the enhancement of the extinction coefficient. However, it is noticeable that the lumen output can be grown after adding $Y_2O_3{:}Eu^{3+}$ with low weight range. Finally, the paper proves the implications of $Y_2O_3{:}Eu^{3+}$ phosphor application for developing the pc-WLEDs of MCW-LEDs manufacturing.

References

[1] ZHENG, H., X. B. LUO, R. HU, B. CAO, X. FU, Y. WANG and S. LIU. Conformal phosphor coating using capillary microchannel for controlling color deviation of phosphor-converted white light-emitting diodes. *Optics Express*. 2012, vol. 20, iss. 5, pp. 5092–5098. ISSN 1094-4087. DOI: 10.1364/OE.20.005092.

[2] LIU, S. and X. B. LUO. *LED Packaging for Lighting Applications: Design, Manufacturing and Testing*. Hoboken: John Wiley&Sons, 2011. ISBN 978-0-470-82840-3.

[3] LIU, Z. Y., S. LIU, K. WANG and X. B. LUO. Optical Analysis of Color Distribution in White LEDs with Various Packaging Methods. *IEEE Photonics Technology Letters*. 2008, vol. 20, iss. 24, pp. 2027–2029. ISSN 1041-1135. DOI: 10.1109/LPT.2008.2005998.

[4] LIU, Z. Y., S. LIU, K. WANG and X. B. LUO. Analysis of Factors Affecting Color Distribution of White LEDs. In: *International Conference on Electronic Packaging Technology & High Density Packaging*. Shanghai: IEEE, 2008, pp. 386–393. ISBN 978-1-4244-2740-6. DOI: 10.1109/ICEPT.2008.4607013.

[5] SOMMER, C., F. REIL, J. R. KRENN, P. HARTMANN, P. PACHLER, H. HOSCHOPF and F. P. WENZL. The Impact of Light Scattering on the Radiant Flux of Phosphor-Converted High Power White Light-Emitting Diodes. *Journal of Lightwave Technology*. 2011, vol. 29, iss. 15, pp. 5145–5150. ISSN 0733-8724. DOI: 10.1109/JLT.2011.2158987.

[6] CHIEN, W. T., C. C. SUN and I. MORENO. Precise optical model of multi-chip white LEDs. *Optics Express*. 2007. vol. 15, iss. 12, pp. 7572–7577. ISSN 1094-4087. DOI: 10.1364/OE.15.007572.

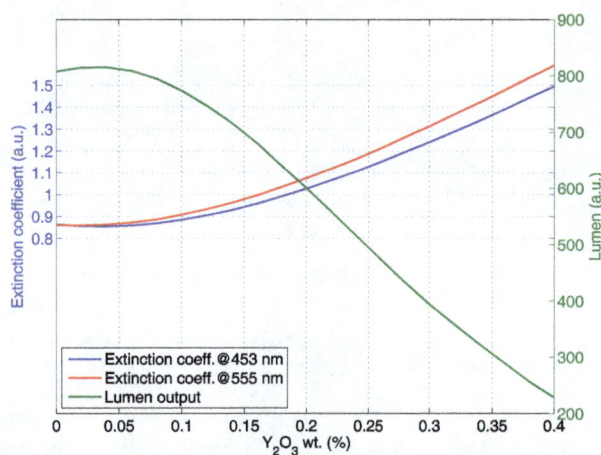

(a) IPP@7000 K

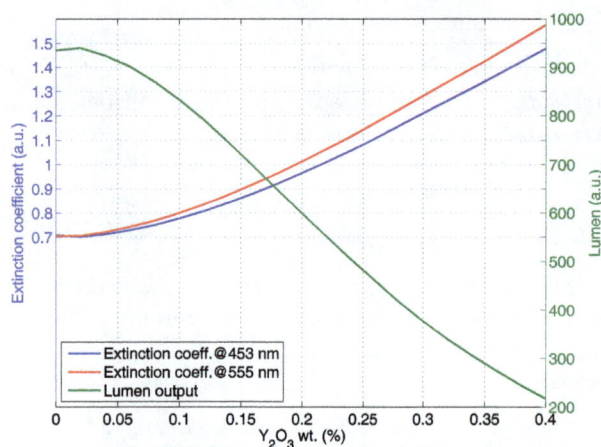

(b) IPP@8500 K

Fig. 4: Luminous flux according to computed extinction coefficient of IPPs.

[7] WANG, K., S. LIU, X. B. LUO, Z. Y. LIU and F. CHEN. Optical Analysis of A 3W Light-Emitting Diode (LED) MR16 Lamp. In: *International Conference on Electronic Packaging Technology & High Density Packaging*. Shanghai: IEEE, 2008, pp. 1–5. ISBN 978-1-4244-2740-6. DOI: 10.1109/ICEPT.2008.4607028.

About Authors

Tran Hoang Quang MINH defended his Ph.D. thesis at Tomsk Polytechnic University, Tomsk City, Russian Federation. The author's major fields of study are High-Voltage Power System, Relay Protections and Optoelectronics. He is working as Lecturer in Faculty of Electrical and Electronic Engineering, Ton Duc Thang University, Ho Chi Minh City, Vietnam.

Nguyen Huu Khanh NHAN defended his Ph.D. thesis at Institute of Research and Experiments for Electrical and Electronic Equipment, Moscow, Russian Federation. He is working as Lecturer in Faculty of Electrical and Electronic Engineering, Ton Duc Thang University, Ho Chi Minh City, Vietnam. His research interests include VLSI, MEMS and LED driver chips.

Thoai Phu VO was born in An Giang province, Vietnam. He has been working at the Faculty of Electrical and Electronics Engineering, Ton Duc Thang University. He received his Ph.D. degree from Dayeh University, Taiwan in 2015. His research interests are optoelectronics, fuzzy control, image processing.

Nguyen Doan Quoc ANH was born in Khanh Hoa province, Vietnam. He has been working at the Faculty of Electrical and Electronics Engineering, Ton Duc Thang University. Quoc Anh received his Ph.D. degree from National Kaohsiung University of Applied Sciences, Taiwan in 2014. His research interest is optoelectronics (such as Multi-chip white light LEDs, free-form lens, optical material).

4

Hybrid Solar Cell with TiO$_2$ Film: BBOT Polymer and Copper Phthalocyanine as Sensitizer

Saptadip SAHA, Priyanath DAS, Ajay Kumar CHAKRABORTY,
Ruchira DEBBARMA, Sharmistha SARKAR

Department of Electrical Engineering, National Institute of Technology Agartala,
Jirania, West Tripura 799046, India

saptadip.saha@gmail.com, priyanath70@gmail.com, akcall58@gmail.com,
ruchira.debbarma85@gmail.com, sharmisthabsl@gmail.com

Abstract. *An organic-inorganic hybrid solar cell was fabricated using Titanium dioxide (TiO$_2$): 2,5-bis(5-tert-butyl-2-benzoxazolyl) thiophene (BBOT) film and Copper Phthalocyanine (CuPc) as a sensitizer. BBOT was used in photodetector in other reported research works, but as per best of our knowledge, it was not implemented in solar cells till date. The blend of TiO$_2$: BBOT blend was used to fabricate the film on ITO-coated glass and further a thin layer of CuPc was coated on the film. This was acted as photoanode and another ITO coated glass with a platinum coating was used as a counter electrode (cathode). An optimal blend of acetonitrile (solvent) (50–100 %), 1,3-dimethylimidazolium iodide (10–25 %), iodine (2.5–10 %) and lithium iodide, pyridine derivative and thiocyanate was used as electrolytes in the hybrid solar cell. The different structural, optical and electrical characteristics were measured. The Hybrid solar cell showed a maximum conversion efficiency of 6.51 %.*

Keywords

BBOT, CuPc, electrolyte, organic, solar cell.

1. Introduction

Earth contains a finite amount of fossil fuels and continuous use of these fossil fuels in huge amount is resulting in the depletion and high cost of these resources. Fossil fuels are a non-renewable source of energy, which means that these resources once depleted cannot be replenished at a sufficient rate for sustainable economic extraction in meaningful human time-frames. In contrast, the many types of renewable energy resources, such as the wind and solar energy, are constantly replenished.

Solar energy in the direct or indirect form, is the source of most of the renewable energy, which can be used for heating homes, generating electricity and a variety of commercial and industrial uses. Therefore, to shift the dependency for energy from non-renewable to renewable resources, renewable energy has become an important topic for research [14]. Out of all the renewable energy sources, solar energy has the most advantages because it cuts down the need for a distribution network since it is possible to place the supply at or near the consumption area [11]. Photovoltaic effect was first observed by Becquerel [15]. The conventional solar cell or first generation based solar cell is made of crystalline silicon. With further research and modification of the first generation solar cell, the thin film solar cell came into picture. This was based on thin films of silicon and other materials which reduced the costs, normally associated with conventional semiconductor wafer production. Presently, the third generation solar is still a newly emerging field of research which is based on solar cell made of organic materials. This organic based solar cell has the advantage over the previous generation solar cell as far as the cost of materials and manufacturing is concerned [1] and [2]. The downside of the organic solar cell is the low efficiency. To overcome the major disadvantages of the pure organic solar cell, many significant changes have been made in the device structure, and also a new choice of materials consisting of both organic and inorganic materials was incorporated in the device. A solar cell based on organic and inorganic known as Hybrid solar cell has attracted attention due to its potential of reaching an efficiency of about 10 % [2], [3], [4], [5], [6] and [7].

In recent research investigation of polymer solar cell, a power conversion efficiency of ~5 % was obtained [8] and [9]. Copper phthalocyanine is chosen as sensitizer and 2,5-bis(5-tert-butyl-2-benzoxazolyl) thiophene (BBOT) for the organic blend of BBOT and TiO_2 in the active region of hybrid solar cell in this research because they have high optical stability, chemical stability and photovoltaic property [13].

2. Materials and Methods

The fabrication of hybrid solar cell involved following steps. Two pieces of ITO coated glass of the dimension of 2 cm × 1 cm were cleaned with Ethanol (C_2H_5OH) and de-ionized (DI) water (Fig. 1(a)). The resistance of the conducting side of the glass was measured with a multimeter to be 19–25 Ω on average. Then the glass slides were allowed for natural drying. A fine solution of nanoporous TiO2 powder (Global NanoTech), 2,5-Bis(5-tert-butyl-benzoxazo-2-yl) thiophene (BBOT) (Sigma Aldrich) (2:1) and acetic acid (CH_3COOH) (Nice chemicals) was prepared. A small portion of the conductive side of an ITO coated glass piece was covered with tape to avoid generation of TiO_2 film: BBOT blend during fabrication. A film of the blend was coated on the conductive side by using tape casting method, the colour of the film was light yellow (Fig. 1(b)). The dimension of the film was 2 cm × 1 cm. When the liquid part of the solution was evaporated naturally, it left a film of nanoporous TiO2 and BBOT (Fig. 1(c)). After the growth of the film the tape was removed. The sample was annealed for 25 minutes at 400 °C (Fig. 1(d)). After annealing the film colour changed to dark yellow (Fig. 1(e)). Now, a film of Copper Pthalocyanine (CuPc) was generated on the TiO_2 and BBOT film and was allowed for natural drying (Fig. 1(f)). For the fabrication of counter electrode, another glass slide was coated with platinum solution (Platisol, Solaronix) on the conductive

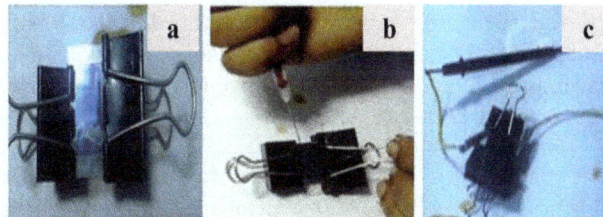

Fig. 2: (a) The solar cell is clamped between two binding clips, (b) electrolyte is being injected, (c) electrical characteristics are being measured.

side. Once both of the samples dried completely, they were clamped together facing two conductive sides using binder clips such that a conductive portion in both the slides was available to connect to the measurement probes (Fig. 2(a)). A small amount of electrolyte (HI-30, solaronix) was injected in the junction of the two slides (Fig. 2(b)). The electrolyte solution contained acetonitrile (50–100 %), 1,3-dimethylimidazolium iodide (10–25 %), iodine (2.5-10 %). Thus the device was ready to analyse different electrical characteristics. Finally the device was ready for electrical measurements (Fig. 2(c)). The schematic diagram is shown by Fig. 3.

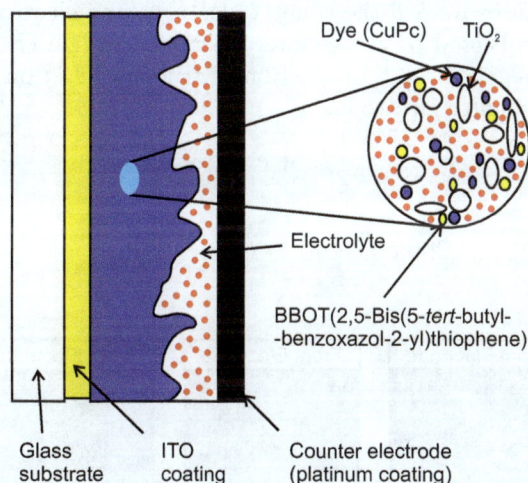

Fig. 3: Schematic diagram of the TiO_2 film, BBOT polymer and Copper Phthalocyanine based hybrid solar cell.

3. Working Principle

CuPc, which is a photo sensitizer, was fabricated on the top layer of the solar cell. CuPc absorbs photon from incident light. The BBOT particles, which were present next to the CuPc layer, are fluorescent in nature [17]. BBOT particles receive the photons donated by the CuPc layer and radiate the photons to the active material [18]. These radiated photons are absorbed and are trapped by the nanoporous TiO_2 particles, which are present in the blend. Some researchers have been

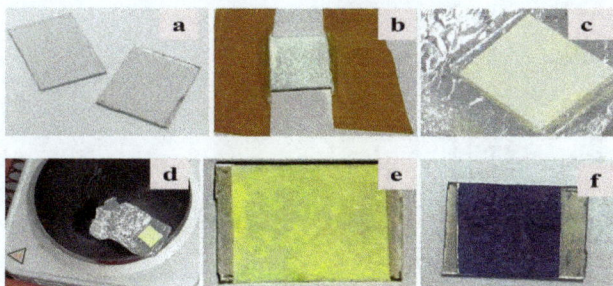

Fig. 1: (a) Cleaned ITO coated glass pieces, (b) fabricated film of TiO2 and BBOT blend on ITO coated glass by the tape casting method, (c) the film after drying (light yellow) (d) The sample is being annealed on a hot plate at 400?C, (e) The sample after annealing (color turned to bright yellow), (f) CuPc coated TiO2: BBOT film.

reported regarding fabrication of photodetector using BBOT [16]. The following mechanism was involved in conversion of electrical energy from the light energy [19], [20] and [21].

- Absorption: The incident photon is absorbed by the CuPc on the TiO_2 surface. After absorbing photon, the photosensitizer is excited from the ground state (S) to the excited state (S^*).

$$S \overset{h\gamma}{\to} S^*. \qquad (1)$$

- Electron injection: The excited electrons are injected into the conduction band of the TiO_2 and BBOT blend. This oxidizes the photosensitizer (S^+).

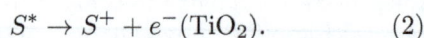

$$S^* \to S^+ + e^-(TiO_2). \qquad (2)$$

- Regeneration: The I^- ion redox mediator, present in the electrolyte, donates electrons to the oxidized photosensitizer (S^+) and again regenerates the ground state (S), and the I^- is oxidized to I_3^-.

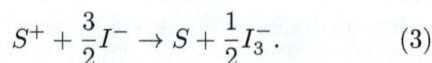

$$S^+ + \frac{3}{2}I^- \to S + \frac{1}{2}I_3^-. \qquad (3)$$

- Collection: The oxidized redox mediator (I_3^-), diffuses toward the counter electrode and electron is collected by the counter electrode. Iodide (I^-) is regenerated by reduction of triiodide (I_3^-) on the counter electrode.

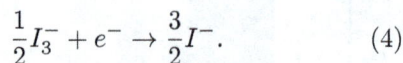

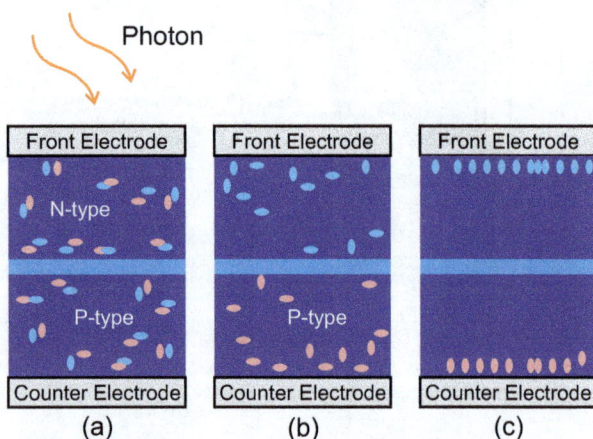

$$\frac{1}{2}I_3^- + e^- \to \frac{3}{2}I^-. \qquad (4)$$

Fig. 4: Working mechanism of the fabricated solar cell: (a) light absorption and exciton formation, (b) charge separation and (c) charge collection.

4. Results and Discussion

4.1. Structural Characterizations

Atomic Force Microscopy (AFM) study was carried out to analyse the surface morphology of the TiO_2: BBOT.

Figure 5(a) and Fig. 5(b) show the 2D and 3D AFM images of the film of TiO_2: BBOT blend. An average of 10 ∼ 15 nm TiO_2 and BBOT nanoparticles were observed in the images.

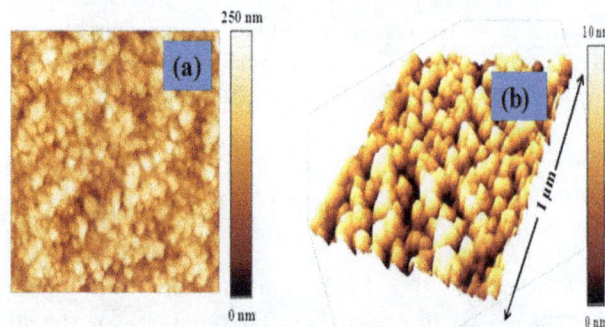

Fig. 5: (a) 2D and (b) 3D AFM images of the TiO_2: BBOT sample.

The crystalline structures of the nanoporous TiO_2 film (Fig. 6(a)) and TiO_2: BBOT (Fig. 6(b)) were con-

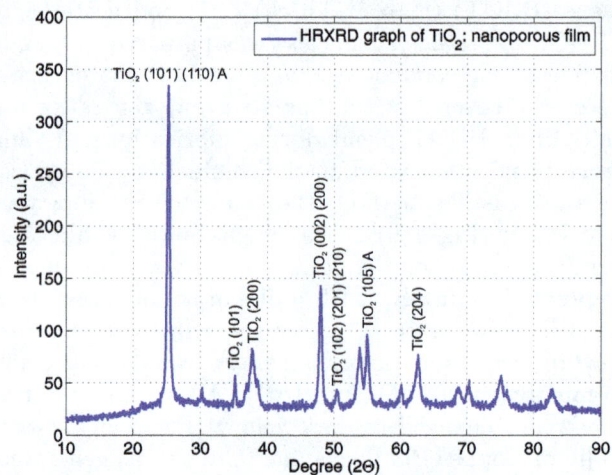

(a) TiO_2 nanoporous film.

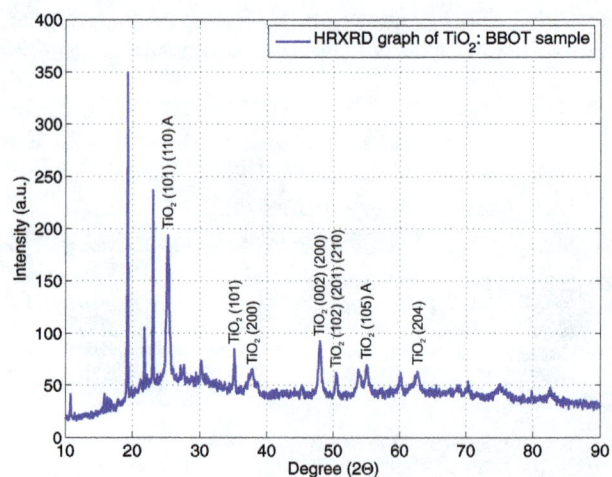

(b) TiO_2: BBOT film.

Fig. 6: HRXRD image.

firmed by the High Resolution X-Ray Diffraction analysis (HRXRD) (Panalytical, X'Pert Pro X-ray diffractometer). The TiO$_2$: BBOT film showed new peaks due to the presence of the BBOT (Fig. 6(b)). These peaks were not present in the pattern of pure TiO$_2$ (Fig. 6(a)). All the diffraction peaks refer to anatase and rutile phases (JCPDS No. 84-1285 for anatase and 87-0920 for rutile) [7] and [12]. The pattern shows an average size for both the phases.

4.2. Optical Characterizations

The UV-VIS optical absorption measurement (250 − 1000 nm) was done on the nanoporous TiO$_2$, BBOT, CuPc and CuPc: BBOT: TiO$_2$ sample at room temperature by a UV- visible near infrared spectrophotometer (Lambda 950, Perkin Elmer) (Fig. 7). The nanoporous TiO$_2$ film shows a peak in UV region (375 nm) (Fig. 7(a)). The BBOT showed large absorption spectra in the UV-visible region with a peak at 352 nm ((Fig. 7(b)). CuPc also had absorption spectra in visible region with main peaks at 618 nm and 722 nm and a small hump at 575 nm (Fig. 7(c)). CuPc: BBOT: TiO$_2$ sample had absorption peaks at 562 nm, 643 nm and 726 nm. The $(\alpha h \nu)^2$ versus eV curves of the nanoporous TiO$_2$ film and CuPc:BBOT:TiO$_2$ sample are shown by Fig. 8. In case of nanoporous TiO$_2$, the main band gap is shown for ~ 3.29 eV, which was also reported by other authors [20]. However, TiO$_2$ film exhibits the band gap at ~ 2.01 eV, which may correspond to the CuPc. Another major hump is obtained at ~ 3.67 eV which may be due to sub band gap transition of the nanoporous TiO$_2$. The band gap shifting was reported in other journals as well [20].

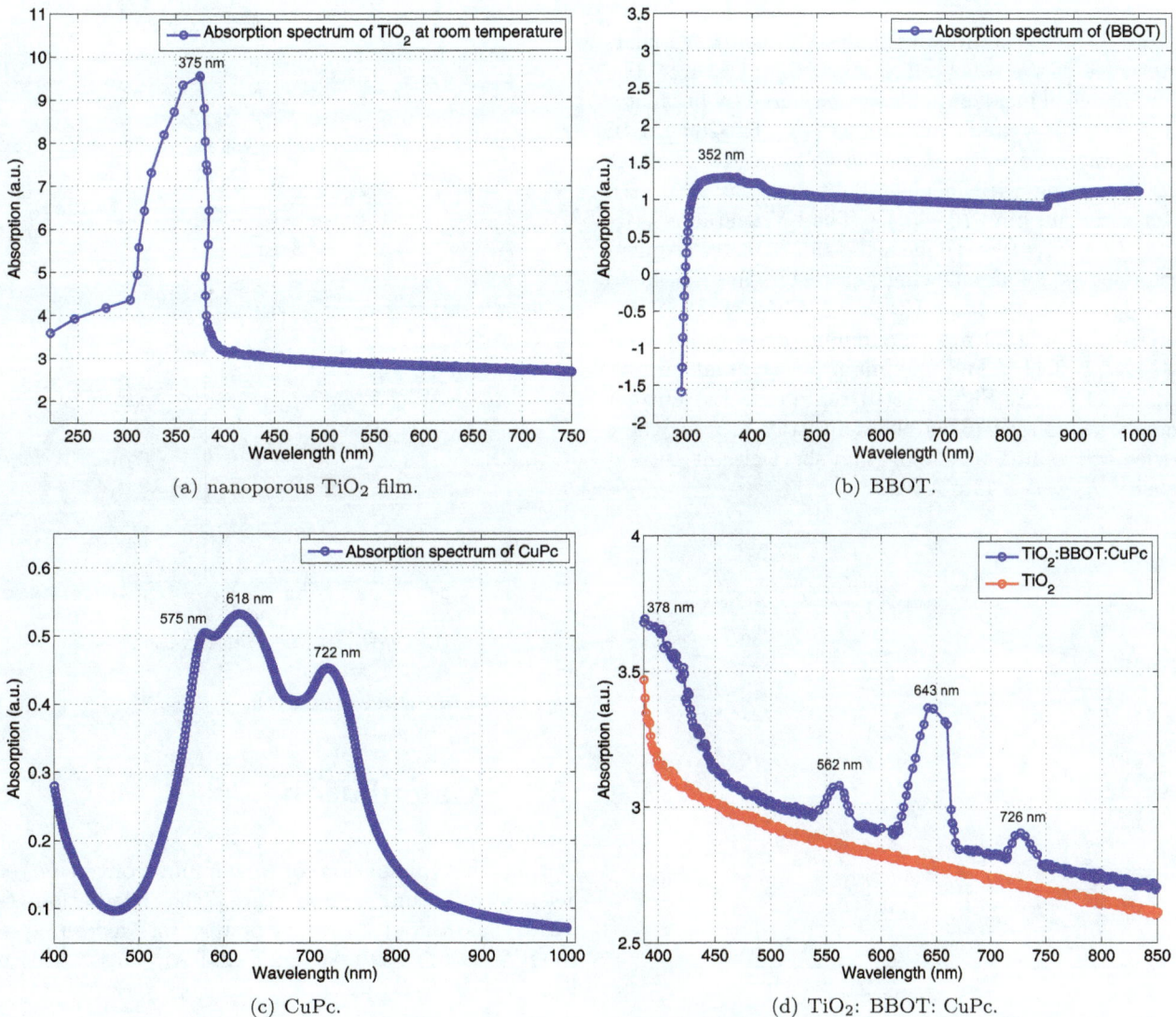

(a) nanoporous TiO$_2$ film.

(b) BBOT.

(c) CuPc.

(d) TiO$_2$: BBOT: CuPc.

Fig. 7: Absorption spectra.

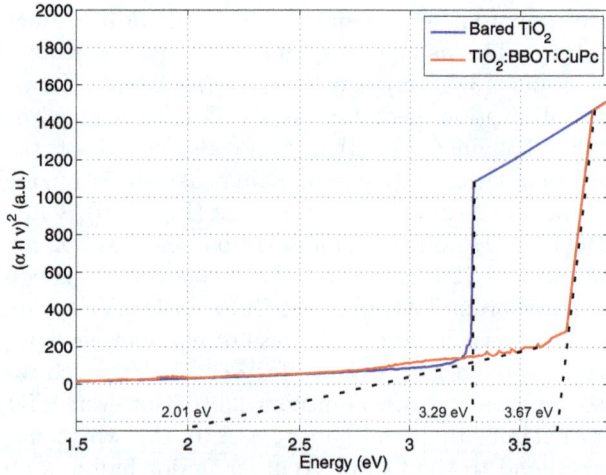

Fig. 8: $(\alpha h\nu)^2$ versus eV graph of nanoporous TiO_2 and TiO_2: BBOT: CuPc hybrid sample.

4.3. Electrical Characteristics

Figure 9 shows the Illuminance vs Voltage (L-V) characteristics of the solar cell at room temperature. The illuminance of incident light was measured by lux meter (HTC LX-101A) and a multimeter (FLUKE 289 TRUE RMS) was used to measure the voltage. The voltage was increasing with incident light and was saturated after a certain limit (0.645 V). The I-V characteristics study (Fig. 11(a)) was done using an I-V meter (Agilent technologies). The following parameters are measured as: open circuit voltage $V_{oc} = 0.6047$ V, short circuit current $I_{sc} = 16.9$ mA, maximum power point voltage $V_{mp} = 0.44$ V and maximum power point current $I_{mp} = 14.8$ mA. Figure 11(b) represents the current density vs voltage (J-V) characteristics graph, where device active-area was 2 cm^2 and short circuit current density J_{sc} was 8.45 mA·cm^{-2}.

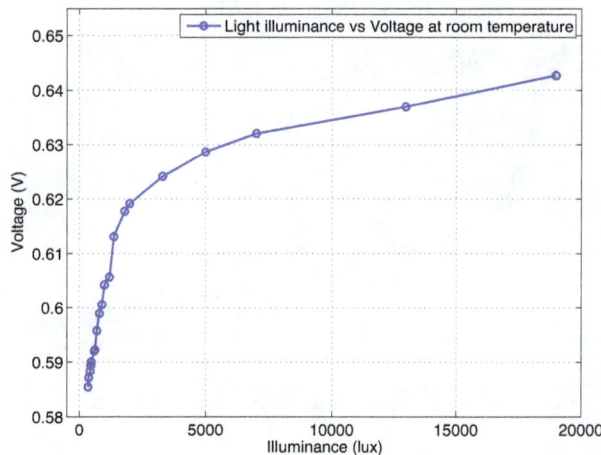

Fig. 9: Illuminance vs. output voltage characteristics of hybrid solar cell.

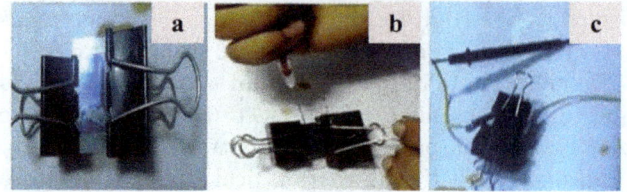

Fig. 10: (a) the solar cell is clamped between two binding clips, (b) electrolyte is being injected, (c) electrical characteristics are being measured.

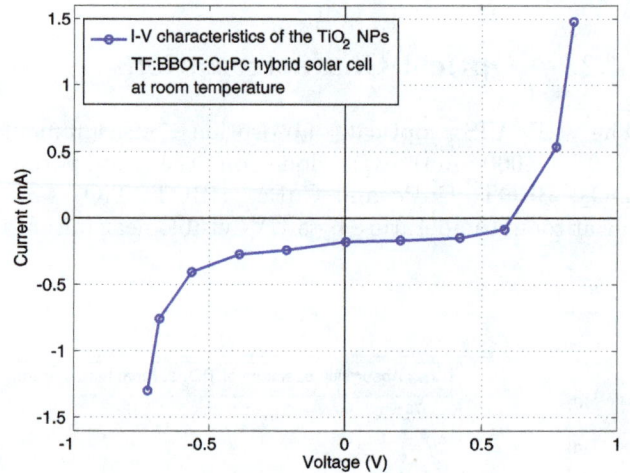

(a) Current I to voltage V.

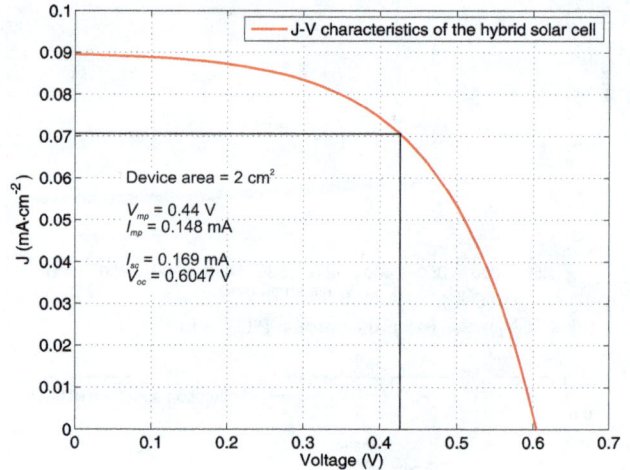

(b) Current density J to voltage V.

Fig. 11: Characteristics of the cell.

5. Calculation

During the calculation of maximum conversion efficiency at room temperature, the irradiation G, which was served as input power, was measured as 1000 W·m^{-2} i.e. 0.1 W·cm^{-2} and Air Mass (AM) of 1.5.

Fill Factor: $FF = \dfrac{V_{mp} \cdot I_{mp}}{V_{oc} \cdot I_{sc}} = \dfrac{0.44 \cdot 14.8}{0.6047 \cdot 16.9} = 0.6372.$

Efficency: $\eta = \dfrac{V_{oc} \cdot I_{sc} \cdot FF}{P_{in}} \cdot 100 =$

$$= \frac{0.6047 \cdot 16.9 \cdot 0.6372}{0.1 \cdot 1000} \cdot 100 = 6.51 \ \%.$$

6. Conclusion

In this paper, we have demonstrated the fabrication of nanoporous TiO_2: BBOT: CuPc film based hybrid solar cell. A layer of Copper Pthalocyanine (CuPc) was used as sensitizer and optimal blend of electrolytes was used to maximize the cell efficiency. ITO coated glass was used for both - the front and counter electrodes and a platinum solution coating was used on the counter electrode to maximize the collection of electrons and to reduce the current loss. The AFM and HRXRD studies confirmed the nanoporous morphology of the cell and the crystalline structure of the TiO_2 film respectively. The optical absorption study revealed the absorption properties of the TiO_2, CuPc and BBOT individually and their absorption contributions in the blend. The I-V characteristics graph defined the values $V_{oc} = 0.6047$ V, $I_{sc} = 16.9$ mA, $V_{mp} = 0.44$ V and $I_{mp} = 14.8$ mA. The maximum conversion efficiency of the cell was measured to be 6.51 %, which is much higher than in other research works reported earlier for this type of solar cell.

Acknowledgment

The authors thankfully acknowledge the financial support provided by The Institution of Engineers (India) for carrying out Research & Development work in this subject.

References

[1] CHAMBERLAIN, G. A. Organic solar cells: a review. *Solar Cells*. 1982, vol. 8, iss. 1, pp. 47–83. ISSN 1878-2655. DOI: 10.1016/0379-6787(83)90039-X.

[2] SPANGGAARD, H. and F. C. KREBS. A brief history of the development of organic and polymeric photovoltaics. *Solar Energy Materials and Solar Cells*. 2004, vol. 83, iss. 2–3, pp. 125–146. ISSN 0927-0248. DOI: 10.1016/j.solmat.2004.02.021.

[3] PEUMANS, P., A. YAKIMOV and S. R. FORREST. Small molecularweight organic thin-film photodetectors and solar cells. *Journal of Applied Physics*. 2003, vol. 425, iss. 7, pp. 158–162. ISSN 0021-8979. DOI: 10.1063/1.1534621.

[4] WINDER, C. and N. S. SARICIFTCI. Low Bandgap polymers for photonharvesting in bulk heterojunction solar cells. *Journal of Materials Chemistry*. 2004, vol. 14, iss. 1, pp. 1077–1086. ISSN 0959-9428. DOI: 10.1039/b306630d.

[5] SERVAITES, D. J., S. YEGANEH, T. J. MARKS and M. A. RATNER. Efficiency Enhancement in Organic Photovoltaic Cells: Consequences of Optimizing Series Resistance. *Advanced Functional Materials*. 2010, vol. 20, iss. 1, pp. 97–104. ISSN 1616-3028. DOI: 10.1002/adfm.200901107.

[6] ARICI, E., D. MEISSNER, F. SCHAEFFLER and N. S. SARICIFTCI. Core/shell nanomaterials in photovoltaics. *International Journal of Photoenergy*. 2003, vol. 5, iss. 1, pp. 199–208. ISSN 1687-529X. DOI: 10.1155/S1110662X03000333.

[7] BRABEC, C. J., V. DYAKONOV, J. PARISI and N. S. SARICIFTCI. *Organic Photovoltaics: Concepts and Realization*. New York: Springer, 2003. ISBN 978-3-642-05580-5.

[8] PADINGER, F., R. S. RITTBERGER and N. S. SARUCIFTCI. Effect of Postproduction Treatment on Plastic Solar Cells. *Advanced Functional Materials*. 2003, vol. 13, iss. 1, pp. 85–88. ISSN 1616-3028. DOI: 10.1002/adfm.200390011.

[9] WRIGHT, M., C. YANG, X. GONG, K. LEE and A. J. HEEGER. Thermally Stable, Efficient Polymer Solar Cells with Nanoscale Control of the Interpenetrating Network Morphology. *Advanced Functional Materials*. 2005, vol. 15, iss. 10, pp. 1617–1622. ISSN 1616-3028. DOI: 10.1002/adfm.200500211.

[10] HAUCH, J. A., P. SCHILINSKY, S. A. CHOULIS, R. CHILDERS, M. BIELE and C. J. BRABEC. Flexible Organic P3HT:PCBM Bulk-Heterojunction Modules with More than 1 Year Outdoor Lifetime. *Solar Energy Materials and Solar Cells*. 2008, vol. 92, iss. 1, pp. 727–731. ISSN 0927-0248. DOI: 10.1016/j.solmat.2008.01.004.

[11] PATHAK, D., T. WAGNER, T. ADHIKARI and J. M. NUNZI. Photovoltaic performance of $AgInSe_2$-conjugated polymer hybrid system bulk heterojunction solar cell. *Synthetic Metals*. 2015, vol. 199, iss. 1, pp. 87–92. ISSN 0379-6779. DOI: 10.1016/j.synthmet.2014.11.015.

[12] CHEN, H. Y., J. HOU, S. ZHANG, Y. LIANG, G. YANG, Y. YANG, L. YU, Y. WU and G. LI. Polymer solar cells with enhanced open-circuit voltage and efficiency. *Nature Photonics*. 2009, vol. 3, iss. 1, pp. 649–653. ISSN 1749-4893. DOI: 10.1038/nphoton.2009.192.

[13] YANG, J., K. C. GORDON, A. J. MCQUIL-LAN, Y. ZIDON and Y. SHAPIRA. Photoexcited carriers in organic light emitting materials and blended films observed by surface photovoltage spectroscopy. *Physical Review B*. 2005, vol. 71, iss. 1, pp. 155–209. ISSN 0556-2805. DOI: 10.1103/PhysRevB.71.155209.

[14] WRIGHT, M. and A. UDDIN. Organic-inorganic Hybrid solar cells: A comparative review. *Solar Energy Materials and Solar Cells*. 2012, vol. 107, iss. 1, pp. 87–111. ISSN 0927-0248. DOI: 10.1016/j.solmat.2012.07.006.

[15] BECQUEREL, A. E. Memoire sur les effets electriques produits sous l'influence des rayons solaires. *Comptes Rendus des Seances Hebdomadaires*. 1839, vol. 9, iss. 1, pp. 561–567. ISSN 0001-4036.

[16] KUROSAWA, K., W. WATANABE, A. TANAKA, Y. KOJIMA, K. FUJII and M. YAMADA. *Light detecting device*. US Patent. US5585915 A, 1996.

[17] FOURATI, M. A., T. MARIS, W. G. SKENE, C. G. BAZUIN and R. E. PRUD'HOMME. Photophysical, electrochemical and crystallographic investigations of the fluorophore 2,5-bis(5-tert-butyl-benzoxazol-2-yl)thiophene. *Journal Physical Chemistry B*. 2011, vol. 115, iss. 43, pp. 12362–12369. ISSN 1520-6106. DOI: 10.1021/jp207136k.

[18] LEWINSKA, G., A. PUSZYNSKI and J. SANETRA. BBOT for applications in photovoltaic cells devices and organic diodes. *Synthetic Metal*. 2015, vol. 199, iss. 1, pp. 335–338. ISSN 0379-6779. DOI: 10.1016/j.synthmet.2014.11.013.

[19] NARAYAN, M. R. Review: Dye sensitized solar cells based on natural photosensitizers. *Renewable and Sustainable Energy Reviews*. 2011, vol. 16, iss. 1, pp. 208–215. ISSN 1879-0690. DOI: 10.1016/j.rser.2011.07.148.

[20] HUG, H., M. BADER, P. MAIR and T. GLATZEL. Biophotovoltaics: Natural pigments in dye-sensitized solar cells. *Applied Energy*. 2013, vol. 115, iss. 1, pp. 216–225. ISSN 0306-2619. DOI: 10.1016/j.apenergy.2013.10.055.

[21] LUDIN, N. A., A. M. A. A. MAHMOUD, A. B. MOHAMAD, A. A. H. KADHUM, K. SOPIAN and N. S. A. KARIM. Review on the development of natural dye photosensitizer for dye-sensitized solar cells. *Renewable Sustainable Energy Reviews*. 2013, vol. 31, iss. 1, pp. 386–396. ISSN 1879-0690. DOI: 10.1016/j.rser.2013.12.001.

About Authors

Saptadip SAHA was born in Agartala, India. He received his M.Tech. from National Institute of Technology Agartala in 2014. His research interests include fabrication of photodector, organic and inorganic solar cells, DSSC, different studies of PV system.

Priyanath DAS was born in Agartala, India. He received his Ph.D. from Jadavpur University in 2012. His research interests include PV systems, PV faults and grid connected PV systems.

Ajoy Kumar CHAKRABORTY was born in Jhargram, India. He received his Ph.D. from Jadavpur University in 2007. His research interests include PV systems, Smart Metering and grid connected PV systems.

Ruchira DEBBARMA was born in Agartala, India. She received her B.E. from Hindustan Institute of Technology And Science in 2013. Her research interest includes Hybrid Solar Cell, Organic Solar Cell.

Sharmistha SARKAR was born in Bishalgarh, India. She received her B.E. from Tripura Institute of Technology in 2014. Her research interest includes Dye Sensitized Solar Cell, Organic Solar Cell.

5

Perimeter System Based on a Combination of a Mach-Zehnder Interferometer and the Bragg Gratings

Marcel FAJKUS, Jan NEDOMA, Petr SISKA, Lukas BEDNAREK,
Stanislav ZABKA, Vladimir VASINEK

Department of Telecommunications, Faculty of Electrical Engineering and Computer Science,
VSB–Technical University of Ostrava, 17. listopadu 15/2172, 708 33 Ostrava-Poruba, Czech Republic

marcel.fajkus@vsb.cz, jan.nedoma@vsb.cz, petr.siska@vsb.cz, lukas.bednarek@vsb.cz,
stanislav.zabka@centrum.sk, vladimir.vasinek@vsb.cz

Abstract. *Fiber optic interferometers and Bragg gratings belong to the group of very precise and sensitive devicesthat allow measuring very small deformation, temperature or vibration changes. The described methodology presentsthe useof a Mach-Zehnder interferometer and Bragg gratings together as a sensor system for detecting and monitoring movement within thedefined perimeter of 2.5×1 m. Analyses of the dynamic changes in interferometric patterns were a basis for this method. Also the signal maximum amplitude was measured and compared with the noise background. Perimeter disruptions can be detected by Bragg gratings due to its large deformation sensitivity in transversal or perpendicular directions. The result is then evaluated in the spectral domain. In terms of detected persons it showed very good results. The combination of these sensors was chosen for monitoring both the static and dynamic phenomena. Author's aim is to take advantage of both devicespositive properties. Thus, the system has the abilityto identify people due to frequency analysis in case of interferometers as well as dynamic weighting thanks to Bragg gratings.*

Keywords

Bragg gratings, Mach-Zehnder interferometer, monitoring of movement, perimeter system.

1. Introduction

Security and safety systems are a set of technical and organizational measures. These measures stand between a protected interest and the danger. It may be a property or other values that we protect against theft, damage, destruction or disruption by any other way. As for standard electrical systems, the installation of detection elements is limited to certain parts of the object, due to a reduction of the financial costs. These parts are mostly entrances into objects or rooms with valuable things, or into places with dangerous material etc. In addition, the detection elements must be either hidden or resistant to mechanical deactivation by the intruder or through electromagnetic jammers. All disadvantages of standard systems can be eliminated by using the security system based on fiber-optic principles. Optical fiber can be used either for detection of the intrusion or for transmission of information about the state of the monitored object. Optical fibers can be installed easily into floors, windows or walls. These fibers can be installed so as to be invisible, undetectable or untraceable due to the very small dimensions. Moreover, these elements cannot be decommissioned by the use of jammers due to the immunity against electromagnetic interference. The price of an optical fiber is low, therefore fiber-optic security systems can be implemented in each object and building.

2. Operating Principles

Perimeter detection systems based on the FBG generally consist of a network of FBGs. The network has ten or more FBGs within a single detection system. Thus created networks are capable of detecting disruptions of static and dynamic processes. Disruption of the object causes the formation of the vertical force.

The vertical force will act on one or more FBG sensors, depending on the density of the sensor network. The disadvantage of such systems is the necessity to use more FBGs. Other disadvantages are decreasing of the effectiveness of the detection system and more complex evaluation part in case of large detection surface. Examples of embodiments show articles [1], [2], [3] and [4].

Perimeter detection systems based only on the use of interferometric measurements utilize the known types of interferometric sensors. These systems use the combination of two or more interferometers which form a closed circuit. The resolving capability of such systems is both dependent on the size of the detection area and requires a highly coherent laser source and a more complex evaluation part for the processing and synchronization. Amplitude response can be monitored in dependence on time to evaluate the signal. We can detect the distortion using the frequency analysis in the case of using an FFT transform. A disadvantage arises in detecting static processes when these sensors based on the operating principle of interferometric measurements are ineffective. Examples of embodiments show articles [5] and [6].

Perimeter systems were realized mainly through the use of one type of sensors where we encounter disadvantages that are mentioned above. The combination of FBG gratings and fiber-optic interferometers was chosen with regard to the possibility of monitoring of static and dynamic processes simultaneously. Bragg gratings (hereinafter "FBG") are a typical example of a fiber-optic sensor for use in security systems. FBG are formed by the periodic structure changes of the refractive index within of the core of the optical fiber (Fig. 1). There is a partial spectral reflection of the transmitted light on these interfaces. If we bring wide-spectrum light into an optical fiber with FBG, then a certain part of the spectrum is reflected, and other wavelengths are transmitted through Bragg grating without loss.

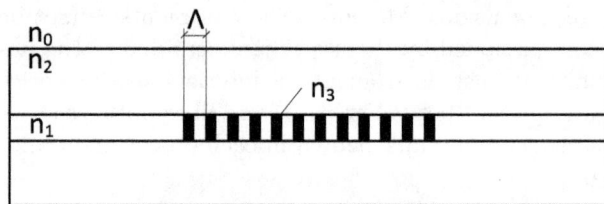

Fig. 1: Structure of the FBG.

The reflected wavelength is called the Bragg wavelength and it is given by:

$$\lambda_B = 2n_{eff}\Lambda, \qquad (1)$$

where n_{eff} is the effective refractive index, and Λ is the period of changes of the refractive index in the

optical fiber. A Bragg grating utilizes the temperature and deformation sensitivity of said parameters on the surrounding imulses in sensorial applications. Respective evaluation is performed in the spectral domain. The temperature sensitivity is 10.1 pm·$°C^{-1}$ for the FBG with Bragg wavelength of 1550 nm, and tensile deformation is 1.01 pm·$\mu strain^{-1}$ for same Bragg wavelength. FBG sensors belong to the group of single-point sensors, but they can be very easily connected to achieve the multipoint measurement using multiplexing techniques. The wavelength division multiplex is the simplest technique which uses the spectral separation of the signal f^{-1}rom the individual FBG sensors. Time division multiplex can be used instead of wavelength division multiplex. Time division multiplexing offers up to several hundred sensors per one optical fiber, but the implementation of the evaluation is significantly more complex [7].

The use of fiber optic interferometers is the other very suitable possibility for solving problems. These devices enable measurement with high sensitivity over a long distance. The reason is obvious -the light passes through the fiber with low attenuation in comparison with electrical cables, having metallic conductivity. Other advantages of fiber optic interferometers are a large dynamic range, their resistance to electrostatic and electromagnetic interference, and the fact that they are relatively less affected by aging components from which they are composed. Sensing of the physical or chemical values is manifested by phase change of the received light beam. This type of sensor requires a single mode optical fiber and a coherent radiation source. It offers the possibility to achieve maximum sensitivity within the fiber-optic sensors. Typical sensitivity can be achieved in the order of 10^{-8} (the wavelength of light is about 1 μm in the optical fiber). The design of the reference arm of the interferometer is the basis for fiber interferometry. The arrangement must be designed to a maximum of the elimination of unwanted signals. Noise (background signal) has the origin in the thermal phenomena, mechanical changes, or changes in the refractive index of the material. It is also necessary to consider different non-specific reactions that we do not want to detect (physical and chemical reactions). It is desirable that the reference arm is positioned as close as possible to the measuring arm. The reference and measuring arms should have the same length, structure and other properties. The only difference must be sensitivity to a specific value, which we want to detect [8] and [9].

Interferometry is an optical method that monitors the phase difference between two optical beams which pass through similar (if possible identical) optical paths. A phase shift arises in the interferometer. Interferometry is able to detect three parameters. These parameters affect the optical beam propagating along the optical path:

- change of the propagation speed,

- change of the wavelength,

- change of the route length.

If the change occurs in any of these parameters, then a change also occurs in a wave-phase. This change depends on the length of the path L, the refractive index n, and the wavelength λ according to the equation:

$$\Phi = 2\pi L \frac{n}{\lambda} = kLn, \qquad (2)$$

where L is the length of used fiber, n is the refractive index of the core, λ is the wavelength of the radiation source and k is the size of the wave vector. Fluctuations of phase delay of the interferometer can be described as:

$$\frac{d\phi}{\phi} = \frac{dL}{L} + \frac{dn}{n} + \frac{dk}{k}. \qquad (3)$$

Variable V (contrast, visibility) is introduced to evaluate the degree of interference provided by the interferometer:

$$V = \frac{I_{max} - I_{min}}{I_{max} + I_{min}}. \qquad (4)$$

Visibility depends on the relative intensity of the signal and reference beams, their relative states of polarization and mutual coherence.

I_{max} is the maximum observed intensity for linearly polarized waves in the same direction:

$$I_{max} = I_1 + I_2 + 2A_1A_2, \qquad (5)$$

and I_{min} is the minimum observed intensity:

$$I_{min} = I_1 + I_2 - 2A_1A_2, \qquad (6)$$

where I_1, I_2 indicate the intensity of interference waves, A_1, A_2 indicate the maximum amplitudes of the waves:

$$I_1 = I_0 \left[1 - V \cos\left(\phi_r - \phi_s\right)\right], \qquad (7)$$

$$I_2 = I_0 \left[1 + V \cos\left(\phi_r - \phi_s\right)\right], \qquad (8)$$

where I_0 is the mean signal value, ϕ_r is the phase of the reference beam, and ϕ_s is the phase of the signal beam.

The output intensity of the interferometer can be expressed as:

$$I = \frac{I_0 \cdot \alpha}{2} \left(1 + \cos \Lambda\phi\right), \qquad (9)$$

where α indicates the optical losses of the interferometer, and $\Lambda\phi$ is difference between phases in both arms and is given by $\Lambda\phi = \phi_r - \phi_s$. The output intensity is converted into electric current using the photodetector. Differential combination of those currents produces the output:

$$i = \varepsilon \cdot l_0 \cdot \alpha \cdot \cos\left(\phi_d + \phi_s \sin \omega t\right), \qquad (10)$$

where ε is the sensitivity of the photodetector, ϕ_d is changing the phase shift, ϕ_s is the phase modulation amplitude, and ω is frequency. The output current from the interferometer is derived from two-phase shifts. The first ϕ_d member (slow changes over time) is the mean value of the variation in the intensity of random influences. The second member $(\phi_s \cdot \sin \omega t)$ is a phase change caused by an external source of vibrations with the frequency omega [10] and [11].

Optical fibers containing FBG can be installed, in the security technology for example, into the floor in combination with fiber-optic interferometers. The sensory network with FBG sensors must be installed to cover large objects or surfaces. Disruption of the object causes the formation of vertical force. This force will act on one or more FBG sensors, depending on the density of the sensory network. This information is detectable in the evaluation unit. Implementation of only one sensor can be used for fiber optic interferometers due to the very high sensitivity. The combination of two or more interferometric sensors can be used to cover a larger surface of an object. The amplitude response of the signal can be monitored for evaluation in dependence on time. In the case of FFT transform, detecting the intrusion can be monitored by the implementation of frequency analysis. The combination of FBG and optical interferometers was chosen with regard to the possibility of monitoring both the static and dynamic processes.

3. Experimental Setup

Practical measurements were divided into the following four phases.

- Testing to achieve maximum possible signal to noise ratio (SNR).

- Achieving a situation where the implemented sensor does not detect impulses outside the test perimeter of 2.5×1 m.

- Testing the detection system with the interferometric sensor and with FBG.

- Testing the combination of both sensors.

The test area was chosen to maximize the ability of detection of static and dynamic processes in the 2.5×1 m range.

Mach-Zehnder interferometer is the solutionbasis of the interferometric sensor operating with the optical fiber withstandard specification G.652D. The interferometer is implemented with respect to the greatest possible sensitivity to the low frequency. Results of the frequency analysis showed that vibration caused by the test persons are low frequencies.The tunable laser source was used as the excitation source operating at the wavelength of 1550 nm. The output power was set to there ference value of 1 mW. This value was constant for all experimental measurements. Isolator was inserted for filtering unwanted back reflections to the source of radiation. A part of the signal processing (electronic evaluation part) includes the PbSe photodetector which detects the signal resulting from the interference of optical beams from the reference (L2) and measuring (L1) interferometerarms. The output intensity is converted into a measurable electric current. The high pass filter is used to ensure the zero offset voltage, amplifier and analog-digital converter (NI USB 6210) as well. The actual evaluation software works for the interferometric sensor with the signal in the time domain. The application displays the progress of the sensed signal as voltage in dependence on time (amplitude spectrum of the input signal). The block scheme of the measurement is shown in Fig. 2.

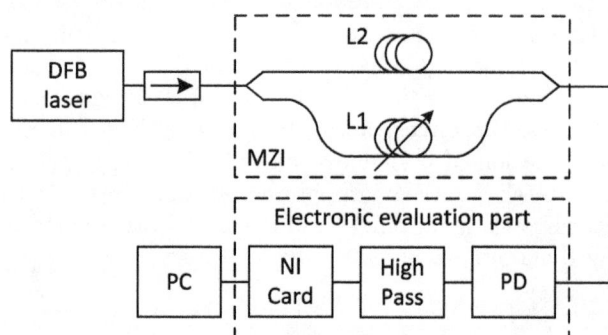

Fig. 2: A scheme of the measurement with the interferometric sensor.

The total of 250 repeated walkthroughs of 10 test persons were performed around the detection pad of 2.5×1 m for testing in order to obtain the maximum possible signal regarding noise ratio (SNR). Typical values of amplitude response did not exceed 0.02 V at the distances of 1, 2 and 3 m (Fig. 3). SNR level 0.1 V was determined on the basis of the obtained values. Application assessed the response as sufficient to confirm the passage of the subject from this value

The total of 100 repeated walkthroughs of 10 test persons were performed through the perimeter system (the responses caused by passing through the detection pad of 2.5×1 m) to test the detection system with an in-

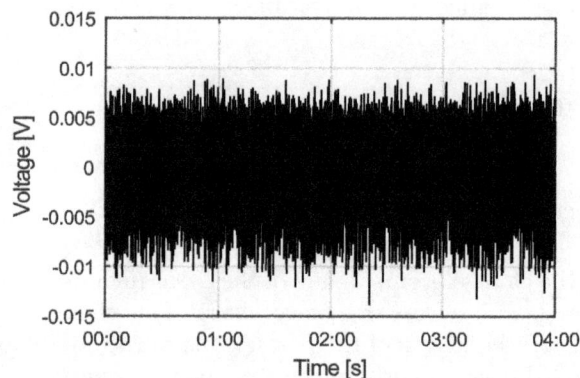

Fig. 3: A typical record of the passage of a person outside the detection pad-detection by interferometric sensor.

terferometric sensor. Each test subject performed four steps, and the response was subsequently evaluated. Individual persons were detected with 100% efficiency. The typical results are shown below (Fig. 4) for two test persons in the time domain. Typical values of the amplitude response varied from 0.5 to 2 V according to the test persons with weight, age as well as gender differences.

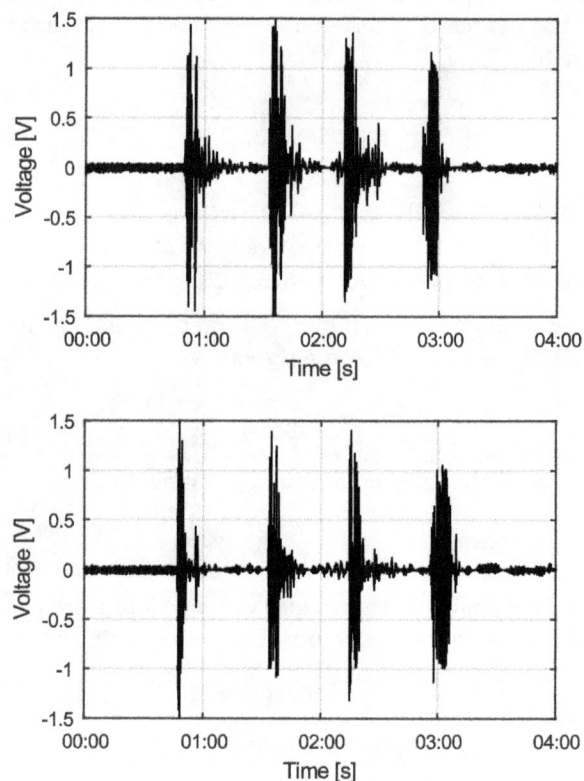

Fig. 4: A typical record of the walkthroughs of two different test persons over the detection pad-detection by the interferometric sensor.

A tunable laser was used for experimental verification of the perimeter system with FBG sensor(Fig. 5). Output radiation was spectral swept in the range from

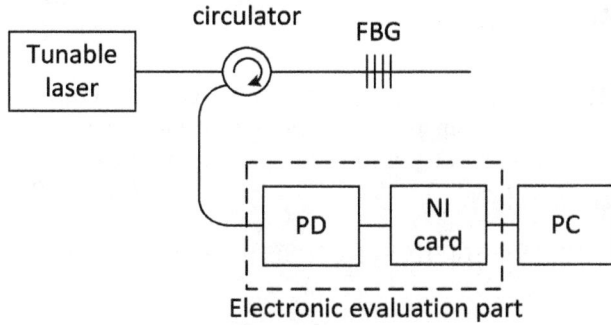

Fig. 5: A scheme of the measurement with the FBG sensor.

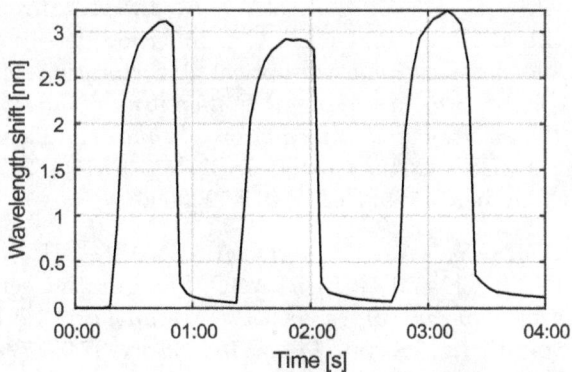

Fig. 7: A typical record of the repeated walkthroughs of a test person over the detection pad (00:00–30:00) versus stopping and staying on the spot (30:00–45:00) detected by the FBG sensor.

1548 to 1557 nm with the period of 250 ms. FBG sensor was tuned to the Bragg wavelength of 1550.104 nm. The reflected light was both detected by a photodetector in time periods and digitized by the measuring card NI USB 6210. The application was implemented on the PC in LabView that shows the resulting spectral characteristic.

The total of 100 repeated walkthroughs of 10 test persons were performed through the perimeter system (the responses caused by passing through the detection pad of 2.5×1 m) to test the detection system with the FBG sensor. Individual persons were detected with 100% efficiency. Each test person performed three steps within one repetition, the response was subsequently evaluated (Fig. 6). The progress of the repeated walkthroughs of the test persons was tested at the end of the measurement after walkthrough over the detection pad versus stopping and staying on the spot (Fig. 7).

sensor. Three steps and stopping are monitored with the FBG sensor (Fig. 9).

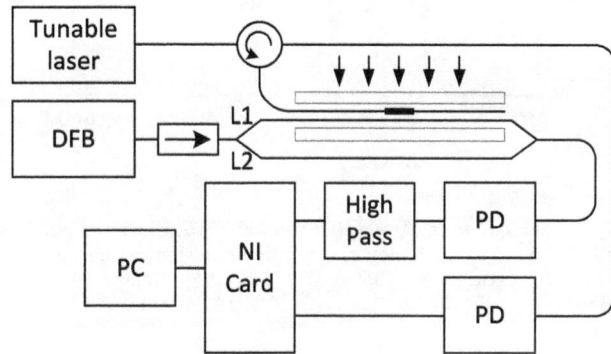

Fig. 8: A scheme of the measurement with the FBG sensor and the Mach-Zehnder interferometer sensor.

4.   Conclusion

The reported results showed that the combination of the fiber optic interferometer and Bragg grating may find its application in safety and security systems. The methodology of the sensors was chosen with a view to the use of the advantageous properties of both types of sensors (detection of the static as well as dynamic phenomena). The test was focused on the maximizing ability of detect testing persons within a detection pad with the dimensions of 2.5×1 m. The results showed that the individual objects were detected with 100% efficiency in cases using both an interferometric sensor and the sensor with FBG. For both sensors, there was a total of 200 repetitions performed (10 different test persons). Each testing person performed steps within one measurement. Tests showed the combination of these sensors is able to monitor both static and dynamic phenomena simultaneously, and eliminates the risk of failure of our perimeter system. A further re-

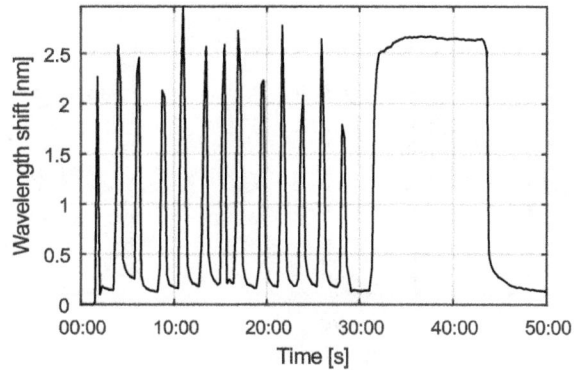

Fig. 6: A typical record of the walkthrough of a test person over the detection pad detected by the FBG sensor.

The testing scheme with the combination of both sensors is shown in Fig. 8. The combination of these sensors is able to monitor both static and dynamic phenomena simultaneously. This combination reduces the risk of failure of the perimeter system of dimensions 2.5×1 m to a minimum. The graph describes a dynamic phenomenon monitored by an interferometric

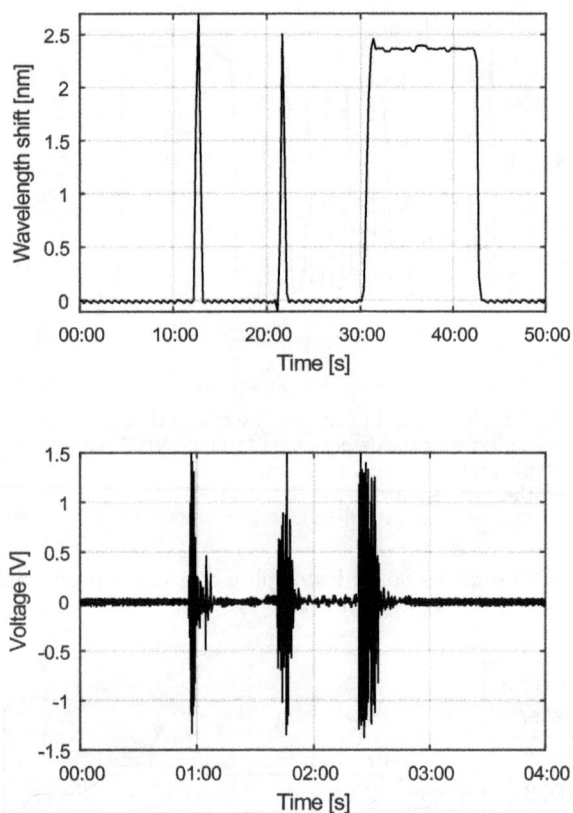

Fig. 9: An example of record of the walkthrough of test persons over the detection pad-the combination of the interferometric sensor (bottom) and the FBG sensor (top).

search will focus on expanding the proposed prototype. The interferometric sensor detects the intrusion of the monitored perimeter using frequency analysis, and the FBG sensor expands with the possibility of dynamic weighing of the persons.

Acknowledgment

This article was supported by the projects of the Technology Agency of the Czech Republic TA03020439 and TA04021263, and by Ministry of Education of the Czech Republic within the project no. SP2016/149. The research has been partially supported by the Ministry of Education, Youth and Sports of the Czech Republic through the grant project no. CZ.1.07/2.3.00/20.0217 within the frame of the operation programme Education for competitiveness financed by the European Structural Funds and from the state budget of the Czech Republic. This article was supported too by the project VI20152020008 (GUARDSENSE II).

References

[1] WU, H., Y. QIAN, W. ZHANG, H. LI and X. XIE. Intelligent detection and identification in fiber-optical perimeter intrusion monitoring system based on the FBG sensor network. *Photonic Sensors.* 2015, vol. 5, iss. 4, pp. 365–375. ISSN 1674-9251. DOI: 10.1007/s13320-015-0274-8.

[2] ALLWOOD, G., S. HINCKLEY and G. WILD. Optical Fiber Bragg grating based intrusion detection systems for homeland security. In: *IEEE Sensors Applications Symposium.* Galveston: IEEE, 2013, pp. 66–70. ISBN 978-1-4673-4636-8. DOI: 10.1109/SAS.2013.6493558.

[3] LI, S., J. MA and J. HU. Rockfall hazard alarm strategy based on FBG smart passive net structure. *Photonic Sensors.* 2015, vol. 5, iss. 1, pp. 19–23. ISSN 1674-9251. DOI: 10.1007/s13320-014-0203-2.

[4] CATALANO, A., F. A. BRUNO, M. PISCO, A. CUTOLO and A. CUSANO. An intrusion detection system for the protection of railway assets using fiber bragg grating sensors. *Sensors.* 2014, vol. 14, iss. 10, pp. 18268–18285. ISSN 1424-8220. DOI: 10.3390/s141018268.

[5] JIANG, J., Y. JIANG, D. LIU, S. WANG and X. LI. Fiber-optic perimeter security system based on dual Mach-Zehnder interferometer structure. *Guangxue Jishu/Optical Technique.* 2015, vol. 41, iss. 3, pp. 193–196. ISSN 1002-1582.

[6] ALI, T. H., H. MEDJADBA, L. M. SIMO-HAMED and R. CHEMALI. Intrusion detection and classification using optical fiber vibration sensor. In: *3rd International Conference on Control, Engineering and Information Technology.* Tlemcen: IEEE, 2015, pp. 1–6. ISBN 978-1-4799-8212-7. DOI: 10.1109/CEIT.2015.7233060.

[7] KERSEY, A. D., M. A. DAVIS, H. J. PATRICK and M. LEBLANC. Fiber grating sensors. *Journal of Light wave Technology.* 1997, vol. 15, iss. 8, pp. 1442–1463. ISSN 0733-8724. DOI: 10.1109/50.618377.

[8] LOPEZ-HIGUERA, J. M. *Handbook of Optical Fibre Sensing Technology.* 1st ed. Dublin: Wiley, 2002. ISBN 978-0-471-82053-6.

[9] GOODWIN, E. P. and J. C. WYANT. *Field guide to interferometric optical testing.* 1st ed. Bellingham: SPIE, 2006. ISBN 978-0819465108.

[10] UDD, E. and W. B. SPILLMAN. *Fiber optic sensors: an introduction for engineers and scientists.*

2nd ed. Hoboken: Wiley, 2011. ISBN 978-0-470-12684-4.

[11] SALEH, B. E. and M. C. TEICH. *Zaklady fotoniky*. 1st ed. Praha: Matfyzpress, 1995. ISBN 80-85863-05-7.

About Authors

Marcel FAJKUS was born in 1987 in Ostrava. In 2009 he received a Bachelor's degree from VSB–Technical University of Ostrava, Faculty of Electrical Engineering and Computer Science, Department of Telecommunications. Two years later, he received a Master's degree in the field of Telecommunications in the same workplace. He is currently a Ph.D. student, and he works in the field of optical communications and fiber optic sensor systems.

Jan NEDOMA was born in 1988 in Prostejov. In 2012 he received a Bachelor's degree from VSB–Technical University of Ostrava, Faculty of Electrical Engineering and Computer Science, Department of Telecommunications. Two years later, he received his Master's degree in the field of Telecommunications in the same workplace. He is currently a Ph.D. student, and he works in the field of optical communications and fiber optic sensor systems.

Petr SISKA was born in 1979 in Kromeriz. In 2005 he finished M.Sc. study at VSB–Technical University of Ostrava, Faculty of Electrical Engineering and Computer Science, Dept. of Electronic and Telecommunications. Three years later, he finished Ph.D. study in Telecommunication technologies. Currently he is an employee of the Department of Telecommunications. He is interested in Optical communications, Fiber optic sensors and Distributed Temperature Sensing systems.

Lukas BEDNAREK was born in 1988 in Frydek-Mistek. In 2011 he received a Bachelor's degree from VSB–Technical University of Ostrava, Faculty of Electrical Engineering and Computer Science, Department of Computer Science. Two years later he received a Master's degree in the field of Telecommunications at the Department of Telecommunications. He is currently a Ph.D. student, and he works in the field of optical communications and aging of the optical components.

Stanislav ZABKA was born in 1988 in Myjava, Slovakia. In 2010 he received a Bachelor's degree from VSB–Technical University of Ostrava, Faculty of Electrical Engineering and Computer Science, Department of Information and communications technologies. In 2013, he received a Master's degree in the field of Information technologies in the same workplace. In 2016, he will start as a Ph.D. student, and he is currently working as a software developer in an international IT company in the area of energy utilities, distribution processes a communication.

Vladimir VASINEK was born in Ostrava. In 1980 he graduated in Physics, specialization in Optoelectronics, from the Science Faculty of Palacky University. He was awarded the title of RNDr. At the Science Faculty of Palacky University in the field of Applied Electronics. The scientific degree of Ph.D. was conferred upon him in the branch of Quantum Electronics and Optics in 1989. He became an associate professor in 1994 in the branch of Applied Physics. He has been a professor of Electronics and Communication Science since 2007. He pursues this branch at the Department of Telecommunications at VSB–Technical University of Ostrava. His research work is dedicated to optical communications, optical fibers, optoelectronics, optical measurements, optical networks projecting, fiber optic sensors, MW access networks. He is a member of many societies - OSA, SPIE, EOS, Czech Photonics Society; he is a chairman of the Ph.D. board at the VSB–Technical University of Ostrava. He is also a member of habitation boards and the boards appointing to professorship.

6

On the Highly Stable Performance of Loss-Free Optical Burst Switching Networks

Milos KOZAK[1], Brigitte JAUMARD[2], Leos BOHAC[1]

[1]Department of Telecommunication Engineering, Faculty of Electrical Engineering, Czech Technical University in Prague, Technicka 2, 16000 Prague, Czech Republic

[2]Department of Computer Science & Software Engineering, Faculty of Engineering and Computer Science, Concordia University, 1515 Rue Sainte-Catherine, H3G 2W1 Montreal, Canada

milos.kozak@fel.cvut.cz, bjaumard@cse.concordia.ca, bohac@fel.cvut.cz

Abstract. *Increase of bandwidth demand in data networks, driven by the continuous growth of the Internet and the increase of bandwidth greedy applications, raise the issue of how to support all the bandwidth requirements in the near future. Three optical switching paradigms have been defined and are being investigated: Optical Circuit Switching (OCS); Optical Packet Switching (OPS); and Optical Burst Switching (OBS). Among these paradigms, OBS is seen as the most appropriate solution today.*

However, OBS suffers from high burst loss as a result of contention in the bufferless mode of operation. This issue was investigated by Coutelen et al., 2009 who proposed the loss-free CAROBS framework whereby signal convertors of the optical signal to the electrical domain ensure electrical buffering. Convertors increase the network price which must be minimized to reduce the installation and operating costs of the CAROBS framework. An analysis capturing convertor requirements, with respect to the number of merging flows and CAROBS node offered load, was carried out. We demonstrated the convertor location significance, which led to an additional investigation of the shared wavelength convertors scenario. Shared wavelength convertors significantly decrease the number of required convertors and show great promise for CAROBS. Based on this study we can design a CAROBS network to contain a combination of simple and complex nodes that include none or some convertors respectively, a vital feature of network throughput efficiency and cost.

Keywords

CAROBS, hypothesis testing, merging flows, Optical Burst Switching, routing, stationary signal.

1. Introduction

Not as long ago, twisted pairs were replaced by optical fibers in order to ensure higher bandwidth and reach greater distances. Further, Wavelength Division Multiplex (WDM) currently uses already deployed fibers more efficiently supporting wavelengths at 100 Gbps. However, network node power consumption increases as wavelength bandwidth increases which is the result of the switching paradigm residing in the electrical domain using electronic cross-connects [1] where the optical signal must be converted to the electrical domain to be routed. This approach is recognized as the point-to-point network topology relying on Optical-Electrical-Optical (OEO) conversion. The OEO conversion allows reaching long distances and designing mesh topologies but the network performance is limited as it does not allow groom different optical signals [1], which is necessary to decrease operational cost. Therefore, a new strategy incorporating all-optical bypass started to be used as the way of decreasing network operational cost [1] and [2]. This strategy opened a new area of optical networking: All-optical networks.

All-optical networks can be realized with different switching granularities. Optical Circuit Switching (OCS), which is currently deployed as a part of SDH/Sonet and IP over WDM networks, is characterized by switching at the level of wavelengths, with lightpaths usually being long-term settings. OCS paradigm uses bypass nodes, however, grooming is carried out electronically. On the other side of the spectrum, Optical Packet Switching (OPS) allows all-optical grooming while providing sub-wavelength switching of small packets. However, OPS relies on fast switching cross-connects that are not affordable for production (at least as for today). In a nutshell, Optical Burst Switching provides sub-wavelength granularity in the optical domain thanks to so-called bursts where a burst is defined as a set of continuous packets destined to a common egress point, so that the optical cross-connect does not need to change frequently or quickly. OBS allows very fast switching without any burst OEO, and only the burst control message [3] undergoes OEO. OBS is a promising and mature technology that is currently being tested by some ISPs in the field [4], [5] and [6].

However, the perennial always mentioned weakness of OBS is the high burst loss caused by the combination of burst contention and congestion: when burst contention occurs, only one burst is switched, others are dropped, and the dropped bursts are sentenced by the burst priorities or arrival times. Various concepts minimizing burst loss have emerged, with various approaches for resolving contention, e.g., various routing strategies [7] and [8], wavelength conversion [10] and time-slots [10] and [11], as well as some zero burst loss concepts with the addition of electrical buffering [12] and [13], or for particular topologies [13]. The CAROBS framework proposed by Coutelen et al. [13] underpins these concepts, combines Core and Edge node architectures, brings all optical grooming, allows wavelength conversion with recourse to OEO conversion, and uses electrical buffering to provide a loss-free mode of operation. From our perspective, the CAROBS framework is very promising framework for future deployments of asynchronous OBS networks.

The motivation of this paper is to evaluate the viability of CAROBS in terms of CAPEX, OPEX and performance parameters. We carry out comprehensive set of simulations in order to enumerate the number of optical to electrical (O/E) conversion blocks that are essential for loss-free paradigm, but the O/E blocks are recognized as the main parameter influencing CAPEX, OPEX of CAROBS networks. These O/E blocks are used for optical signal conversion which is essential for burst buffering and most importantly for potential optical signal regeneration. Optical signal regeneration is crucial when a certain number of the optical signal amplification is met. When certain distance is crossed the optical signal is amplified otherwise it is difficult to convert it from optical to electrical domain. However, every amplification increases optical signal noise so at some point it is not possible the optical signal only amplify, but it must be regenerated. In terms of CAROBS deployment, a heuristic approach on the optical signal regeneration was investigated by Kozak et al. [15]. We suggested the CAROBS control plane redefinition such that we used the electrical buffering for the purpose of optical regeneration, OEO conversion. This paper disclosed issue of standard dimensioning approach based on the shortest path routing and LAUC-VF [16] for burst scheduling in CAROBS. Consequently, an in-depth analysis of CAROBS behaviour to quantify requirements on electrical buffering is carried out in this paper. The results of this study are formulated in a way that could be used in future research for an OBS deployment integrating the CAROBS framework.

The paper is organized as follows. Section 2. reviews recent work on OBS networks with a focus on buffering and the OBS loss-free paradigm. Section 3. briefly recalls the features of the CAROBS architecture and buffering behaviour, and redefines the architecture in terms of Queueing theory to capture the significant aspects of model traffic behaviour when buffering is assumed. In Section 4. , issues arising as a result of buffering are discussed in great details. In Section 5. , we describe the configuration of the simulator and the simulation approaches. Section 6. discusses the results of our simulations. Conclusions are drawn in the last Section.

2.   Literature Review

The prime motivation of all-optical frameworks, including OBS, is to avoid an OEO grooming bottleneck, which would prevent high transmission speed. OBS uses sub-wavelength scheduling, so burst contentions causing performance deterioration might occur, even with optimized routing, reaffirming the need to minimize them. Recent studies designed time slotted OBS architectures as one way to avoid burst contention, see, e.g. [17] and [18]. Other studies returned to the principals of all-optical networks, with wavelength routing for bursts [19]. In most cases, these architectures rely on a ring topology or a mesh topology with a global in advance signalling to reserve a channel in an OCS-like manner, and some studies tackle synchronous transmissions over OBS network [20]. These solutions are either less scalable or less efficient from the perspective of network performance compared to dynamic just-in time signalling used in OBS and OPS [1]. All these architectures are compliant with the original OBS definition [3] and do not introduce any buffering.

Pavon-Marin et al. [21] carried out an in-depth analysis of buffer-less OBS architectures, and concluded that buffer-less OBS architectures are not viable for mesh topologies. According to [21], the limiting factors of OBS are an inter-burst gap, a separate control wavelength and optical contention resolution. In order to overcome OCS for bursty traffic, the OBS paradigm must be changed [21]. The separate control channel and inter-burst gap cannot be omitted unless a time-slotted approach is used [17]. Therefore, buffering seems to be the only way to increase OBS network performance. Unfortunately, very little work has been done on OBS buffering.

Some early concepts use fiber delay lines (FDL) [22] and subsequent papers deal with dimensioning

FDLs [23]. Utilizing lengthy fibers is problematic on most premises, therefore some authors investigated scenarios utilizing electrical memories [12] and [13]. Among these works [12], [24] and [13], there is the CAROBS framework proposed by Coutelen et al. [13]. Coutelen et al. published a series of papers dealing with node architecture and its performance, but did not devote any attention to the dimensioning of a network consisting of CAROBS nodes. Traditionally, dimensioning of OBS networks has been carried out using $M/M/k/k$ models [25] and [26]. However, this analytical model is not accurate as it does not reflect the streamline effect [27] and [28], which is why the burst blocking probability (BBP) seems to be higher than OBS mathematical models results. The streamline effect is the phenomenon unique to OBS networks wherein bursts traveling in a common link are streamlined and do not content with each other until they diverge [27]. When flows are merged, burst contention arises when there are incoming flows on the same wavelength. The ramification of the streamline effect is that a non-merging flow offering a given load level to a node should result in a BBP level according to $M/M/k/k$, however, thanks to the streamline effect, the BBP is 0. This result is known for classical OBS, but has not been studied for buffering OBS frameworks.

The buffering OBS is unique due to the so-called secondary contention. It occurs when a burst is scheduled from electrical memory back to the optical domain on a given wavelength at the same time as an incoming burst is requesting the same wavelength. In such a case, the new burst is buffered. It means that neither the $M/M/k/k$, nor the streamline effect model work for buffered OBS exclusively. Secondary contention was studied by Delesques et al. [29] using the Engset model. Their main concern was the buffer size dimensioning. However, they tackled buffering probability (BP) as well. BP is a comprehensive parameter of a buffered OBS network. It cannot be easily quantified by $M/M/k/k$ models, with respect to the streamline effect, as for burst loss probability in a regular OBS network. Therefore, the $M/M/k/k$ formulation with streamline effect and secondary contention must be combined when buffering OBS to provide the buffered OBS node model.

In this paper, we focus on a simulation approach to obtain experimental properties of buffering OBS nodes, i.e., the CAROBS node. These results can also be used for mathematical modelling.

In the next Section, we describe the CAROBS node architecture and internal processes that lead to buffering (loss-free) behaviour. We also describe the reformulation of a CAROBS node with respect to the queueing system terminology, in order to incorporate it to the proposed mathematical model described in Section 4.

3. CAROBS Model Description

The focus of this Section is on the CAROBS model with respect to the ways CAROBS resolves contention. CAROBS relies on electrical buffering as a way of avoiding burst loss. There are other concepts using electrical buffers [12] and [24], but all of these architectures are very complex . Our view is that CAROBS incorporates electrical buffering in an efficient way. It extends the classical Core node architecture and adds a buffering property using a software plane. It was defined by Coutelen in 2010 [13]. One of the most significant changes is how it merges Edge and Core node architecture into one CAROBS node. The CAROBS node architecture is depicted in Fig. 2. Thanks to the new architecture, CAROBS can ensure all-optical grooming via a new transmission mechanism called burst train which contains a number of bursts, here called cars, whose destination is along the same flow-path of the most distant destination node as seen in Fig. 1. The term flow-path represents a temporary lightpath, signalled by the JET mechanism, during the burst train. The burst train concept preserves the mandatory inter-car gap, to allow all-optical grooming. CAROBS all-optical grooming is achieved through the head drop as depicted in Fig. 1. Effective transmission in the form of burst trains is ensured by the Curbet Train Algorithm (CTA) [13] that optimally justifies car length so that the gap between two consecutive cars is equal to the mandatory inter-car gap. The burst train is signalled by one CAROBS header which contains the same information as the original Burst Header Packet [3] and adds the section containing information for each car [13]. A CAROBS car supports transmission of the same traffic as a burst in OBS. Each car contains aggregated data for only one destination edge node. In short, the burst trains concept improves OBS network performance [13].

Fig. 1: An example of the CAROBS train structure containing cars for each intermediate node along the longest flow path. All optical grooming removes the head car(s) at intermediate nodes.

The CAROBS concept relies on the node architecture as depicted in Fig. 2. The WDM demultiplexer and multiplexer were omitted to make the illustration easier to read. The horizontal lines represent the input and output ports, and ports are labelled by λ. λ_1 represents the dedicated wavelength for the control channel and λ_x represents the wavelengths used for optical transmission. The CAROBS node architecture spans three logical layers. On the top there is the control plane that contains the *SOA Manager* which reads CAROBS headers and determines further node actions. The CAROBS node may either switch the whole burst train, groom-out the first car and switch the remaining part of the burst train, or, buffer the whole burst train because of contention. Based on this decision, the *SOA Manager* creates a set of instructions for the *SOA Switching Matrix* (MX) and forwards the CAROBS header to the next CAROBS node. The middle layer contains the aggregation and disaggregation ports for user traffic. The term user traffic represents traffic in the neighbourhood of the location of the CAROBS node. The most important block is the Media Access Control (MAC) on the middle layer. The MAC uses the CTA algorithm for car alignment in the burst train, stores the contenting burst trains in the internal electrical memory and re-aggregates the buffered cars. The bottom layer represents the physical layer where all cars are switched. It contains the MX that ensures the switching of the optical signal. If contention occurs the contenting burst trains are switched to the port dedicated for electrical buffering. These ports are directly connected to the O/E that convert the optical signal to the electrical domain where the contenting burst is buffered in MAC memory.

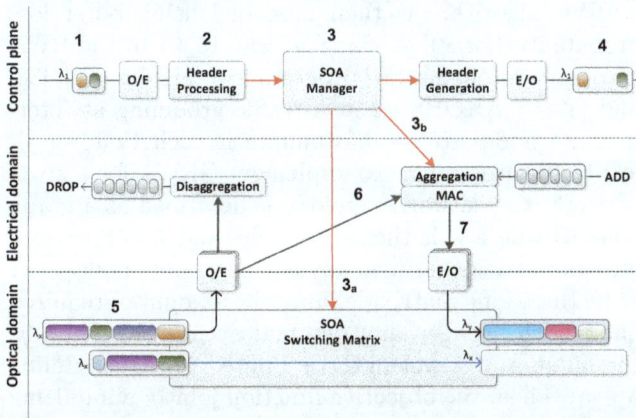

Fig. 2: CAROBS node resolves burst contention with recourse to an electrical buffer. The contenting burst train is sent to the MAC and stored until the output direction is available. The CAROBS header of the contenting burst is also modified as a result of delay buffering.

The contention resolution process is showed in Fig. 2. When a new CAROBS header reaches an input port of the CAROBS node (1) it is detected and processed (2). All relevant information is used by the *SOA Manager* which creates the MX configuration. If the burst train overlaps with a previously scheduled burst train, the burst train is buffered in order to avoid burst contention. In this case, the *SOA Manager* first calculates the buffering delay using the LAUC-VF (Latest Available Unused Channel with Void Filling) algorithm [16] then creates two instructions: one for the MX and the second for the MAC. The first instruction switches the contenting burst train to the dedicated port used

for buffering (3_a). The second instruction informs the MAC for how long the burst train is to be buffered (3_b). Immediately following these two set instructions, the CAROBS header is scheduled and sent toward the next node. The CAROBS header is delayed by the same amount of time as the one by which the burst train is buffered in the MAC. When the contenting burst train arrives at the input port (5) it is deflected to the MAC and stored there (6). When the buffering time is up, the burst train is re-aggregated to the MX (7) and sent toward the next node.

3.1. Dimensioning Model

The description of the buffering process leads to the dimensioning problem of the CAROBS node. At first glance, the optimization of the CAROBS node might seem to be easy using the Erlang C formula to calculate the BBP, which is the same as BP. However, there are two perennial shortcomings: the Erlang C formula only works for systems with buffering before the service [30]. Moreover, the contenting burst cannot be buffered unless there are enough O/E blocks. It means that there is service before the buffering and this service must be a priori optimized before buffering can be optimized. Therefore, for the purpose of further discussion, we have reformulated the CAROBS node architecture, see Fig. 2, using tools of Queueing theory [30], see Fig. 3. Such a redefinition is vital to separate the buffering problem and the O/E block availabilities into two systems that can be tackled individually.

We define three building blocks, the *SOA Switching Matrix* that ensures optical signal switching and two Queueing systems (QS). For the sake of simplicity, we define them as the Input QS (IQS) and Output QS (OQS). The IQS tackles the contenting burst trains through a limited number of O/E blocks. The number of required O/E blocks depends on the IQS offered load. The offered load of the IQS can be quantified as $\alpha_{BUF} = BP \cdot \alpha$, where α represents the total node offered load, α_{BUF} represents the offered load to the IQS, and BP represents the buffering probability. Using the α_{BUF} values, the number of O/E blocks can be calculated with respect to the streamline effect which means the streamline effect evaluation must be carried out for less than five merging flows [27].

If there is more than five merging flows, the M/M/k/0 model can be applied [27]. Then we can use the Erlang B formula to obtain the BBP of IQS. The evaluation of the BBP enables us to obtain an approximation of the number of O/E blocks that are necessary to provide the loss-free mode. The OQS provides the burst train buffering and allows the traffic from connected networks α_{AGG} to be aggregated. The behaviour of both QS is driven by the *SOA Manager*. Since the

OQS receives burst trains in the form of an electrical signal it can store them in electrical memory. Current electrical memories provide only limited space, i.e. a limited number of burst trains can be stored. However, for the sake of simplicity and the CAROBS proposal [13] compliance, we assume that electrical memory is unlimited in this paper for all experiments.

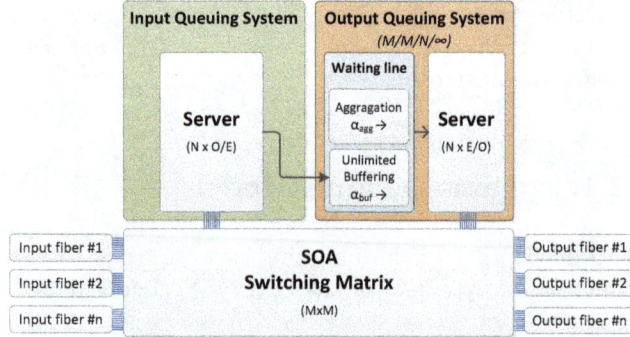

Fig. 3: Simplified block structure of the CAROBS node architecture in terms of QS [30]. It contains two QS and one SOA Switching matrix. The IQS is responsible for the optical signal detection and sends the burst trains to the electrical buffer once they are detected (O/E). The second QS controls the buffered trains and schedules them along the input traffic to the optical layer through a limited number of lasers (E/O) using LAUC-VF.

Every model relies on a number of approximations, the most crucial approximations in this analysis relate to the input traffic characteristics. The input traffic can be modelled using different input arrival processes and distributions of packet size. Some models are applicable only to a specific input traffic distribution or packet size distribution while some of them are more generic. In this analysis, we assume that the packets from the connected network arrive following a Poisson distribution at a rate of α packets per second so that the inter-arrival time between packets is a negative exponential distribution with parameter α. Depending on the car triggering [13], a burst train is created. CAROBS uses both triggering types (time, space) [13], therefore, car assembly tends toward Gaussian distribution asymptotically [26] according to the central limit theorem [31]. Using Kendall's notation [30], we classify the OQS as $M/M/N/\infty$ where N is the number of E/O blocks, see Fig. 3.

4. Dimensioning Problem Formulation

In the previous Section, the buffering process was described with its constraint represented by a limited number of O/E blocks. This Section deals with the traffic routing problem that is bounded to the number of O/E blocks. The key characteristics of the CAROBS

framework is its loss-free mode of operation, i.e., no burst must be dropped.

In other words, there are always enough O/E blocks when contention occurs. On the other hand, O/E blocks are expensive from both an operational and installation perspective, hence it is reasonable to minimize their number. Minimizing the number of O/E blocks implies that the effects of merging flows and secondary contention must be a priori minimized. As long as the secondary contention is the only effect of merging flows, it can be taken care by minimizing the merging flow effects. Therefore, we next focus on the classification of merging flows and their impact on burst buffering.

Currently, traffic in all-optical networks is distributed in a network using both the routing and wavelength assignment (RWA) and grooming RWA (GRWA) approaches [32], [33] and [34]. In OCS networks, RWA is vital because there is no sub-wavelength scheduling; bandwidth sharing is achieved using traffic grooming in the electrical domain, so there is no contention in optical domain. In the design of buffer-less OBS networks, RWA is used extensively as well [8], [35], [36] and [37]. The performance of such a RWA algorithm is then modelled using burst loss probability (BLP). A classical way to write the RWA through a mathematical program is recalled in e.g., [38] and [39]. CAROBS ensures traffic grooming at intermediate nodes so we reformulate it such that we allow traffic grooming. To implement GRWA for a given network, the network topology is described as a graph $G(V, L)$ where V is the set of nodes and L is the set of directional links between any two connected nodes. Let D be the traffic matrix defining the amount of required bandwidth between any two nodes $s, d \in V$. Usually, the number of wavelengths is limited to $|\Lambda|$, assuming $\lambda \in \Lambda$. Then, the objective function jointly minimizing number of used wavelengths and nonprovisioned traffic is as follows:

$$\min \sum_{(s,d) \in V^2 : s \neq d} \sum_{\substack{\ell \in L \\ \lambda \in \Lambda}} y_{\ell,\lambda}^{sd} + \theta \sum_{(s,d) \in V^2 : s \neq d} e_{sd}, \quad (1)$$

where $y_{\ell,\lambda}^{sd}$ is a decision variable: it is equal to 1 if the required bandwidth flow ϕ from s to d is assigned on wavelength λ and link ℓ, and 0 otherwise. The e_{sd} is a variable representing how much of traffic could not be routed because there is not enough wavelengths in Λ to support all traffic D and θ is a objective function parameter. This objective function is subject to: For all $v, s, d \in V$,

$$\sum_{\ell \in \omega^+(v)} \phi_{\ell}^{sd} - \sum_{\ell \in \omega^-(v)} \phi_{\ell}^{sd} =$$

$$= \begin{cases} D_{sd} - e_{sd} & \text{if } v = s \\ -D_{sd} + e_{sd} & \text{if } v = d \\ 0 & \text{otherwise,} \end{cases} \quad (2)$$

where $\omega^+(v)$ is the set of egress links of v and $\omega^-(v)$ is the set of ingress links of node v.

$$\sum_{s,d \in V} \phi_\ell^{sd} \leq y_{\ell,\lambda}^{sd} C, \qquad \lambda \in \Lambda, \ell \in L, \tag{3}$$

$$\sum_{s,d \in V} \phi_\ell^{sd} \leq C, \qquad \ell \in L, \tag{4}$$

$$\sum_{s,d \in V} y_{\ell,\lambda}^{sd} \leq 1 \qquad \lambda \in \Lambda, \ell \in L, \tag{5}$$

$$\phi_\ell^{sd} \geq 0, e_{sd} \geq 0 \qquad \{s,d\} \in V, \ell \in L, \tag{6}$$

$$y_{\ell,\lambda}^{sd} \in \{0,1\} \qquad \{s,d\} \in V, \lambda \in \Lambda, \ell \in L, \tag{7}$$

where C is the wavelength bandwidth (assumed to be the same for all wavelengths).

The only drawback of this formulation is that it does not take care of the merging flows and the sub-wavelength granularity which is allowed by OBS. The number of merging flows should be minimized as much as possible to maximize the occupancy of the simple, already merged flows. Hopefully, it can be done in the online mode using load balancing algorithms maximizing the streamline effect (SLE) [27]. However, such an algorithm corresponds to a heuristic, hence, it does not provide a globally minimal solution from the perspective of the number of O/E blocks. In other words, heuristics do not provide minimal solution which shows the minimal CAROBS requirements of studied network. Additionally, the number of O/E blocks depends on the characteristic of the node offered load, implication of M/M/N/0 model [30]. The incoming burst trains can be classified as a process with random inter-arrival intervals [26], therefore, it may result in infinite waiting in the electronic memory for a certain level of offered load. The infinite waiting is caused by the aforementioned premise of infinite memory, however, in real networks it would result in a burst loss because of limited electronic memory. Consequently, for the minimization of the number of O/E blocks, the offered node load must be engineered to avoid excessive buffering.

The conclusion of this Section is as follows. Both QS must be dimensioned properly, otherwise burst loss can be experienced. Both QS dimensioning relates to the maximal CAROBS node load. It can be formulated as the input traffic intensity ρ, so that $\rho \equiv \alpha/\mu$. The CAROBS node stability condition is $\rho < 1$ which can be written as $\alpha < \mu$ [30]. Here, α stands for total node offered load and μ represents the node intensity of service, i.e., how much traffic a node can transmit. The node offered load α is equal to the sum of offered loads from each tributary flow. It is worth noting that the stability condition applies to the system with merging flows, otherwise applies to the SLE. For the SLE, the stability condition changes to $\rho \leq 1$.

In order to stabilize the GRWA formulation, we must modify Eq. (3) such that it limits the traffic routed to each output link and does not exceed the value of ρ for various numbers of merging flows. The constraint is reformulated as:

$$\sum_{(s,d) \in V^2} \phi_\ell^{sd} \leq y_{\ell,\lambda}^{sd} \cdot C \cdot K \tag{8}$$

$$\lambda \in \Lambda, \ell \in L_{\text{selected},v}, v \in V,$$

where L_{selected} represents the set of outgoing links of the merging node V, e.g., ℓ_0 in Fig. 4, which concentrates traffic from a number of merged flows and K is the coefficient of stationary threshold that ensures that the egress link is used efficiently $C \cdot K \to \rho_{\max}$, i.e., no excessive buffering nor burst loss occurs.

Constraints Eq. (8) imply that the maximum node load is limited. Then the stability condition is valid, additionally the electronic memory is not overloaded.

5. Simulations

Following the discussion on the optimization of the number of O/E blocks issues, BP cannot be estimated using either the M/M/k/k model, or the SLE, because both provide information about BBP but not about BP. There is a clear relation between BP and BBP, but the main difference is that BP is influenced by secondary contention. The BP value can be quantified either mathematically or empirically using simulations. In this paper, an approximation model relying on simulations, prior to the design of a mathematical model in the future, is discussed. We focus on the basic node behaviour under various conditions, using the topology depicted in Fig. 4. The most important node, the node under study, is marked as merging node v, see Fig. 4. There is also destination node d where all traffic flows, from sources $s_\bullet$, are destined. Traffic flows originate in source nodes. Four scenarios with different number of sources are evaluated. Two to five merging flows are evaluated. The maximum of five merging flows was chosen because the maximum node degree that is considered is six, i.e., five merging flows in [40]. Simulations were carried out using OMNeT++ simulator and CAROBS models [15]. Source nodes $s_\bullet$ were supplied with traffic generated according to a Poisson distribution. The generated payload packets of constant size (100 kb) defining the flow were supplied to aggregation queues to generate bursts. It is assumed that electrical storage capacity is unlimited. JET (Just Enough Time) [3] was used as a signalling protocol and LAUC-VF algorithm [16] for burst assembly.

Traffic analysis is quite comprehensive if one wants to reach a specific level of accuracy. In order to obtain accurate results, the number of simulations must

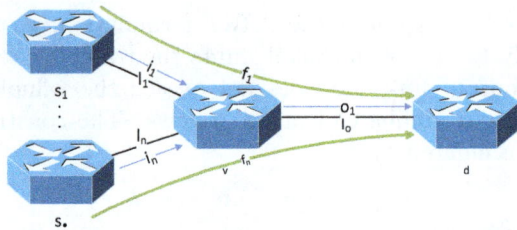

Fig. 4: The elementary topology used for one node behaviour evaluation. The number of sources was changed as is depicted here by $s_1, \ldots, s_\bullet$.

be as high as possible because the results accuracy is achieved through the repetition of identical simulations and by changing the input load patterns. However, the number of repetitions implicate the total simulation time; therefore, a limited number of simulations should be used.

In our simulations, we carry out 25 identical simulations with different patterns of node offered load. The main task of the simulations is to verify the impact of the number of merging flows (MF) and node load α on the buffering probability $BP(\alpha, MF)$ and buffering delay $BD(\alpha, MF)$ of a buffered burst train. The node load α is equal to the sum of loads offered by each tributary flow. Then, the average maximal offered load provided by each source is $1/(MF + 1)$ erl. The term average maximal offered load represents the average value of offered load among the identical simulations for one simulation at the given node offered load α.

In addition to the number of merging flows and the node load, an evaluation for wavelength data rate 1, 10 and 40 Gbps is performed. Wavelength data rate 100 Gbps was simulated, but the difference of results for 40 and 100 Gbps was negligible thus the result for 40 Gbps wavelength data rate are presented. Additionally, only the on-off keying modulation format was implemented into CAROBS model; therefore, results for 100 Gbps are not conclusive and presented. In Subsection 6.2. , we investigate further the O/E block sharing in WDM networks to verify the impact of the wavelength number $|\Lambda|$ on $BP(\alpha, MF) \rightarrow BP(\alpha, MF, |\Lambda|)$ and $BD(\alpha, MF) \rightarrow BD(\alpha, MF, |\Lambda|)$. We vary the number of wavelengths $|\Lambda|$ from 1 up to 60.

6.   Results

First, we deal with simulations restricted to only one wavelength where we vary wavelength data rates. Therein, we first focus on the stability of measured parameters and their confidence. Based on the results we formulate recommendation for the number of O/E blocks in order to ensure loss-free mode of operation. Also, these results open question about viability

of one wavelength systems, i.e., wavelength sensitive O/E blocks.

Subsequently, in Subsection 6.2. , we evaluate the same simulation scenario for multiple wavelengths (WDM mode of operation) and colour-less O/E blocks. These results favors the WDM in CAROBS networks because WDM mode decreases number of O/E blocks that are necessary for the same load as for one wavelength scenario significantly. Additionally these results open a very promising deployment scenario that is unique for CAROBS.

6.1.   One Wavelength Evaluation

One-wavelength transmission systems were common before the emergence of WDM systems. Here we return to the one-wavelength system because of its simplicity, i.e., less degree of variance in the analysis. Less degree of variance means a smaller number of parameters that can change. Such a (lower) number of parameters is vital for the stability discussion of results as other parameters are fixed. The one-wavelength analysis consists of two main steps. In the first step, we remedy the number of identical simulations, so the following results are a trade-off between accuracy and the overall time of running simulations. In the second step, we repeat the predefined simulation n times and then we evaluate the results.

In order to find the trade-off between accuracy and overall simulation running time, simulations for three different node loads $\alpha = \{0.1, 0.5, 0.9\}$ erl are conducted. For each of these three-node loads, 2,000 simulations with different offered load patterns are performed. Additionally, we conduct this analysis for data rates 1, 10, and 40 Gbps. Only the stationary simulations are used, and simulations are tested using the mathematical tool of hypothesis testing. The signal is stationary when the values of BD and BP do not depend on time, i.e., statistical properties do not change during the simulation. This test is applied to the whole simulation time interval and in addition to the sub-intervals. Each sub-interval contains 120 samples. In the test, it is assumed that the time dependent values of BD and BP can be approximated using a linear regression line, which is quantified by time vector X, the coefficients of linear regression b_0, b_1, and the coefficient of linear regression error e_i which must be minimal. Then, we can formulate a null hypothesis $H0$ claiming that the signal is stationary if every sub-interval is stationary. Sub-interval stationarity is characterized by coefficient $b_1 = 0$. If $b_1 \neq 0$, then such a sub-interval is not stationary, so the whole time interval, i.e., the simulation cannot be used for further stability analysis. The core of the linear regression analysis is shown in Eq. (9), Eq. (10), Eq. (11) and Eq. (12).

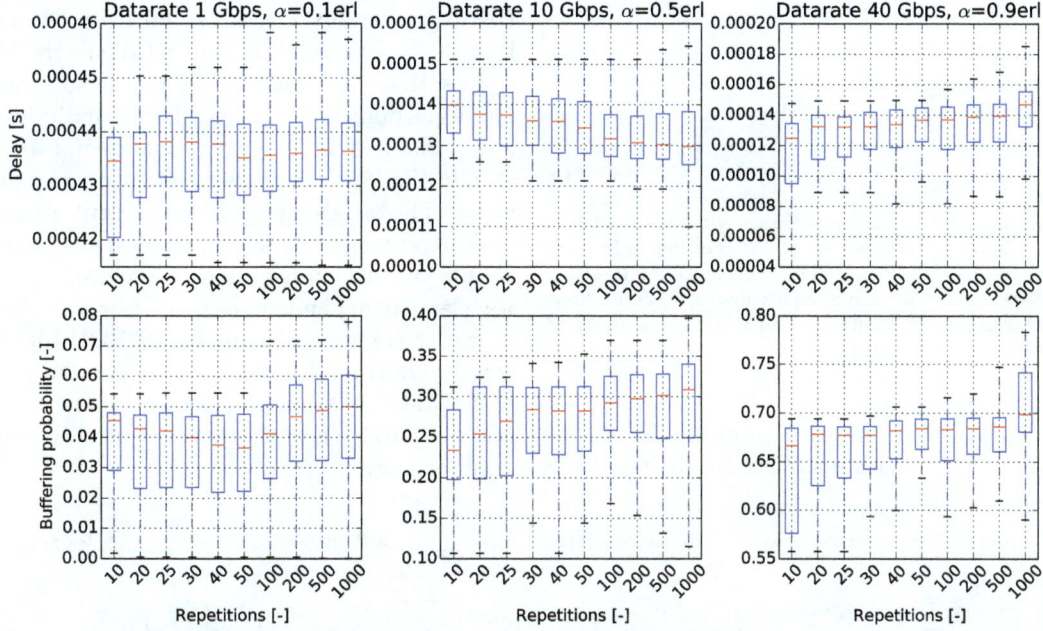

Fig. 5: Statistical properties of BD and BP captured using boxplots for various node loads and various data rates. In the upper row, characteristics of buffering delay are depicted. In the bottom row, characteristics of buffering probability for a different number of identical simulations are shown.

$$Y_i = b_0 + b_1 x_i + e_i \qquad i = 1, 2, \dots, n, \qquad (9)$$

$$b_1 = \frac{n \sum_{i=1}^{n} x_i^2 Y_i - \sum_{i=1}^{n} x_i \sum Y_i}{n \sum_{i=1}^{n} x_i^2 - \left(\sum_{i=1}^{n} x_i \right)^2}, \qquad (10)$$

$$b_0 = \frac{\sum_{i=1}^{n} Y_i - b_1 \sum_{i=1}^{n} x_i}{n}, $$

$$s^2 = \frac{\sum_{i=1}^{n} Y_i^2 - b_0 \sum_{i=1}^{n} Y_i - b_1 \sum_{i=1}^{n} x_i Y_i}{n - 2}, \qquad (11)$$

$$\frac{|b_1| \sqrt{\sum_{i=1}^{n} x_i^2 - n \bar{x}^2}}{s} \geq t_{n-2}(\Psi), \qquad (12)$$

where n is the number of verified samples in the sub-interval and the number of sub-intervals for the hypothesis testing of the whole simulation interval. Then, the decision on stationarity is valid with a level of confidence Ψ. If the critical value Eq. (12) is higher than the coefficient of Student's distribution, the hypothesis $H0$ does not apply, i.e., BD or BP is not stationary and the simulation cannot be used for further evaluations. The valid set of simulations is then used for the evaluation of the number of identical simulations. This approach is also used in the next analysis of the maximal node load.

In order to find the minimal number of identical simulations, which are necessary, we perform 2,000 simulations for each case (combination of load and data rate)

and evaluate these results. The statistical properties of datasets representing various numbers of identical simulations can be seen in Fig. 5. The statistical properties do not change after 1,000 identical simulations; therefore, for the sake of readability, we did not depict cases for more than 1,000 identical simulations in Fig. 5. One can see that in most cases after 25 identical simulations the mean value and variance do not change significantly against their values with 1,000-simulation case. Therefore, we chose to have 25 identical simulations for all our simulations in the paper. As long as there are only 25 identical simulations, the results can be corrected using the coefficients of Student's distribution.

The impact of the number of merging flows to the BD and BP is an extremely important aspect of the analysis which is carried out using the dataset containing 2,000 simulations. For the sake of simplicity, only the case for the node load 0.5 erl is depicted, however, all other simulation schemes led to the same conclusion that BD does not depend on the number of merging flows, see Fig. 6, only on the node load, see Fig. 8(a). This conclusion comes from the Poisson character of the merged flows. Therefore, in the delay analysis, we can assume $BD(\alpha, MF) \equiv BD(\alpha)$.

The same dataset of 2,000 different patterns was used for the BP analysis with the results for node load 0.5 erl depicted in Fig. 7. Other node load results led to the same conclusion. According to the results depicted in Fig. 7, it can be seen that above 2 MF the difference of BP is negligible, therefore, we can only define BP for two or more MF as two different parameters in

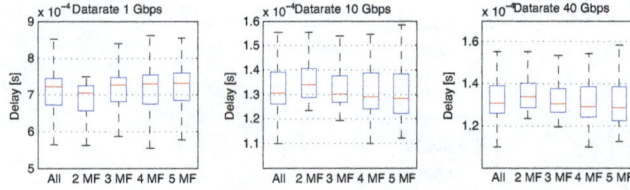

Fig. 6: Buffering delay test of dependance for node load $\alpha =$ 0.5 erl showing Buffering delay does not depend on the number of merging flows. The routing policy does not avoid scenarios with a higher number of merging flows for the same level of offered load when the end-to-end delay is the main concern.

the next analysis. This is indirect contradiction to the $M/M/k/k$ with an inclination to the SLE for small MF scenarios.

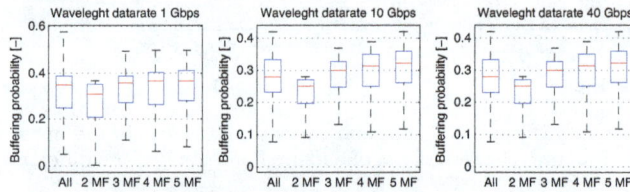

Fig. 7: Buffering probability dependance test on the number of merging flows. The test was carried out for the node load $\alpha =$ 0.5 erl. This test shows that the Buffering probability does not depend on the number of merging flows for $MF \geq 3$.

In the performance analysis of the CAROBS system using one wavelength, 25 schemes were performed with a variety of offered load patterns for load $\alpha =$ $[0.1; 1.02]$ erl with equidistant step 0.02 erl for $\alpha =$ $[0.1; 0.95]$, and with step 0.01erl for $\alpha = [0.95; 1.02]$ erl. The scheme using 2-5 MF and three data rates 1, 10 and 40 Gbps of a wavelength was retained and BD and BP stationary tests were evaluated in order to obtain the stationary threshold K which defines the maximal load when buffering is bounded. The values determining the stationary thresholds are captured in Tab. 1. Therein, a column F_{BW} is added to expresses the stationary threshold in terms of the bandwidth that must be free to avoid excessive buffering.

Tab. 1: Table of load thresholds and spare bandwidth. F_{BW} represents the spare bandwidth, which cannot be used to keep the node stable in a long-term perspective.

MF	1 Gbps		10 Gbps		40 Gbps	
	K	F_{BW}	K	F_{BW}	K	F_{BW}
2	0.89	110 Mbps	0.93	700 Mbps	0.96	1.6 Gbps
3	0.88	120 Mbps	0.89	1.1 Gbps	0.89	4.4 Gbps
4	0.88	120 Mbps	0.89	1.1 Gbps	0.89	4.4 Gbps
5	0.87	130 Mbps	0.88	1.2 Gbps	0.88	4.8 Gbps

The values of BD(α) and BP(α, MF) are seen in Fig. 8. The differences among BD(α) for various wavelength data rates in Fig. 8(a) are shown. When the end-to-end delay is a routing concern, it can be seen that the node load cannot exceed ≈ 0.9 erl, otherwise

BD exponentially increases. Further, the value of the load ≈ 0.9 erl is equal to the stationary threshold which means that the node could not ensure the loss-free mode permanently. It would eventually lead to burst loss. The second monitored parameter BP(α, MF) is seen in Fig. 8(b) where we depict only two and five MF cases. The notably high BP is evident even for the low node load from this figure. Such a situation is not vital for production networks; therefore, we assume that wavelength dependent optical detector (O/E) blocks do not pave the path to the CAROBS in WDM networks. On the other hand, this situation provides very useful data that can be used for further analysis as an upper bound when the number of O/E blocks is the main concern.

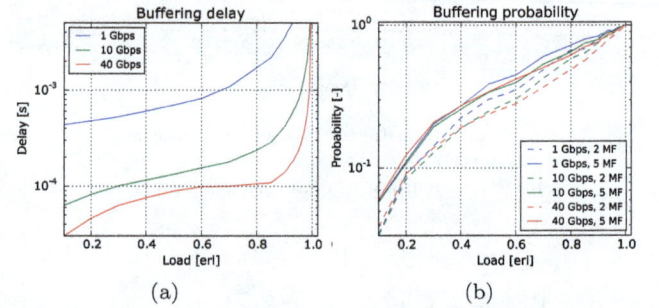

(a)   (b)

Fig. 8: The BD and BP rely on the value of the offered load. Below 0.9 erl the BD is in scale of μs, however, above 0.9 erl, it significantly increases. The BP is more proportional to the value of the offered load α. Therefore, when engineering the number of regenerators, BP is the main objective.

The accuracy and results of BP are crucial when estimating O/E blocks. Each O/E block can be used by only one burst at a time and only one burst train can come at a given time period because wavelength can carry only one single burst train at a moment. Subsequently, the number of buffered bursts depend on the offered load α and the number of MF, BP(α, MF). The BP(α, MF) specifies the probability a new incoming burst train will be blocked by another burst train (burst contention); in the worst-case scenario, the new incoming burst train can be blocked by a rescheduled burst train (secondary contention) [41]. The worst-case scenario results in the corner case of equal numbers of O/E blocks and MF, and the O/E block measurement is depicted in Fig. 9. Notice, it is necessary to install at least $MF - 1$ O/E blocks even for a low load. These graphs quantify the O/E block requirements so they can be used in further studies of CAROBS GRWA. The O/E block measurement reveals a high demand of the number of O/E blocks, even for a low load, therefore, it is not viable for real CAROBS deployment in WDM networks where deployment requires a specific number of O/E blocks per wavelength depending on the α at the wavelength.

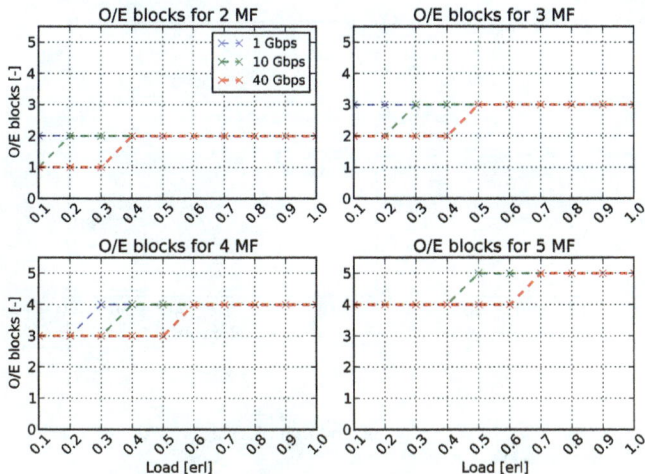

Fig. 9: Dependance of the number of O/E blocks for various MF scenarios and wavelength data rates. The most resilient case is for 2 MF. This scenario is also the cheapest, though sometimes at the expense of wavelength greediness. The evaluation of $R(\alpha, MF)$ is calculated from Fig. 8(b).

The total number of O/E blocks required for the CAROBS network and given traffic D can be calculated through the CAROBS GRWA extension which gives information about virtual routing (different routing for different wavelengths). Each wavelength can have a different number of merging flows (MF_λ) and a different node offered load α_λ. The evaluation of the number of O/E blocks can be formulated as follows:

$$\sum_{\substack{v \in V \\ \lambda \in \Lambda}} R(\alpha_{v,\lambda}, \mathrm{MF}_{v,\lambda}), \qquad (13)$$

where function R represents the graphs depicted in Fig. 9, i.e., the requirement on the O/E blocks to deliver the loss-free mode of operation.

Such a formulation can be used for CAPEX or OPEX studies where it can define the part of objective function in order to minimize monetary sources while maintaining an appropriate quality of transmission.

6.2. Multiwavelength Evaluation

Following the results obtained in the previous Section, the focus shifts to only the BD, BP, and colourless O/E blocks. First, the stationary threshold of the CAROBS WDM system with a various number of wavelengths $|\Lambda|$ will be evaluated the the required number of O/E blocks will be enumerated. The term "Load" will be used in all figures, however, the meaning is slightly modified compared to the previous Section where it meant the total utilization of one wavelength. From now on, the term "Load" represents the overall utilization of all wavelengths in a specific link – link utilization. For example, a link supporting 10

wavelengths, where only one wavelength is utilized by a 1 erl traffic, according to the notation used in the previous Section, means that Load is equal to 0.1 erl. K is redefined the same way as Load.

The stationary threshold of CAROBS WDM was evaluated based on the dataset of the CAROBS WDM system where the $|\Lambda|$ was varied, and the number of merging flows MF and node offered load α was changed. The accuracy of each step, as defined in Subsection 6.1. , is evaluated using the mean value analysis (MVA) approach. The MVA was carried out so the final value is uncertain with less than 5% of probability. The dependance of the coefficient $K(|\Lambda|)$ is illustrated in Fig. 10.

The inclination of coefficient $K(|\Lambda|)$ to the value one is seen; however, in the studied range of wavelengths it does not meet it. It results into the gap of bandwidth F_{BW} that cannot be used for static traffic, but this gab of bandwidth can be used for frequently bursting short term flows which cannot result in excessive buffering. The values of coefficient $K(|\Lambda|)$ delimit the working area where the CAROBS WDM system can be provisioned. Subsequently, the graphs of BD and BP are depicted in Fig. 11. BP is captured for two and five MF in Fig. 11a), Fig. 11b) Fig. 11c), $BD(\alpha, |\Lambda|)$ is depicted in Fig. 11d) Fig. 11e) Fig. 11f) without respect to the number of MF, because of the Poisson character of merging flows, see Subsection 6.1. Both BP and BD improved significantly as the number of wavelengths $|\Lambda|$ increased. This improvement in values of BP and BD is an excellent indicator that further research on colour-less O/E blocks is the right direction.

The gap between the two and five MF scenarios is worth studying as it takes on importance as the wavelength data rate increases. The gap is the direct result of SLE, i.e., suppressed secondary contention. Presumably, it can be used to minimize the number of deployed O/E blocks in the network. Then the objective can be formulated as minimization of MF and maximization of the stream lining. The contribution of this Section is the prove that SLE is highly important for the CAROBS node and its performance.

The number of O/E blocks relates to BP, and the graphs are depicted in Fig. 12. The one-wavelength scenario is captured here to depict the upper bound described in the previous Section. We can see that O/E blocks sharing among wavelengths can significantly reduce their necessity for the same level of offered load. It is significant to note that it is not necessary to install any O/E blocks for buffering up to a specific level of node offered load. Such a threshold can be used in order to design simple CAROBS nodes with minimal requirements, and such a node could be deployed into distant areas. On the other hand, this approach al-

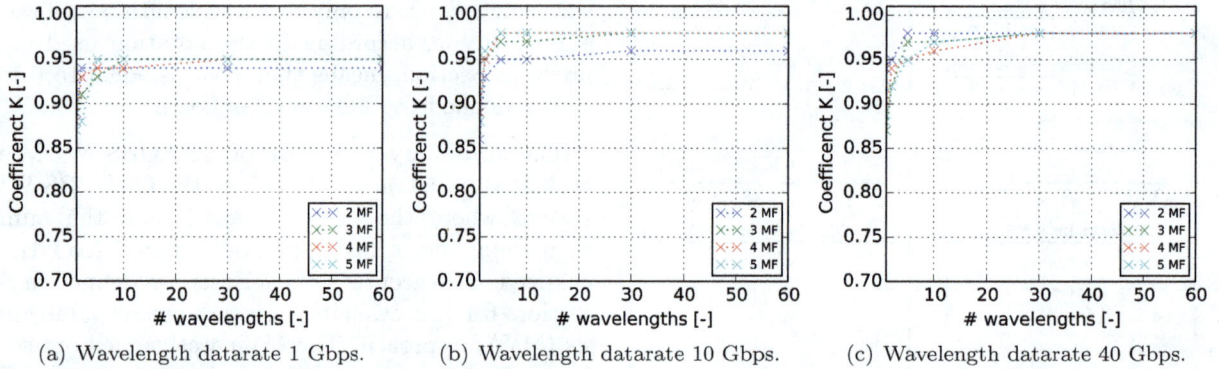

(a) Wavelength datarate 1 Gbps.   (b) Wavelength datarate 10 Gbps.   (c) Wavelength datarate 40 Gbps.

Fig. 10: Dependance of reduction coefficient K on the number of wavelengths. The value of coefficient $K_{bps}(\lambda, |\Lambda|) \to 1$ is not equal to one, i.e., there is some free capacity necessary to keep the system stationary.

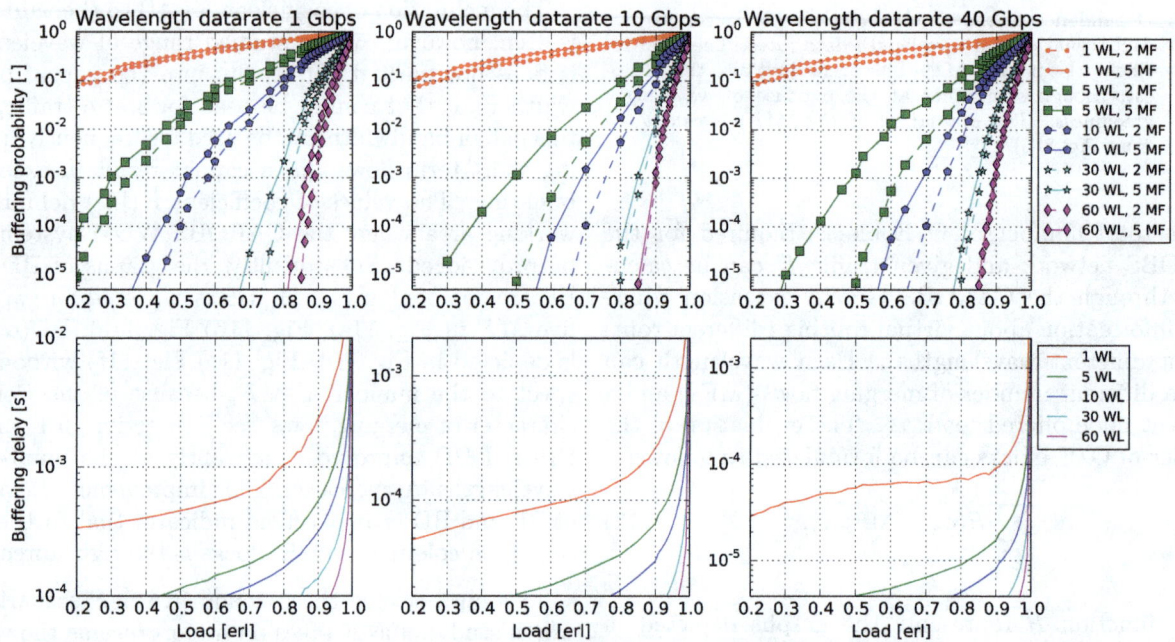

Fig. 11: A comparison of CAROBS using the WDM system. The upper set of figures depict the buffering probability for six different wavelength sets and two and five MF. The lower set of figures captures the average buffering delay with no respect to the number of MF. Both sets offer clear evidence for shared O/E blocks deployment.

lows designing CAROBS nodes which tackle most contentions.

The shared O/E blocks approach allows extension of CAROBS nodes deployment with no O/E blocks which is promising for networks with centralized buffering, as it allows traffic routing without any O/E blocks for contention resolution at a particular node. Geographically extensive deployments, with some nodes low loaded nones can be deployed more cheaply and easily. That notwithstanding, more powerful nodes can be installed in data centers allowing contention resolution through O/E blocks. It is important in future studies to return to the regeneration of optical signal because of optical impairments and these results give us a good starting point. This Section showed that the CAROBS WDM with shared O/E blocks has minimal requirements on

O/E blocks for contention resolution. Therefore, it is worth investing more research into optimal routing and O/E block installation in further work.

6.3.   Highlight of Results

In this Section, the analysis based on OMNeT++ simulations was split into two Subsection 6.1. and Subsection 6.2. In the first Subsection we mainly focused on the simulation condition in order to ensure valid outputs and defined the CAROBS stability condition. Satisfying these conditions, we carried out information on CAROBS behaviour for wavelength selective O/E blocks. Based on these results we concluded that CAROBS WDM relying on wavelength selective O/E blocks will present very high CAPEX and OPEX, i.e.,

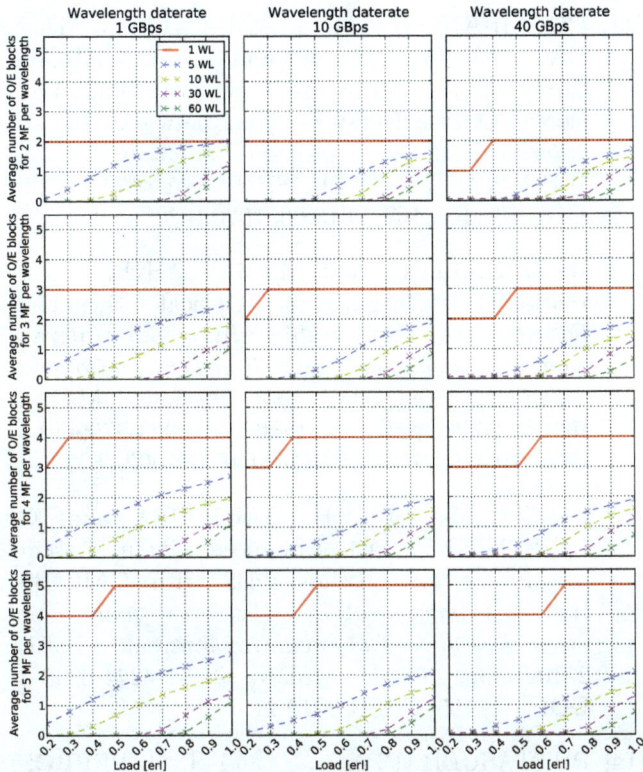

Fig. 12: Estimation of the average number of O/E blocks per wavelength with respect to the egress port load. The higher the number of wavelengths shared by one O/E block, the lower the total number of O/E blocks. The upper boundary for the system containing a higher number of wavelengths is lowered because it is over the threshold K, i.e., a load that we do not want to reach.

it raises motivation for CAROBS WDM using colourless O/E blocks.

CAROBS WDM based on colour-less blocks analysis was carried out in Subsection 6.2. This analysis shows very significant O/E blocks amount reduction. Very interesting property of CAROBS networks is that it allows traffic routing without any O/E block at some nodes thus make the CAROBS network maintenance easier.

7. Conclusion

The feasibility of an OBS deployment has been proven by recent prototypes developed by vendors [4], but, still, there is no large scale deployment of OBS networks. The initial reason, related to burst loss, was addressed by the CAROBS framework of Coutelen et al. [13] and resolved by the loss-free paradigm. Nevertheless, geographically extensive deployment for burst traffic as we can see in access or metropolitan networks is not possible since the optical signal degrades before it is received by the end node. This problem

was outlined in CAROBS [15], but a thorough study was not carried out thus we went through these lacking experiments and brought detail analysis of CAROBS behaviour which we plan to use for CAROBS GRWA formulation.

GRWA formulation is very useful for CAROBS. CAROBS is the only OBS framework that allows traffic disaggregation at intermediate node, so it would be inefficient to use RWA strategy as is usually done for OBS. Consequently the OBS seems to be inefficient when compared to OCS; however, both paradigms are good for different traffic characteristics, i.e., network topologies thus it is not fair to compare them under the assumption that one can replace the other one. CAROBS can perform very well in access networks with a lot of bandwidth granularities that are merged. Also CAROBS can be used for construction of very simple nodes that can be placed in outlaid areas with almost no maintenance.

As we believe in OBS and CAROBS networks, we would like to focus our future work on GRWA for CAROBS WDM using colour-less O/E blocks in order to design optimal traffic routing that allows to cross geographically significant installations as well as complex metropolitan mesh networks.

Acknowledgment

The first author was supported by student grant at Czech Technical University in Prague SGS16/158/OHK3/2T/13, by grants ICT systems analytical risk processing, and Telecommunications infrastructure operational risks analysis. The second author has been supported by a Concordia University Research Chair (Tier I) and by an NSERC (Natural Sciences and Engineering Research Council of Canada) grant.

References

[1] SALEH, A. A. M. and J. M. SIMMONS. Technology and architecture to enable the explosive growth of the internet. *IEEE Communications Magazine*. 2011, vol. 49, iss. 1, pp. 126–132. ISSN 0163-6804. DOI: 10.1109/MCOM.2011.5681026.

[2] SALEH, A. A. M. and J. M. SIMMONS. All-Optical Networking-Evolution, Benefits, Challenges, and Future Vision. *Proceedings of the IEEE*. 2012, vol. 100,

iss. 5, pp. 1105–1117. ISSN 0018-9219. DOI: 10.1109/JPROC.2011.2182589.

[3] QIAO, C. and M. YOO. Optical Burst Switching (OBS) - A New Paradigm for an Optical Internet. *Journal of High Speed Networks.* 1999, vol. 8, iss. 1, pp. 69–84. ISSN 1875-8940.

[4] Optical Packet Switched Transport. *Intune* [online]. Available at: `http://www.intunenetworks.com`.

[5] DE LEENHEER, M., P. THYSEBAERT, B. VOLCKAERT, F. DE TURCK, B. DHOEDT, P. DEMEESTER, D. SIMEONIDOU, R. NEJABATI, G. ZERVAS, D. KLONIDIS and M. J. O'MAHONY. A View on Enabling-Consumer Oriented Grids through Optical Burst Switching. *IEEE Communications Magazine.* 2006, vol. 44, iss. 3, pp. 124–131. ISSN 0163-6804. DOI: 10.1109/MCOM.2006.1607875.

[6] LIU, L., H. GUO, T. TSURITANI, Y. YIN, J. WU, X. HONG, J. LIN and M. SUZUKI. Dynamic Provisioning of Self-Organized Consumer Grid Services Over Integrated OBS/WSON Networks. *Journal of Lightwave Technology.* 2011, vol. 30, iss. 5, pp. 734–753. ISSN 0733-8724. DOI: 10.1109/JLT.2011.2180508.

[7] THACHAYANI, M. and R. NAKKEERAN. Combined probabilistic deflection and retransmission scheme for loss minimization in OBS networks. *Optical Switching and Networking.* 2015, vol. 18, iss. 1, pp. 51–58. ISSN 1573-4277. DOI: 10.1016/j.osn.2015.03.004.

[8] KLINKOWSKI, M., J. PEDRO, D. CAREGLIO, M. PIORO, J. PIRES, P. MONTEIRO and J. SOLE-PARETA. An overview of routing methods in optical burst switching networks. *Optical Switching and Networking.* 2010, vol. 7, iss. 2, pp. 41–53. ISSN 1573-4277. DOI: 10.1016/j.osn.2010.01.001.

[9] LIU, Y., E. TANGDIONGGA, Y. LI, H. DE WAARDT, A. M. J. KOONEN, G. D. KHOE, H. J. S. DORREN, X. SHU and I. BENNION. Error-free 320 Gb/s SOA-based Wavelength Conversion using Optical Filtering. In: *Optical Fiber Communication Conference and the National Fiber Optic Engineers Conference.* Anaheim: IEEE, 2006, pp. 1–3. ISBN 1-55752-803-9. DOI: 10.1109/OFC.2006.216061.

[10] MOUNTROUIDOU, X., V. PUTTASUBBAPPA and H. PERROS. A Zero Burst Loss Architecture for star OBS Networks. In: *Network Control and Engineering for Qos, Security and Mobility, V.* Santiago: Springer US, 2006, pp. 1–13. ISBN 978-0-387-34827-8. DOI: 10.1007/978-0-387-34827-8.

[11] COULIBALY, Y., M. S. A. LATIFF, S. MANDALA, A. M. UMARU and N. M. GARCIA. Study on the performance of slotted and nonslotted Optical Burst Switched networks. In: *4th International Conference on Photonics.* Malaka: IEEE, 2013, pp. 69–71. ISBN 978-1-4673-6073-9. DOI: 10.1109/ICP.2013.6687070.

[12] RAFFAELLI, C. and M. SAVI. Hybrid Contention Resolution in Optical Switching Fabric with QoS traffic. In: *6th International Conference on Broadband Communications, Networks, and Systems.* Madrid: IEEE, 2009, pp. 1–8. ISBN 978-963-9799-49-3. DOI: 10.4108/ICST.BROADNETS2009.7833.

[13] COUTELEN, T., B. JAUMARD and G. HEBUTERNE. An Enhanced Train Assembly Policy for Lossless OBS with CAROBS. In: *Communication Networks and Services Research Conference.* Montreal: IEEE, 2010, pp. 61–68. ISBN 978-1-4244-6248-3. DOI: 10.1109/CNSR.2010.21.

[14] PUTTASUBBAPPA V. S. and H. G. PERROS. Quality of Service in an Optical Burst Switching Ring. *Photonic Network Communications.* 2005, vol. 9, iss. 3, pp. 357–371. ISSN 1572-8188. DOI: 10.1007/s11107-004-6438-x.

[15] KOZAK, M., B. JAUMARD and L. BOHAC. On regenerator placement in loss-less optical burst switching networks. In: *36th International Conference on Telecommunications and Signal Processing.* Rome: IEEE, 2013, pp. 311–315. ISBN 978-1-4799-0402-0. DOI: 10.1109/TSP.2013.6613942.

[16] WANKHADE, S. V. and S. B. KAMBALE. An Evolutionary Approach for LAUC Scheduler in Optical Burst Switching Networks. *International Journal of Applied Information Systems.* 2012, vol. 2, iss. 8, pp. 1–4. ISSN 2249-0868. DOI: 10.5120/ijais12-450383.

[17] TRIKI A. A., P. GAVIGNET, B. ARZUR, E. LE ROUZIC and A. GRAVEY. Bandwidth allocation schemes for a lossless Optical Burst Switching. In: *17th International Conference on Optical Network Design and Modeling.* Brest: IEEE, 2013, pp. 205–210. ISBN 978-1-4799-0491-4.

[18] YUAN, J., X. ZHOU, J. WANG, X. LI and F. LIN. An irregularly slotted ring scheme for contention-free optical burst switching. *Optical Switching and Networking.* 2014, vol. 12, iss. 1, pp. 45–55. ISSN 1573-4277. DOI: 10.1016/j.osn.2014.01.002.

[19] GONZALES DE DIOS, O., J. P. FERNANDEZ-PALACIOS, I. DE MIGUEL, J. C. AGUADO, N.

MERAYO, R. J. DURAN, P. FERNANDEZ, R. M. LORENZO and E. J. ABRIL. Experimental demonstration of a PCE for Wavelength-Routed Optical Burst-Switched (WR-OBS) networks. In: *17th International Conference on Optical Network Design and Modeling*. Brest: IEEE, 2013, pp. 269–274. ISBN 978-1-4799-0491-4.

[20] HAQ, I. U., H. M. SALGADO and J. C. S. CASTRO. Resource reservation schemes for synchronous traffic in cooperative clustered OBS networks. In: *High Capacity Optical Networks and Emerging/Enabling Technologies*. Magosa: IEEE, 2013, pp. 1949–4092. ISBN 978-1-4799-2568-1. DOI: 10.1109/HONET.2013.6729779.

[21] PAVON-MARIO, P. and F. NERI. On the Myths of Optical Burst Switching. *IEEE Transactions on Communications*. 2011, vol. 59, iss. 9, pp. 2574–2584. ISSN 0090-6778. DOI: 10.1109/TCOMM.2011.063011.100192.

[22] WANG, B. and N. LELLA. Dynamic contention resolution in optical burst switched networks with partial wavelength conversion and fiber delay lines. In: *GLOBECOM Global Telecommunications Conference*. Dallas: IEEE, 2004, pp. 1862–1866. ISBN 0-7803-8794-5. DOI: 10.1109/GLOCOM.2004.1378312.

[23] DUTTA, M. K. and V. K. CHAUBEY. Contention resolution in optical burst switching (OBS). In: *International Conference on Fiber Optics and Photonics*. Chennai: IEEE, 2012, pp. 1–3. ISBN 978-1-4673-4718-1. DOI: 10.1364/PHOTONICS.2012.WPo.18.

[24] PEDROLA, O., D. CAREGLIO, M. KLINKOWSKI and J. SOLE-PARETA. Regenerator Placement Strategies for Translucent OBS Networks. *Journal of Lightwave Technology*. 2011, vol. 29, iss. 22, pp. 3408–3420. ISSN 0733-8724. DOI: 10.1109/JLT.2011.2168806.

[25] VU, H. L. and M. ZUKERMAN. Blocking probability for priority classes in optical burst switching networks. *IEEE Communications Letters*. 2002, vol. 6, iss. 5, pp. 214–216. ISSN 1089-7798. DOI: 10.1109/4234.1001668.

[26] VENKATESH, T. and C. S. R. MURTHY. *An Analytical Approach to Optical Burst Switched Networks*. New York: Springer, 2010. ISBN 978-1-4419-1510-8.

[27] PHUNG, M. H., K. C. CHUA, G. MOHAN, M. MOTANI and T. C. WONG. The streamline effect in OBS networks and its application in load balancing. In: *2nd International Conference on Broadband Networks*. Boston:

IEEE, 2005, pp. 283–290. ISBN 0-7803-9276-0. DOI: 10.1109/ICBN.2005.1589625.

[28] CHEN, B., G. N. ROUSKAS and R. DUTTA. On Hierarchical Traffic Grooming in WDM Networks. *IEEE/ACM Transactions on Networking*. 2008, vol. 16, iss. 5, pp. 1226–1238. ISSN 1063-6692. DOI: 10.1109/TNET.2007.906655.

[29] DELESQUES, P., T. BONALD, G. FROC, P. CIBLAT and C. WARE. Enhancement of an optical burst switch with shared electronic buffers . In: *17th international conference on Optical Network Design and Modeling*. Brest: IEEE, 2013, pp. 137–142. ISBN 978-1-4799-0491-4.

[30] DAIGLE, J. *Queueing Theory with Applications to Packet Telecommunication*. New York: Springer, 2005. ISBN 978-0-387-22859-4.

[31] VELASCO, L., M. KLINKOWSKI, M. RUIZ and J. COMELLAS. Modeling the routing and spectrum allocation problem for flexgrid optical networks. *Photonic Network Communications*. 2012, vol. 24, iss. 3, pp. 177–186. ISSN 1572-8188. DOI: 10.1007/s11107-012-0378-7.

[32] MUKHERJEE, B. *Optical WDM networks*. New York: Springer, 2006. ISBN 978-0-387-29188-8.

[33] VIGNAC, B., B. JAUMARD and F. VANDERBECK. Hierarchical optimization procedure for traffic grooming in WDM optical networks. In: *International Conference on Optical Network Design and Modeling*. Braunschweig: IEEE, 2009, pp. 1–6. ISBN 978-1-4244-4187-7.

[34] HU, J. Q. and B. LEIDA. Traffic grooming, routing, and wavelength assignment in optical WDM mesh networks. In: *23rd AnnualJoint Conference of the IEEE Computer and Communications Societies*. Hong Kong: IEEE, 2004, pp. 495–501. ISBN 0-7803-8355-9. DOI: 10.1109/INFCOM.2004.1354521.

[35] LE, Z., M. FU and W. DONG. Gradient projection based RWA algorithm for OBS network. In: *7th International Symposium on Communication Systems Networks and Digital Signal Processing*. Newcastle upon Tyne: IEEE, 2010, pp. 272–277. ISBN 978-1-86135-369-6.

[36] TRIAY, J. and C. CERVELLO-PASTOR. Topology analysis of auto load-balancing RWA in optical burst-switched networks. In: *20th Annual Wireless and Optical Communications Conference*. Newark: IEEE, 2011, pp. 1–6. ISBN 978-1-4577-0453-6. DOI: 10.1109/WOCC.2011.5872289.

[37] COUTELEN, T., G. HEBUTERNE and B. JAUMARD. An OBS RWA formulation for

asynchronous loss-less transfer in OBS networks. In: *International Conference on High Performance Switching and Routing.* Paris: IEEE, 2009, pp. 1–6. ISBN 978-1-4244-5174-6. DOI: 10.1109/HPSR.2009.5307423.

[38] JAUMARD, B., C. MEYER and B. THION-GANE. Comparison of ILP Formulations for the RWA Problem. *Optical Switching and Networking.* 2007, vol. 4, iss. 3–4, pp. 157–172. ISSN 0711-2440. DOI: 10.1016/j.osn.2007.05.002.

[39] JAUMARD, B., C. MEYER and B. THION-GANE. On column generation formulations for the RWA problem. *Discrete Applied Mathematics.* 2009, vol. 157, iss. 6, pp. 1291–1308. ISSN 0166-218X. DOI: 10.1016/j.dam.2008.08.033.

[40] BETKER, A., C. GERLACH, R. HULSER-MANN, M. JAGER, M. BARRY, S. BODAMER, J. SPATH, C. GAUGER and M. KOHN. Reference Transport Network Scenarios. *Technical report German Ministry of Education and Research within the MultiTeraNet.* 2004.

[41] KOZAK, M., B. JAUMARD and L. BOHAC. On the efficiency of stream line effect for contention avoidance in optical burst switching networks. *Optical Switching and Networking.* 2015, vol. 18, iss. 1, pp. 35–50. ISSN 1573-4277. DOI: 10.1016/j.osn.2015.03.002.

About Authors

Milos KOZAK received the M.Sc. and Ph.D. degrees in electrical engineering from the Czech Technical University in Prague, in 2009 and 2015, respectively. Since 2009 until 2012, he tought optical communication systems and data networks with the Czech Technical University in Prague. His research interest is on the application of high-speed optical transmission systems in a data network. Particularly regenerators placement in all optical networks.

Brigitte JAUMARD holds a Concordia University Research Chair, Tier 1, on the Optimization of Communication Networks in the CIISE - Concordia Institute for Information Systems and Engineering - Institute at Concordia University. She was previously awarded a Canada Research Chair - Tier 1 - in the Department of Computer Science and Operations Research at Universite de Montreal. She is an active researcher in combinatorial optimization and mathematical programming, with a focus on applications in telecommunications and artificial intelligence. Recent contributions include the development of efficient methods for solving large-scale mathematical programs, and their applications to the design and the management of optical and wireless, access and core networks. In Artificial Intelligence, contributions include the development of efficient optimization algorithms for probabilistic logic (reasoning under uncertainty) and for automated mechanical design. B. Jaumard has published over 150 papers in international journals in Operations Research and in Telecommunications.

Leos BOHAC received the M.Sc. and Ph.D. degrees in electrical engineering from the Czech Technical University, Prague, in 1992 and 2001, respectively. Since 1992, he has been teaching optical communication systems and data networks with the Czech Technical University, Prague. Since 2014, he holds position of associate professor with Czech Technical University, Prague. His research interest is on the application of high-speed optical transmission systems in a data network. He has also participated in the optical research project CESNET - the academic data network provider to help implement a long-haul high-speed optical research network. Currently, he has been actually involved in and led some of the projects on optimal protocol design, routing, high speed optical modulations and industrial network design.

7

Analysis of the Applicability of Singlemode Optical Fibers for Measurement of Deformation with Distributed Systems BOTDR

Marcel FAJKUS, Jan NEDOMA, Lukas BEDNAREK, Jaroslav FRNDA, Vladimir VASINEK

Department of Telecommunications, Faculty of Electrical Engineering and Computer Science, VSB–Technical University of Ostrava, 17. listopadu 15, 708 33 Ostrava, Czech Republic

marcel.fajkus@vsb.cz, jan.nedoma@vsb.cz, lukas.bednarek@vsb.cz, jaroslav.frnda@vsb.cz, vladimir.vasinek@vsb.cz

Abstract. *Distributed optical fiber sensors allow monitoring physical effects across the whole cable. The paper presents results obtained from the performed tests and shows that single mode fibers can provide analyses of the deformation changes, when distributed optical systems BOTDR used. We used standard optical fiber G.652.D with primary and secondary protected layers and specialized cable SMC-V4 designed for this purpose. The aim was to compare the deformation sensitivity and determine which fiber types are the best to use. We deformed the fiber in the longitudinal and transverse directions and mechanically stressed in orthogonal directions to find how to localize optical fibers. They could be deployed in real use. For achieving optimal results of mechanical changes and acting forces, sensor fibers have to be located carefully.*

Keywords

Deformation, distributed system, sensor, special cable, standard telecommunication fiber.

1. Introduction

Distributed fiber optic sensors is a group of fiber-optic sensors. Distributed sensors enable the measurement of quantity along the entire length of optical fiber. The general principle of these systems is the phenomenon called light scattering. Based on the light scattering analysis, we distinguish several types of such systems. Rayleigh scattering is used for the measurement of attenuation profile of the optical fiber [1]. Raman scattering is possible to use for monitoring the temperature along the optical fiber [2] and Brillouin scattering can be used for measuring the temperature and deformation [3] and [4]. This article is focused on the measurement of deformation utilizing Brillouin Time Domain Reflectometry (BOTDR). The principle is based on the measurement of the stimulated Brillouin scattering. The frequency shift of Brillouin scattering is linearly dependent on the temperature and deformation. Distributed sensors are used in many areas of measurement, for example, for monitoring of the condition of building structures, the temperature distribution along the electrical cables, etc. Special optical cable SMC-V4 is designed for measurement of deformation with BOTDR system. However, this cable is more expensive, and implementation of the cable is considerably expensive for the measurement of long distance. The alternative possibility is the use of standard telecommunications optical fiber, which is not recommended for these purposes. The disadvantage of standard optical fiber is a non-tight bond between the optical fiber itself and the secondary protection in the case of measurement of the deformation. For this reason, it is considered that the applied deformation is weakly transmitted to the optical fiber itself. Standard telecommunications optical fibers G.652.D can be used for the measurement of deformation, where there is not a big emphasis on accuracy of measurement, for example, in security systems. This combination offers less accuracy, but the main requirement is knowledge about a disruption of perimeter system. This solution represents a possible low-cost alternative solution of perimeter systems using BOTDR systems. The aim of this article is the comparison of sensitivity to deformation of these types of optical fiber and determining the suitability of these standard optical fibers for the measurement of deformation with distributed systems.

2. Distributed System BOTDR

In practical applications, optical fiber is sticked on the surface of an analyzed structure. The optical fiber is mounted directly inside of the building structure for the analysis of construction. In this paper, we use of a special cable SMC-V4, because the emphasis is placed on the accuracy of measurement. Special cables, with higher tensile strength and better protection, are also more suitable for measurement of the terrain and soil layers [5]. The authors [6] and [7] show that it is possible to use different types of cables for measurement of deformation.

BOTDR (Brillouin Optical Time Domain Reflectometry) operates on the principle of measurement of stimulated Brillouin scattering. Brillouin scattering arises due to the interaction of acoustic waves and pump of the light beam, under the condition of supercritical power of light passing through the optical fiber. The transmitted light is diffused according to the changes in the refractive index. The scattered light is shifted due to Doppler effect on frequency by an amount v_B. This value is given by:

$$v_B = \frac{2nV_a}{\lambda_0}, \qquad (1)$$

where λ_0 is the wavelength of transmitted light, n represents the refractive index of the core of the optical fiber and V_a denotes the propagation velocity of acoustic waves within the optical fiber. The resultant value is given by:

$$V_a = \sqrt{\frac{K}{\sigma}}, \qquad (2)$$

where σ is the density of the material and K express module of volume compressibility. The value σ is determined by the magnitude of deformation and temperature on the optical fiber itself.

The strain response is defined as follows:

$$v_B(\varepsilon, n) = C_{\varepsilon 1}\varepsilon + C_{\varepsilon 0}, \qquad (3)$$

where coefficient $C_{\varepsilon 1}$ is 0.5 GHz/% and coefficient $C_{\varepsilon 0}$ is 10.87 GHz [8] for standard ITU-G.652 optical fiber at wavelength 1550 nm.

Figure 1 shows a Brillouin frequency shift, which is linearly dependent on the applied deformation and temperature. The value of the magnitude of applied deformation and temperature is obtained by scan Brillouin frequency shift using the probing of light. The light is directed into the optical fiber from the opposite end.

Experimental measurements were performed using a distributed system for the measurement of deformation and temperature DiTEST STA-R from Omnisens,

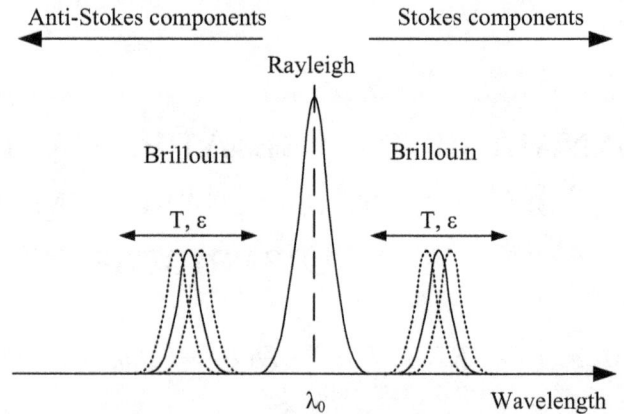

Fig. 1: Rayleigh and Brillouin scattering.

which exploits the sensitivity of the Brillouin frequency shift for sensing of temperature and strain. This technique uses standard low-loss single mode optical fiber. The optical fiber offers the longest distance range with unrivaled performances and a compatibility with standard telecommunication components. The functionality principle consists of the frequency measurement of the Brillouin scattered light. The spatial resolution of 1 m is a distance of 20 km in the optical fiber, resolution 2 m is a distance 30 km, and value of resolution 3 m is a maximum length of optical fiber 50 km. The system contains two independent channels with a reach of 50 km. A step of measurement of 10 cm can be set for short segments of optical fibers in hundreds of meters. A maximum of measuring steps is limited to a value of 100 000 measurement points across the length of optical fiber.

In Fig. 2, we can see how to measure the Brillouin frequency shift with a BOTDR. Pulsed light is introduced into the optical fiber (from one end of the optical fiber), and the power of spontaneous Brillouin backscattered light is then measured in the time domain using heterodyne detection. The frequency of the incident light slightly changes. Therefore, the same measurements

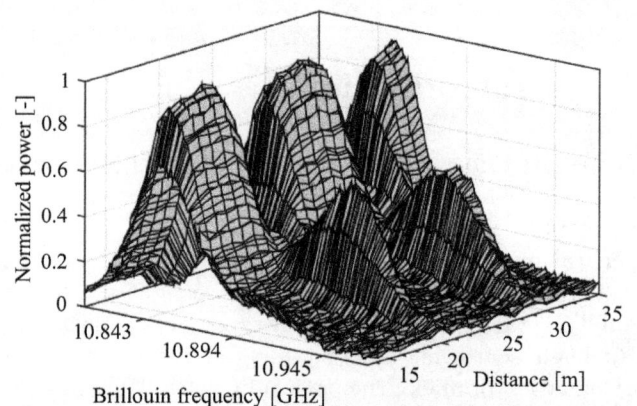

Fig. 2: Principle of BOTDR strain measurement.

are carried out to obtain the Brillouin spectrum repeatedly at many frequencies. The specific frequency, which indicates the peak power, is calculated by fitting the spectrum to a Lorentzian curve. It must be calculated at each point of the optical fiber. The strain is obtained from this frequency. The distance D defines a place, where the pulsed light is released into the position, where the scattered light is generated.

$$D = \frac{cT}{2n}, \qquad (4)$$

where n is the refractive index, c is the light velocity in a vacuum and T is the time interval between receiving the scattered light and launching the pulsed light.

3. Experimental Setup

The constructed preparation was used to deform of the optical fiber for experimental measurement (Fig. 3). The preparation consists of galvanized sheet with dimensions 100×100 cm. Sheet with glued optical fiber is affixed to the wooden frame. Subsequently, this sheet is bending from the unloaded state '0' ($0°$) to the maximum bend '10' ($180°$).

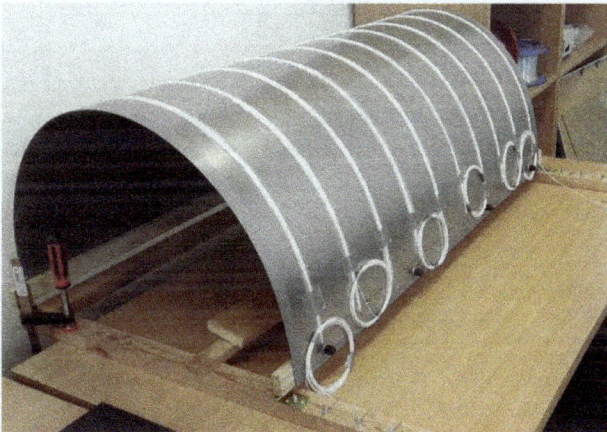

Fig. 3: Equipment for deformation measurement.

Bragg grating was used for the deformation assessment of glued optical fiber at preparation in Fig. 3 for various size of radius bending in the individual positions '1'–'10'. Deformation was zero in position '0'. Figure 4 shows the size of deformation at each position.

Two types of standard telecommunications fibers G.652.D (in primary protection and in a tight secondary protection) were used for experimental measurements, and special optical cable SMC-V4 was also used. This cable is designed for the deformation

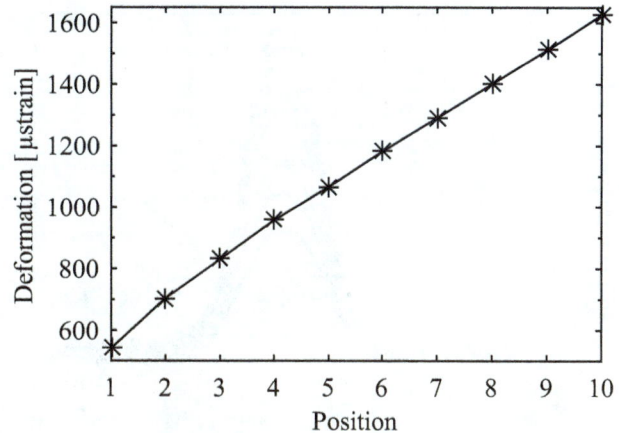

Fig. 4: Deformation of optical fiber with the Fiber Bragg grating glued on preparation for various size of radius bending in the individual positions '1'–'10'.

measurement. It provides the maximum sensitivity on deformation and the ability to detect very small deformations. Individual types of optical fibers were glued on the test sheet in the longitudinal and the transverse direction (Fig. 5) along its entire length. The directions were chosen based on previous research [9]. Polymer adhesive MAMUT Glue was used for gluing. Then measurements were carried out for each arrangement and each condition and each fiber, thus a total 600 repetitions.

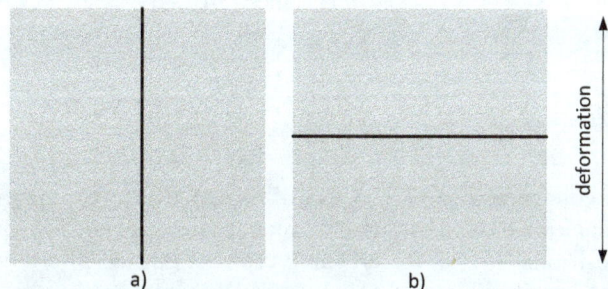

Fig. 5: Arrangement of optical fiber.

Distributed system BOTDR is based on measurement of Brillouin scattering, whose frequency is dependent on applied deformation. Figure 6 shows individual waveforms of Brillouin frequency depending on the distance of optical fiber for deformation at positions '1' to '10'. Figure 6(a) shows the waveforms of Brillouin frequency for longitudinally mounted optical fiber G.652.D in the primary protection. Figure 6(b) shows the waveforms of Brillouin frequency for longitudinally mounted optical cable SMC-V4. In this cable, the optical fiber is characterized by the shifted Brillouin frequency towards lower frequencies (see waveform between 3 and 5 m in Fig. 6(b)).

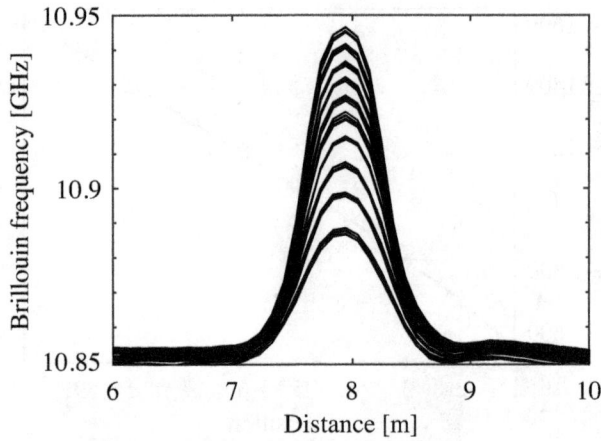

(a) G.652.D in primary protection.

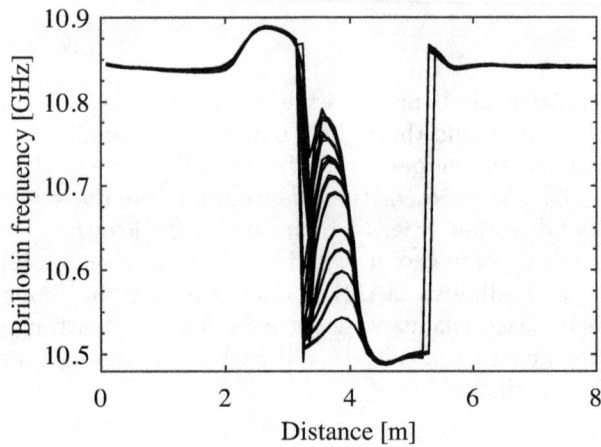

(b) SMC-V4.

Fig. 6: Dependence of Brillouin frequency on applied deformation.

Following Fig. 7 (sef of 6 images) displays dependencies of the Brillouin frequency on deformation in positions '1'–'10' for all three types of optical fibers in the longitudinal and transverse configuration.

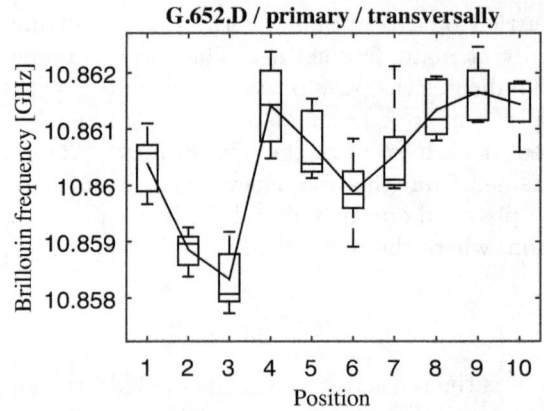

G.652.D / primary / transversally

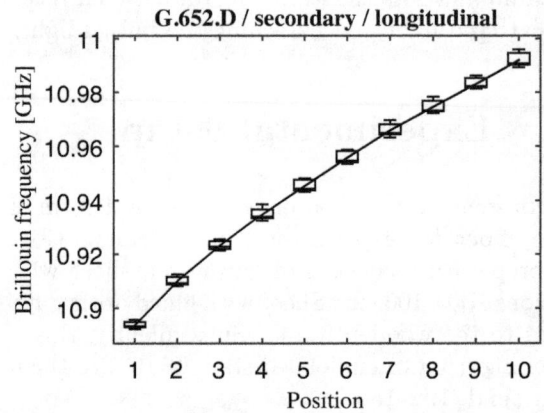

G.652.D / secondary / longitudinal

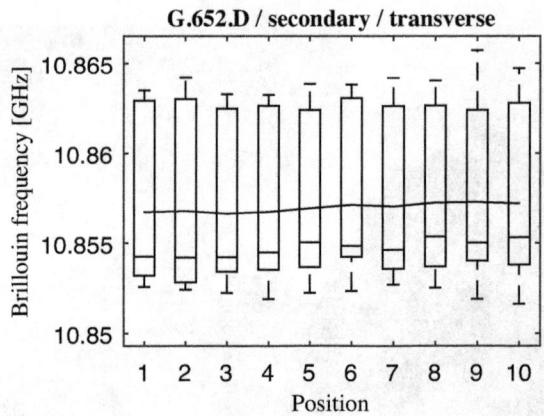

G.652.D / secondary / transverse

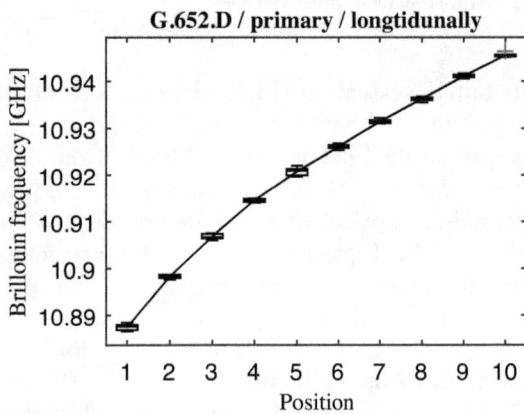

G.652.D / primary / longtidunally

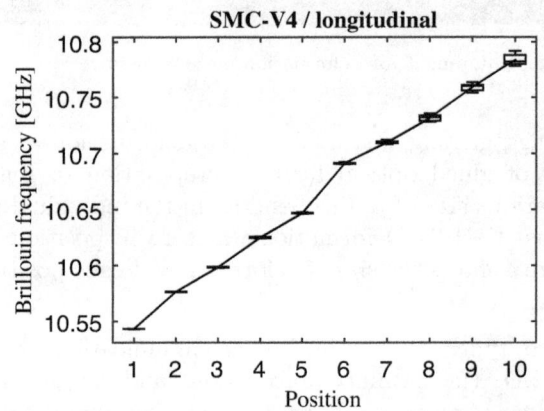

SMC-V4 / longitudinal

SMC-V4 / transverse

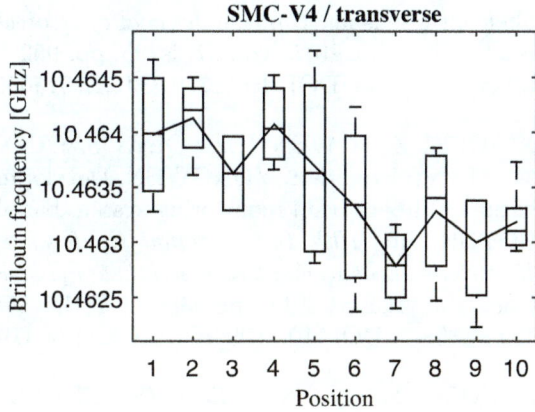

Fig. 7: The dependence of Brillouin frequency on deformation in positions '1' to '10' for all types of optical fibers and the arrangement.

Figure 8 shows the comparison of different configurations. The individual curves represent the average value of the measured data for each configuration and the type of optical fibers.

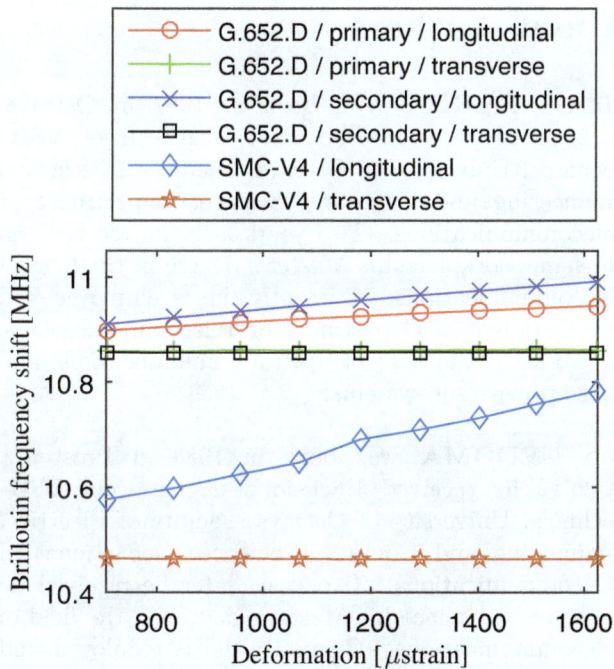

Fig. 8: Comparison of individual configuration and types of optical fibers.

A sensitivity of Brillouin frequency on deformation and the maximum change of deformation were calculated for shown arrangement (Tab. 1).

4.   Conclusion

The presented results show that the tested optical fibers have a small sensitivity to transverse deformation. This sensitivity reaches lower values than

Tab. 1: The sensitivity of Brillouin frequency on the type of arrangement of the optical fiber during measurement of deformation and the maximum change of deformation.

Configuration	Sensitivity	Max Brillouin change frequency
	(kHz/µstrain)	(MHz)
G.652.D/ primary/ longitudinal	53.561	58.09
G.652.D/ primary/ transverse	0.981	1.064
G.652.D/ secondary/ longitudinal	90.688	98.356
G.652.D/ secondary/ transverse	0.459	0.497
SMC-V4/ longitudinal	221.952	240.72
SMC-V4/ transverse	-0.733	-0.7995

1 kHz/µstrain. It can be stated that the tested types of optical fibers are not suitable for measurement of transverse deformations. The sensitivity of SMC-V4 is 4.15 times greater than the sensitivity of G.652.D in the primary protection, and 2.45 times greater than the sensitivity of G.652.D in tightly secondary protection in the case of longitudinal deformation. The ratio of sensitivity is 1.69:1 between the optical fiber G.652.D with tight secondary protection and primary protection. An optical cable SMC-V4 is characterized by shifted of typical Brillouin frequency to lower frequencies about 350 MHz.

This article confirms further the possibility of using standard single-mode fiber for the measurement of deformation using apparatus DiTEST STA-R, which is based on measurement of the Brillouin frequency. However, there is the significant difference in sensitivity between the longitudinal and transverse effect of deformation, which is a hundred times larger in a longitudinal action than in a transverse action. The article confirms further the use of standard telecommunication fibers in security systems, where there is not a big emphasis on the measurement accuracy of deformation. According to the results, this combination offers less accuracy, but the primary requirement is knowledge about a disruption of perimeter system. This solution represents a possible low-cost alternative solution of perimeter systems using BOTDR systems.

Acknowledgment

This article was supported by projects Technology Agency of the Czech Republic TA03020439 and TA04021263 and Ministry of Education of the Czech Republic within the project no. SP2016/149. The Ministry of Education has partially supported the research, Youth and Sports of the Czech Republic through grant project no. CZ.1.07/2.3.00/20.0217 within the frame of the operation program Education for competitiveness financed by the European Structural Funds and from the state budget of the Czech Republic. This article was supported too by project VI20152020008 (GUARDSENSE II).

References

[1] HAGEMANN, H. J., J. UNGELENK and D. U. WIECHERT. Optical time-domain reflectometry (OTDR) of diameter modulations in single mode fibers. *Journal of Lightwave Technology*. 1990, vol. 8, no. 11, pp. 1641–1645. ISSN 0733-8724. DOI: 10.1109/50.60559.

[2] LIU, Y. and Z. ZONGJIU. Design of distributed fiber optical temperature measurement system based on Raman scattering. In: *International Symposium on Signals, Systems and Electronics*. Nanjing: IEEE, 2010, pp. 1–4. ISBN 978-1-4244-6352-7. DOI: 10.1109/ISSSE.2010.5607025.

[3] TUR, M., A. MOTIL, I. SOVRAN and A. BERGMAN. Recent progress in distributed Brillouin scattering fiber sensors. In: *IEEE SENSORS*. Valencia: IEEE, 2014, pp. 138–141. ISBN 978-1-4799-0162-3. DOI: 10.1109/IC-SENS.2014.6984952.

[4] PRADHAN, H. S. and P. K. SAHU. Spontaneous Brillouin scattering based distributed fiber optic temperature sensor design and simulation using phase modulation and optimization technique. In: *Sixth International Conference on Sensing Technology (ICST)*. Kolkata: IEEE, 2012, pp. 300–304. ISBN 978-1-4673-2246-1. DOI: 10.1109/IC-SensT.2012.6461691.

[5] WU, J., H. JIANG, J. SU, B. SHI, Y. JIANG and K. GU. Application of distributed fiber optic sensing technique in land subsidence monitoring. *Journal of Civil Structural Health Monitoring*. 2015, vol. 5, iss. 5, pp. 587–597. ISSN 2190-5452. DOI: 10.1007/s13349-015-0133-8.

[6] LEE, C. C. and S. CHI. Measurement of stimulated-Brillouin-scattering threshold for various types of fibers using Brillouin optical-time-domain reflectometer. *IEEE Photonics Technology Letters*. 2000, vol. 12, no. 6, pp. 672–674. ISSN 1041-1135. DOI: 10.1109/68.849080.

[7] ZHANG, D., P.-S. ZHANG, B. SHI, H.-X. WANG and C.-S. LI. Monitoring and analysis of overburden deformation and failure using distributed fiber optic sensing. *Chinese Journal of Geotechnical Engineering*. 2015, vol. 37, iss. 5, pp. 952–957. ISSN 1000-4548. DOI: 10.11779/CJGE201505023.

[8] XIAOFEI, Z., H. WENJIE, Z. QING, S. YANXIN, M. XIANWEI and H. YONGWEN. Development of optical fiber strain monitoring system based on BOTDR. In: *10th International Conference on Electronic Measurement & Instruments (ICEMI)*. Chengdu: IEEE, 2011, pp. 38–41. ISBN 978-1-4244-8158-3. DOI: 10.1109/ICEMI.2011.6037942.

[9] FAJKUS, M., J. NEDOMA, S. KEPAK, J. JAROS, J. CUBIK, O. ZBORIL, M. NOVAK, and V. VASINEK. Effect of the geometric deformations on the Brillouin scattering in the standard single-mode optical fiber. In: *Proceedings of SPIE 9889: Optical Modelling and Design IV*. Bellingham: SPIE, 2016, pp. 1–6. ISBN 978-151060134-5. DOI: 10.1117/12.2239550.

About Authors

Marcel FAJKUS was born in 1987 in Ostrava. In 2009 he received Bachelor's degree from VSB–Technical University of Ostrava, Faculty of Electrical Engineering and Computer Science, Department of Telecommunications. Two years later he received on the same workplace his Master's degree in the field of Telecommunications. Currently, he is employee and Ph.D. student of Department of Telecommunications. He works in the field of optical communications and fiber optic sensor systems.

Jan NEDOMA was born in 1988 in Prostejov. In 2012 he received Bachelor's degree from VSB–Technical University of Ostrava, Faculty of Electrical Engineering and Computer Science, Department of Telecommunications. Two years later he received on the same workplace his Master's degree in the field of Telecommunications. Currently, he is employed and Ph.D. student of Department of Telecommunications. He works in the field of optical communications and fiber optic sensor systems.

Lukas BEDNAREK was born in 1988 in Frydek-Mistek. In 2011 he received Bachelor's degree from VSB–Technical University of Ostrava, Faculty of Electrical Engineering and Computer Science, Department of Computer Science. Two years later he received Master's degree in the field of Telecommunications on Department of Telecommunications. He is currently Ph.D. student, and he works in the field of optical communications and aging of the optical components.

Jaroslav FRNDA was born in 1989 in Martin,

Slovakia. He received his M.Sc. from the VSB–Technical University of Ostrava, Department of Telecommunications, in 2013, and now he is continuing his Ph.D. study at the same place. His research interests include Quality of Triple play services and IP networks.

Vladimir VASINEK was born in Ostrava. In 1980 he graduated in Physics, specialization in Optoelectronics, from the Science Faculty of Palacky University. He was awarded the title of RNDr. at the Science Faculty of Palacky University in the field of Applied Electronics. The scientific degree of Ph.D. was conferred upon him in the branch of Quantum Electronics and Optics in 1989. He became an associate professor in 1994 in the branch of Applied Physics. He has been a professor of Electronics and Communication Science since 2007. He pursues this branch at the Department of Telecommunications at VSB–Technical University of Ostrava. His research work is dedicated to optical communications, optical fibers, optoelectronics, optical measurements, optical networks projecting, fiber optic sensors, MW access networks. He is a member of many societies: OSA, SPIE, EOS, Czech Photonics Society; he is a chairman of the Ph.D. board at the VSB–Technical University of Ostrava. He is also a member of habitation boards and the boards appointing to professorship.

8

Encapsulation of FBG Sensor into the PDMS and its Effect on Spectral and Temperature Characteristics

Jan NEDOMA, Marcel FAJKUS, Lukas BEDNAREK, Jaroslav FRNDA,
Jan ZAVADIL, Vladimir VASINEK

Department of Telecommunications, Faculty of Electrical Engineering and Computer Science,
VSB–Technical University of Ostrava, 17. listopadu 15/2172, 708 33 Ostrava, Czech Republic

jan.nedoma@vsb.cz, marcel.fajkus@vsb.cz, lukas.bednarek@vsb.cz, jaroslav.frnda@vsb.cz,
jan.zavadil@vsb.cz, vladimir.vasinek@vsb.cz

Abstract. *Fiber Bragg Grating (FBG) is the most distributed type of fiber-optic sensors. FBGs are primarily sensitive to the effects of temperature and deformation. By employing different transformation techniques, it is possible to use FBG to monitor any physical quantity. To use them as parts of sensor applications, it is essential to encapsulate FBGs to achieve their maximum protection against external effects and damage. Another reason to encapsulate is increasing of sensitivity to the measured quantity. Polydimethylsiloxane (PDMS) encapsulation appears to be an interesting alternative due to convenient temperature and flexibility of the elastomer. This article describes an experimental proposal of FBG PDMS encapsulation process, also providing an analysis of the FBG spectral characteristics and temperature sensitivity, both influenced by high temperature and the process of polydimethylsiloxane curing itself. As for the PDMS type, Sylgard 184 was employed. Encapsulation consisted of several steps: allocation of FBG to PDMS in its liquid state, curing PDMS at the temperature of 80 °C ± 5 %, and a 50-minute relaxation necessary to stabilize a Bragg wavelength. A broadband light source and an optical spectrum analyzer were both used to monitor the parameters during the processes of curing and relaxation. Presented results imply that such a method of encapsulation does not have any influence on the structure or functionality of the FBG. At the same time, a fourfold increase of temperature sensitivity was monitored when compared to a bare FBG.*

Keywords

Encapsulated, fiber Brag grating, polydimethylsiloxane, spectral characteristic, temperature sensitivity.

1. Introduction

Fiber-optic Bragg gratings are elements whose function is based on periodic changes of refractive index in the core of the optical fiber. Their usage is now widespread within fiber-optic sensor applications. The key areas of concern are physical quantities such as temperature, deformation, pressure, vibration, etc. Optical fibers, which are made of germanium silicate glass, are very strong in tensile, but these fibers have very low resistance to mechanical damage. Currently, we use some types of encapsulation which are designed according to the requirements for the measurements of different physical quantities. The tasks of encapsulation are an additional protection of the fiber with FBG, empowerment of sensitivity on the measured quantity, and minimalized sensitivity to other physical quantities. PDMS elastomer, which exhibits suitable thermal and elastic properties, is a suitable alternative to material for encapsulation of FBG. This elastomer is harmless, nontoxic, non-flammable and electrically nonconductive.

Polyimide, acrylate and ormocer are the basic protections. In addition to these primary protections, we use a number of other encapsulations. For example, the team of authors [1] describes the encapsulation of FBG into a variety of metallic coatings which are used for increasing sensitivity for the measurement of tensile stress. An interesting alternative is the use of nickel. The authors [2] describe achieving an increase of the temperature sensitivity of FBG. The article [3] describes encapsulation of FBG into steel, and the influence of encapsulation on the deformation sensitivity of FBG. Paper [4] describes a 4.2 times increase in the temperature sensitivity of FBG due to the insertion of FBG into PDMS. We have not found the team of authors focused on the issue of encapsulation of FBG into PDMS, and the impact of encapsulation on the parameters of FBG. So the aim of the authors was to analyse the impact of high temperature and to cure PDMS on spectral characteristics and temperature sensitivity of FBG. Sylgard 184 was chosen as a product that exhibits excellent heat resistance, excellent elastic properties, harmlessness, nontoxicity, and electric non-conductivity. The actual encapsulation Bragg grating, within PDMS, extends the application potential of FBG sensors, e.g. in the field of medicine.

2. Operating Principles

PDMS belongs among polymeric organosilicon compounds, and it is often referred to as silicone rubbers. These compounds contain a bond of Si-O in one molecule. The toughness of silicone rubbers is low, but their advantage is that they remain almost unchanged in a broad range of temperature. Conventional temperature applicability is $-60\ ^\circ$C to $+200\ ^\circ$C. PDMS can withstand temperature up to $350\ ^\circ$C for short-term temperature straining. Silicone rubbers can be divided into three groups. These groups are PDMS for general use, the PDMS having phenyl substituents (for improved low-temperature flexibility), and PDMS with 1,1,1- trifluoro propyl substituents (resistant against oils and fuels).

As for its chemical composition, PDMS belongs among optically pure materials. PDMS only contain a small degree of impurities. Therefore, PDMS is not a suitable environment for bacteria. PDMS is a clear liquid which is odorless and tasteless, resistant to chemicals, radiation, UV radiation and high temperatures in hundreds of degrees Celsius ($^\circ$C). The main disadvantage is the expensive and complicated production. PDMS is used in a broad range of fields such as electronics, medicine, astronautics or automotive industry.

PDMS is produced using technical silicon and a combination of hydrochloric acid and methanol. This combination creates the so-called chloromethane. The production goes through four chemical phases (by synthesis, by rectification, by hydrolysis, by polycondensation). The final chemical composition of polydimethylsiloxane can be seen in Fig. 1. The organic substituent is almost always represented by methyl (CH_3).

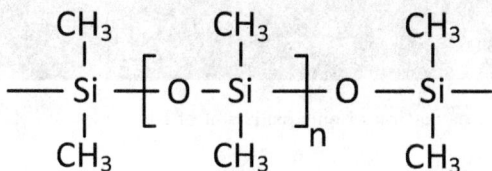

Fig. 1: Chemical composition of PDMS.

Sylgard 184 is a designation for a two-component potting and encapsulating elastomer, on the basis PDMS, which is supplemented by a curing agent. PDMS can be cured at the elevated temperature after addition of the curing agent. Sylgard 184 belongs among moderately viscous liquid elastomers. Temperature range of usability is $-55\ ^\circ$C to $+\ 200\ ^\circ$C. Highlights include an excellent physical resistance to mechanical damage, radiation, and electrical non-conductivity. Sylgard 184 is already cured at room temperature of $25\ ^\circ$C. However, it needs a long period in tens of hours. At the temperature of around $100\ ^\circ$C it is possible to achieve the curing in matter of hours or earlier [5], [6] and [7].

Bragg gratings are the most common type of single-point sensors. They consist of a periodic change of core index in the optical fiber (Fig. 2).

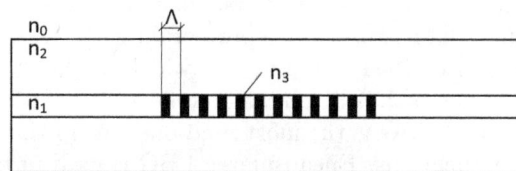

Fig. 2: Structure of fiber Bragg Gratings.

The spectral reflection of the specific wavelength, which is called the Bragg wavelength, occurs at these interfaces. Other wavelengths pass through the structure without attenuation. The important parameter which defines the size of the Bragg wavelength is the period of the changes in the refractive index. Bragg wavelength is given by:

$$\lambda_B = 2n_{eff}\Lambda, \tag{1}$$

where n_{eff} is the effective refractive index, and Λ is the period of changes in the refractive index. The used FBG sensor is based on the temperature and deformation sensitivity. The size of the Bragg wavelength, which is dependent on the operating temperature and mechanical strain, can be expressed as:

$$\frac{\Delta\lambda}{\lambda_0} = k\varepsilon + (\alpha_\Lambda + \alpha_n)\,\Delta T, \tag{2}$$

where α_Λ is the coefficient of thermal expansion, α_n is temperature-optical coefficient and k is the deformation coefficient defined as:

$$k = 1 - p_e. \tag{3}$$

In Eq. (3), p_e is the photo-elastic coefficient. This coefficient takes the value of 0.21 for standard silica optical fiber G.652.D. Coefficient α_Λ is $0.55{\cdot}10^{-6}\ ^\circ$C^{-1} and α_n is in the range 6.4–$8.6{\cdot}10^{-6}\ ^\circ$C^{-1} [8] for optical fiber G.652.D. Temperature and deformation dependence is caused both by the values of parameters and the central Bragg wavelength. Therefore, we state normalized temperature coefficient for determining of these sensitivities:

$$\frac{1}{\lambda_B}\frac{\Delta\lambda_B}{\Delta T} = 6.678 \cdot 10^{-6} \ {}^{\circ}\mathrm{C}^{-1}, \quad (4)$$

and normalized deformation coefficient:

$$\frac{1}{\lambda_B}\frac{\Delta\lambda_B}{\Delta\varepsilon} = 0.78 \cdot 10^{-6} \ \mu\mathrm{strain}^{-1}. \quad (5)$$

Uniform Bragg grating (used in this article) on the wavelength 1554.1203 nm shows temperature sensitivity of 10.378 pm/°C and a strain sensitivity of 1.212 pm/μstrain.

3. Experimental Setup

Uniform FBG was used to implement the sensor. This uniform FBG has polyimide protection with the Bragg wavelength of 1554.1203 nm, the width of the reflecting spectrum of 2.3247 nm, and a reflectivity of 95.6 %. This type of FBG is the most used one within the sensorial applications. Encapsulated FBG is used in medical applications (such as pulse, respiration) because this FBG has both tighter polyimide protection for the optical fiber and better transfer of deformation effect on FBG. The experiment is an innovative type of encapsulation. In the experiment, we investigated the effect of temperature, the curing, and the mechanical stress on spectral characteristics of FBG and temperature sensitivity of FBG. A follow-up research will focus on comparing different types of FBG for verifying and extending the application of such encapsulated FBG.

For the experiment we used a two-component (PDMS and curing agent) elastomer Sylgard 184. The chosen ratio of the mixture was 10:1 where the two-component Sylgard 184 comprises 10 parts and the curing agent forms 1 part. The actual implementation of encapsulation consists of three phases. In all the phases of the encapsulation, we monitored the influence of temperature, the curing and the mechanical stress on the spectral characteristics of FBG. Broad-spectrum LED (Light-Emitting Diode) which has a central wavelength of 1550 nm, and an output power of 1 mW, was used as the radiation source. LED was stabilized by temperature and the current controller labeled LDC 202C and TED 202C made by Thorlabs. Therefore, we obtained a stable optical power. The spectral characteristics were monitored using the optical spectrum analyzer OSA203 by Thorlabs with the Wavelength Meter Resolution about 0.1 pm. Values in Tab. 1 have been rounded to two decimal places. Temperature box has the designation of Concept ET 5050. The optical circulator directed the reflected signal from the FBG to the optical spectrum analyzer within the experiment. The used type is "Polarization Insensitive

Circulator" with the value of insertion loss port 1 to port 2 0.54 dB and port 2 to port 3 0.68 dB. Figure 3 shows a diagram of the experimental measurement.

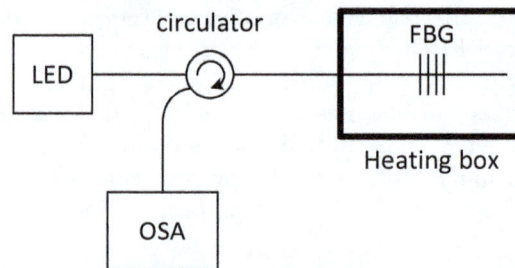

Fig. 3: Scheme of measurement with heating box.

Prepared Sylgard 184 with the volume of 25 ml and in the above-mentioned ratio of 10:1, was placed in an ultrasonic bath Ultrasonic Cleaner for 60 minutes. Therefore, we achieved both maximum homogeneity of the mixture and eliminated air bubbles (Fig. 4).

Fig. 4: Used ultrasonic bath for preparing Sylgard 184 mixture in the ratio 10:1.

In the first phase, we performed the encapsulation of FBG into PDMS in the liquid state. Figure 5 shows a prepared form, in which the FBG was placed and subsequently encapsulated by liquid PDMS.

Fig. 5: Realization of encapsulation of FBG.

The course of Bragg wavelength change is shown in Fig. 6(a) during potting FBG into liquid PDMS. The next step moved forms with FBG into a preheated thermal cabinet with the temperature of 80 °C. The course of the Bragg wavelength is shown in Fig. 6(b). In the second phase, curing PDMS with the FBG in the temperature box was carried out (Fig. 6(c)). Based on the datasheet, the temperature in the box was set to 80 °C. The temperature was monitored by the thermal cell during curing, and the temperature deviation was 80 °C ± 5 % within the measurement. Figure 6(d) represents the evolution of Bragg wavelength within a 50-minute relaxation.

The changes of Bragg wavelength, which are evident in the attached graphs (Fig. 6) can be explained by the influence of temperature, the curing and mechanical stress in handling. The most significant change of Bragg wavelength can be seen in Fig. 6(c), and it represents the time interval 10-15 minutes from insertion FBG in the liquid PDMS into a thermal box with temperature 80 °C. Figure 6(d) represents a relaxation time (50 minutes) at room temperature of 25 °C until stabilization of Bragg wavelength.

During the potting into PDMS and curing we observed rapid increase of Bragg wavelength that was affected by higher temperature (from 25 °C to 80 °C). Initial Bragg wavelength corresponds to the value of 1554.09 nm that was measured after the process of potting had been finished, see Fig. 6(a). Maximal Bragg wavelength (Fig. 6(c)) that we obtained is 1554.66 nm (difference of 570 nm). In this phase only temperature sensitivity for bare FBG is presented with the value of 10.38 pm/°C. This value corresponds to the temperature change of 54.91 °C. Linear growth of Bragg wavelength from 18 min in Fig. 6(c) is caused by low value of thermal conductivity of PDMS material and slow expansion influenced by thermal expansion. In the relaxation phase we indicated sharp reduction of Bragg wavelength because of PDMS material cooling. We observed the high thermal sensitivity of PDMS, which is 4 times bigger in comparison to bare FBG. After the cooling process of PDMS from 80 °C to 25 °C, thermal sensitivity reached 39.44 pm/°C, and the Bragg wavelength reduced to 1552.26 nm. This decline corresponds to the temperature decrease by 57.55 °C.

Figure 7 shows the reflective spectral characteristic of the Bragg grating before and after the curing, including the relaxation time of 50 min. The presented results indicate that this type of encapsulation does not affect the structure of the FBG. However, the encapsulation causes a shift of the reflected spectrum of FBG to lower wavelengths due to the shrinkage of PDMS during curing. A shift of the Bragg wavelength was 1.8652 nm. The spectral width of Bragg grating (Full Width Half Max) increased from the value 2.324 nm to 2.386 nm. However, this shift does not affect the functionality of the FBG sensor.

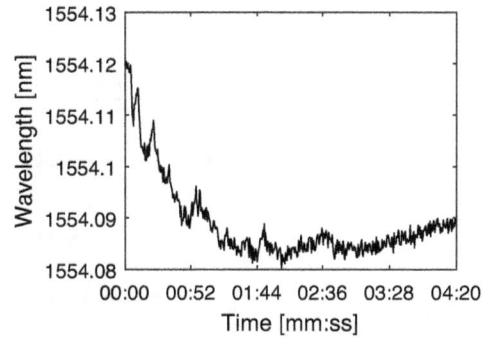

(a) Potting into PDMS.

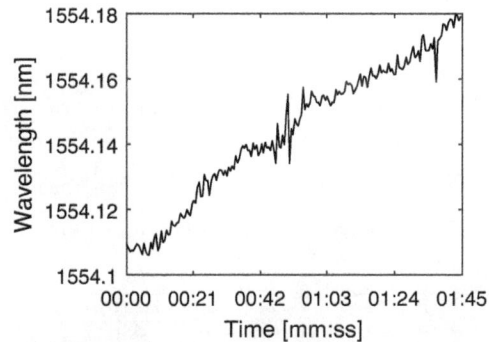

(b) Inserting into thermal box.

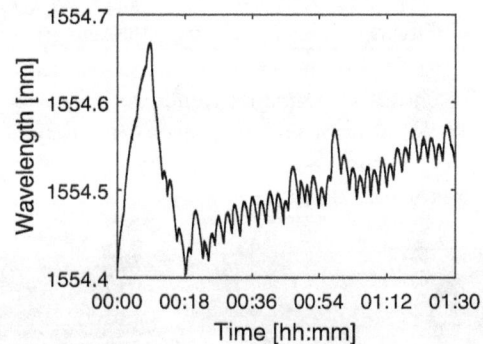

(c) Curing.

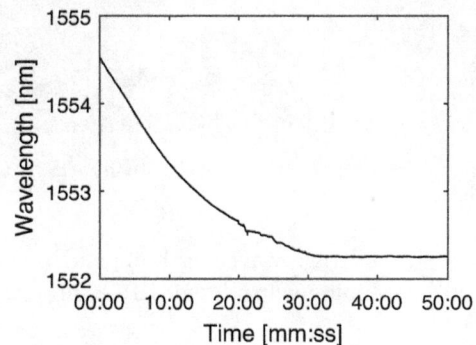

(d) 50-minute relaxation.

Fig. 6: Spectral characteristics of Bragg grating during encapsulation.

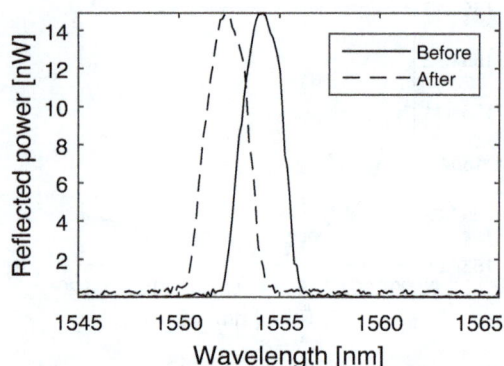

Fig. 7: The spectral characteristic of FBG before and after curing.

Table 1 shows the spectral characteristics of FBG in various stages of encapsulation. There are given values of the parameters before and after each phase.

Tab. 1: Spectral characteristics of FBG during encapsulation of FBG (all 4 phases).

Phases of encapsulation	Wavelength [nm]	
	Before	After
Encapsulating	1554.12	1554.12
Inserting into heating box	1554.11	1554.18
Curing	1554.40	1554.53
Relaxation	1554.52	1552.26

Figure 8 shows implemented encapsulation of FBG into PDMS. Dimensions of the sensor are 60x25x4 mm. The priority of the research was not achieving minimization regarding design.

Fig. 8: The spectral characteristic of FBG before and after curing.

Figure 9 shows Bragg wavelength dependence of encapsulated and non-encapsulated FBG on temperature.

Non-encapsulated Bragg grating at wavelength 1554.1203 achieves temperature sensitivity of 10.378 pm/°C. After encapsulation, the Bragg wavelength was changed to the value 1552.2551 nm and temperature sensitivity was increased to 39.44 pm/°C. Our results correspond to the results of the paper [4].

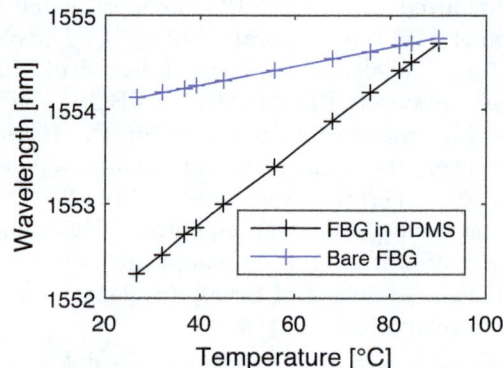

Fig. 9: Bragg wavelength dependence of non-encapsulated (blue line) and encapsulated (black line) FBG on temperature.

4. Conclusion

The aims of authors were both an analysis of the impact of high temperature and itself curing of PDMS on spectral characteristics of FBG and temperature sensitivity of FBG within performed experiment. A secondary motivation is to use this encapsulation of FBG sensor in medical applications. The selected product was Sylgard 184 due to suitable properties. PDMS has very good heat resistance, excellent elastic properties, harmlessness, non-toxicity, non-flammability, and electric non-conductivity. Realization of encapsulation of FBG was split into consecutive phases. At the beginning, we made a reference measurement of spectral characteristics of FBG. The obtained data were compared with the encapsulated FBG, including a 50-min relaxation period, until the stabilization of Bragg wavelength (Tab. 1 and Fig. 7). At all stages of the encapsulation, we monitored the influence of temperature, the curing, and the mechanical stress on spectral characteristics of FBG and temperature sensitivity of FBG (Fig. 6 and Fig. 9). The presented results indicate that this type of encapsulation does not affect the structure of the FBG, it does not affect the functionality, and it represents an alternative method of encapsulation of FBG. The advantage is the fact that we can use the potential properties of PDMS, and we can expand potential application utilization of FBG in the sensor applications including medical ones.

A follow-up research will focus on both the analysis of the influence of different temperatures and duration of curing on the spectral characteristic of the FBG, the reflectivity of FBG, temperature sensitivity, and deformation sensitivity. The reproducibility experiment is also a future goal. This research was not the subject of this article.

Acknowledgment

This article was supported by the projects of Technology Agency of the Czech Republic TA03020439 and TA04021263. This article was also supported by Ministry of Education, Youth and Sports of the Czech Republic within the projects no. SP2016/149 and project no. CZ.1.07/2.3.00/20.0217 within the frame of the operation programme Education for competitiveness financed by the European Structural Funds and from the state budget of the Czech Republic. This article was also supported by the project VI20152020008 (GUARDSENSE II).

References

[1] LUPI, C., F. FELLI, L. IPPOLITI, M. A. CAPONERO, M. CIOTTI, V. NARDELLI and A. PAOLOZZI. Metal coating for enhancing the sensitivity of fibre Bragg grating sensors at cryogenic temperature. *Smart Materials and Structures*. 2005, vol. 14, iss. 6, pp. 71–76. ISSN 0964-1726. DOI: 10.1088/0964-1726/14/6/N02.

[2] LI, X. C., F. PRINZ and J. SEIM. Thermal behavior of a metal embedded fiber Bragg grating sensor. *Smart Materials and Structures*. 2001, vol. 10, iss. 4, pp. 575–579. ISSN 0964-1726. DOI: 10.1088/0964-1726/10/4/301.

[3] WANG, Y., T. G. LIU, L. N. LIU and J. F. JIANG. Study on fiber Bragg grating sensor encapsulated by the alloyed steel. *Guangxue Jishu/Optical Technique*. 2006, vol. 32, iss. 6, pp. 923–925. ISSN 1002-1582.

[4] PARK, C. S., K. I. JOO, S. W. KANG and H. R. KIM. A PDMS-Coated Optical Fiber Bragg Grating Sensor for Enhancing Temperature Sensitivity. *Journal of the Optical Society of Korea*. 2011, vol. 15, iss. 4, pp. 329–334. ISSN 1226-4776. DOI: 10.3807/JOSK.2011.15.4.329.

[5] HOPF, R., L. BERNARDI, J. MENZE, M. ZUNDEL, E. MAZZA and A. E. EHRET. Experimental and theoretical analyses of the age-dependent large-strain behavior of Sylgard 184 (10:1) silicone elastomer. *Journal of the Mechanical Behavior of Biomedical Materials*. 2016, vol. 60, iss. 1, pp. 425–437. ISSN 1751-6161. DOI: 10.1016/j.jmbbm.2016.02.022.

[6] FENDINGER, N. J. *Organosilicon Chemistry Set - Polydimethylsiloxane (PDMS): Environmental Fate and Effects*. Weinheim: Wiley, 2005. ISBN 978-3527620777. DOI: 10.1002/9783527620777.ch103c.

[7] FENDINGER, H. J., R. G. LEHMANN and E. M. MINAICH. *Organosolicon Materials - Polydimethylsiloxane*. New York: Springer, 1997. ISBN 978-3-662-14822-8. DOI: 10.1007/978-3-540-68331-5_7.

[8] KERSEY, A. D., M. A. DAVIS, H. J. PATRICK, M. LEBLANC, K. P. KOO, C. G. ASKINS, M. A. PUTNAM and E. J. FRIEBELE. Fiber grating sensors. *Journal of Lightwave Technology*. 2002, vol. 15, iss. 8, pp. 1442–1463. ISSN 1558-2213. DOI: 10.1109/50.618377.

About Authors

Jan NEDOMA was born in 1988 in Prostejov. In 2012 he received his B.Sc. from VSB–Technical University of Ostrava, Faculty of Electrical Engineering and Computer Science, Department of Telecommunications. Two years later he received his M.Sc. in the field of Telecommunications from the same institution. Currently he is an employee and a Ph.D. student of the Department of Telecommunications. He works in the field of optical communications and fiber optic sensor systems.

Marcel FAJKUS was born in 1987 in Ostrava. In 2009 he received his B.Sc. degree from VSB–Technical University of Ostrava, Faculty of Electrical Engineering and Computer Science, Department of Telecommunications. Two years later he received his M.Sc. in the field of Telecommunications from the same institution. Currently he is an employee and a Ph.D. student of the Department of Telecommunications. He works in the field of optical communications and fiber optic sensor systems.

Lukas BEDNAREK was born in 1988 in Frydek-Mistek. In 2011 he received his B.Sc. on VSB–Technical University of Ostrava, Faculty of Electrical Engineering and Computer Science, Department of Computer Science. Two years later he received his M.Sc. in the field of Telecommunications on Department of Telecommunications. He is currently a Ph.D. student, and works in the field of optical communications and aging of optical components.

Jaroslav FRNDA was born in 1989 in Martin, Slovakia. He received his M.Sc. from the VSB–Technical University of Ostrava, Department of Telecommunications, in 2013, and now he is continuing in his Ph.D. study at the same place. His research interests include Quality of Triple play services and IP networks.

Jan ZAVADIL was born in 1989 in Vyskov, Czech Republic. In 2014 he received his Bachelor's degree from VSB–Technical University of Ostrava, Faculty of Electrical Engineering and Computer Science, Department of Telecommunication Technology. In 2016 he received his M.Sc. in the field of Information technologies from the same institution. He is currently a Ph.D. student.

Vladimir VASINEK was born in Ostrava. In 1980 he graduated in Physics, specialization in Optoelectronics, from the Science Faculty of Palacky University. He was awarded the title of RNDr. at the Science Faculty of Palacky University in the field of Applied Electronics. The scientific degree of Ph.D. was conferred upon him in the branch of Quantum Electronics and Optics in 1989. He became an associate professor in 1994 in the branch of Applied Physics. He has been a professor of Electronics and Communication Science since 2007. He pursues this branch at the Department of Telecommunications at VSB–Technical University of Ostrava. His research work is dedicated to optical communications, optical fibers, optoelectronics, optical measurements, optical networks projecting, fiber optic sensors, MW access networks. He is a member of many societies – OSA, SPIE, EOS, Czech Photonics Society, he is a chairman of the Ph.D. board at the VSB–Technical University of Ostrava. He is also a member of habitation boards and the boards appointing to professorship.

9

Enhancing Lighting Performance of White LED Lamps by Green Emitting Ce,Tb Phosphor

Nguyen Doan Quoc ANH, Tran Hoang Quang MINH, Nguyen Huu Khanh NHAN

Faculty of Electrical and Electronics Engineering, Ton Duc Thang University, 19 Nguyen Huu Tho Street, Tan Phong Ward, District 7, Ho Chi Minh City, Vietnam

nguyendoanquocanh@tdt.edu.vn, tranhoangquangminh@tdt.edu.vn, nguyenhuukhanhnhan@tdt.edu.vn

Abstract. *With the development of high-efficiency and high-power Light-Emitting Diodes (LEDs), it has become possible to use LEDs in lighting and illumination. In last decades, developing a new method for improving lumen output and Angular Color Uniformity (ACU) is the main direction in LED technology. In this paper, an innovative approach for enhancing lighting performance (lumen output and angular color uniformity) of Multi-Chip White LED lamps (MCW-LEDs) was proposed and demonstrated by mixing the green $Ce_{0.67}Tb_{0.33}MgAl_{11}O_{19}$: Ce,Tb (CeTb) phosphor into their phosphor compounding. With varying CeTb concentration, ACU and lumen output with Conformal Phosphor Package (CPP) and In-cup Phosphor Package (IPP) are calculated, displayed and analyzed. The results show that the lumen output and the ACU of 7000 K and 8500 K MCW-LEDs increased remarkably in comparison with the older works. Using green CeTb is a prospective method for improving lighting performance of MCW-LEDs in future.*

Keywords

Angular color uniformity, $Ce_{0.67}Tb_{0.33}MgAl_{11}O_{19}$: CeTb, lighting performance, lumen output, multi-chip white LED lamps.

1. Introduction

Multi-chip white LED lamps have many benefits for LED lighting applications such as long life, compactness, high efficiency, low power consumption [1]. The main lighting properties of MCW-LEDs are lumen output and the Angular Color Uniformity (ACU), which are considered and analyzed in many studies [2] and [3]. The yellow YAG: Ce phosphor material is mixed with silicone glue to form a phosphor compounding which absorbs the blue light of the LED chips to emit the yellow light. The white light with various average Correlated Color Temperatures (CCTs) is achieved by the mixture of the yellow, blue and red light. Light rays go through phosphor particles and scatter in the phosphor compounding. The blue light is compounded with the yellow light during the operation of scattering. The blue light becomes weaker due to the absorption process of the yellow YAG: Ce phosphor particles. Meanwhile, the converted yellow light is amplified through each scattering. Correspondingly, the difference of light intensity distributions can cause a yellow ring phenomenon [4].

In purpose to enhance light quality and lumen output, the multiple phosphors or the optical structures of white LED lamps must be optimized. The substantial influence of phosphor geometry on the lumen output of LED lamps with higher Color Rendering Index (CRI) by adding green $(Ba,Sr)_2SiO_4:Eu_2^+$ and red $CaAlSiN_3:Eu_2^+$ phosphors with various phosphor structures was proposed in [5]. In another way, [6] and [7] provided G-A-R multi-package LED with higher lumen output and good CRI. The CRI of the LED lamps was enhanced by using multi-chromatic phosphor. In general, the above works only focused on improvement of CRI and lumen output without solving ACU enhancement problem. Moreover, these works only focused on single-chip white LED lamps, to this date very few works focused on improvement of the light quality of multi-chip white LED lamps with high CCTs.

The green-emitting CeTb is one type of hexagonal poly-aluminate. Moreover, its structure is similar to magnetoplumbite, which is characterized by the hexagonal symmetry of the space group P63/mmc [5]. Its related compositions include CeO_2, Tb_4O_7, MgO, Al_2O_3, Ce_3^+ and Tb_3^+ ions, all of which were thoroughly mixed in agate mortar [7]. The Ce_3^+ ion plays a role of the sensitizer for Tb$_3^+$ luminescence in the green-emitting CeTb. CeTb is applied particularly for very high-loading and long lifetime fluorescent lamps. Correspondingly, it is one of the popular commercialized oxide phosphors.

In this paper, an innovative application of adding green CeTb phosphor particles into the phosphor compounding of MCW-LEDs with Conformal Phosphor Package (CPP) and In-cup Phosphor Package (IPP) for improving color uniformity and lumen output is presented and demonstrated. Simulation results indicated that the participation of the CeTb particles enhanced the scattering event of the phosphor compounding. Therefore, the light distribution of MCW-LEDs could not depend on their wavelengths. After that, the uniform spatial color distribution of the MCW-LEDs could thus be accomplished. This research work has three parts. Firstly, the physical model of MCW-LEDs using the Light Tool simulation program was simulated. Then, by putting CeTb particles into the phosphor compounding of the MCW-LEDs with the conformal and in-cup package was proposed. Finally, the influence of CeTb particles concentration on the lighting performance of the MCW-LEDs was calculated, analyzed and demonstrated. The researched results showed that the ACU and lumen output of the MCW-LEDs increased significantly after mixing green CeTb with yellow YAG:Ce phosphor particles.

2. Physical and Mathematical Model of Real MCW-LEDs

By using LightTools 8.1.0 software, the physical model of the MCW-LEDs is presented and simulated. The modeling work can be divided into 2 main parts:

- Constructing the mechanical structures and the optical properties of MCW-LED lamps.

- Verifying the optical properties of phosphor compounding through varying CeTb particles concentration.

The 2-D model of MCW-LED lamps by using Light-Tools 8.1.0 software is presented in Fig. 1. In this model, the reflector has a bottom length of 8 mm, a height of 2.07 mm, and a length of 9.85 mm at its top surface. The conformal phosphor compounding, with the fixed thickness 0.08 mm, which covers the nine chips. Each LED chip with a square base of 1.14 mm and a height of 0.15 mm is bonded in the cavity of the reflector shown in Fig. 1(a). The radiant flux of each blue chip is 1.16 W. For the conformal structure as shown in Fig. 1(b), the novel phosphor compounding is coated conformally on the chips. As for the in-cup

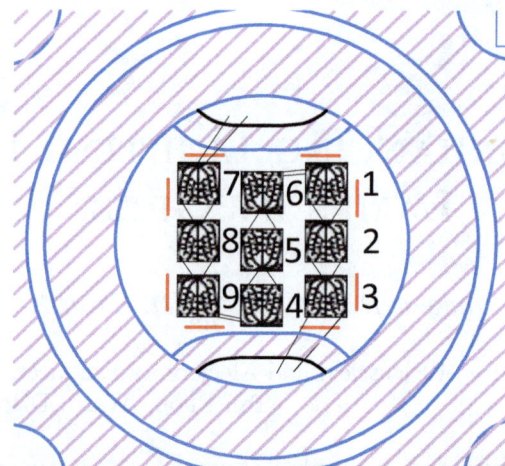

(a)

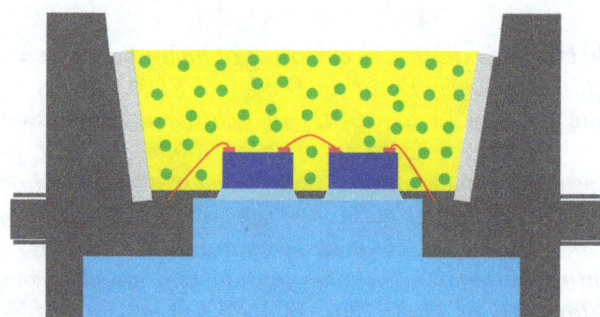

(b)

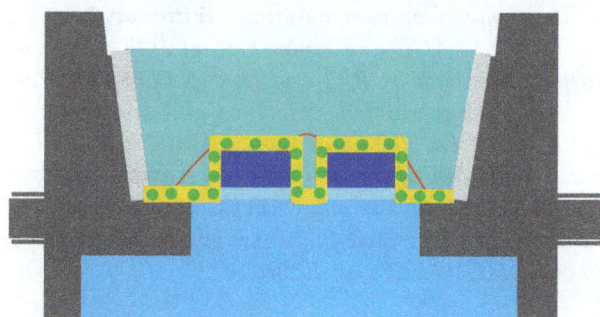

(c)

Fig. 1: (a) is the position diagram of chips, (b) and (c) are the simulated conformal and in-cup phosphor package.

phosphor structure, the novel phosphor compounding is mixed in the silicone lens, as displayed in Fig. 1(c). The scattering of phosphor particles could be analyzed by using Mie-theory. In this research, the average diameters of the phosphor particles are 14.5 nm, which are same as the real parameters [4]. The new phosphor compounding is the mixture of the CeTb and YAG:Ce particles, and the silicone glue. The refractive indexes of CeTb and YAG:Ce phosphors and its silicone glue are in turn 1.85, 1.83 and 1.52, which are same as the real parameters. Besides the refractive index and the size of the phosphor particles, the emission spectra of the novel phosphor compounds was determined. The emission spectra of the conformal phosphor compounding are verified by enhancing CeTb concentration from 0 % to 20 %, as presented in Fig. 2(a). As for Fig. 2(b), the emission spectra of the in-cup phosphor compounding are displafrefyed by varying CeTb concentration from 0 % to 1.0 %. The results demonstrated that the luminous efficiency of MCW-LEDs remarkably enhanced after compounding CeTb particles to phosphor compounds.

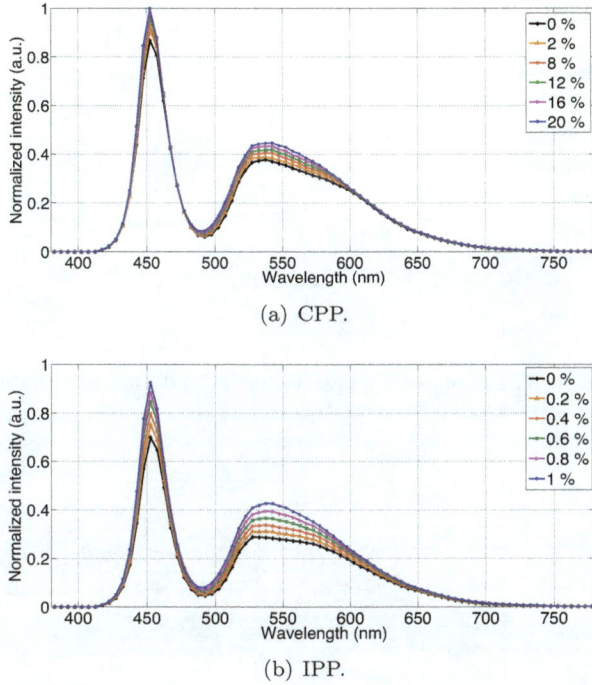

(a) CPP.

(b) IPP.

Fig. 2: Emission spectra of multiple phosphors.

In this section, Mie theory is applied to calculate the scattering of phosphor particles in an angle range from 0° to 360°. The angular light scattering intensities could be calculated by the below equations:

$$i_1 = \sum_{n=1}^{\infty} \frac{2n+1}{n(n+1)} \left[\begin{array}{c} a_n(x,m)\pi_n(\cos\theta) \\ b_n(x,m)\tau_n(\cos\theta) \end{array} \right], \quad (1)$$

$$i_2 = \sum_{n=1}^{\infty} \frac{2n+1}{n(n+1)} \left[\begin{array}{c} a_n(x,m)\tau_n(\cos\theta) \\ b_n(x,m)\pi_n(\cos\theta) \end{array} \right], \quad (2)$$

where m is a refractive index, x is a size parameter, $\pi_n(\cos\theta)$ and $\tau_n(\cos\theta)$ are the angular functions, a_n and b_n are the expansion coefficients. The scattering intensities are calculated for 555 nm and 453 nm wavelengths, which are the emission peaks of the yellow light of YAG:Ce phosphor and the blue light of chips, respectively.

Figure 3 displayed the scattered light intensity distributions of CeTb and YAG:Ce particles with wavelength (Fig. 3(a)) 453 nm, (Fig. 3(b)) 555 nm. The

scattering enhancement of phosphor compounding can be achieved after adding CeTb phosphor. Correspondingly, the angular color distribution of MCW-LEDs may also be reconfigured and get better values with CeTb particles (Fig. 4).

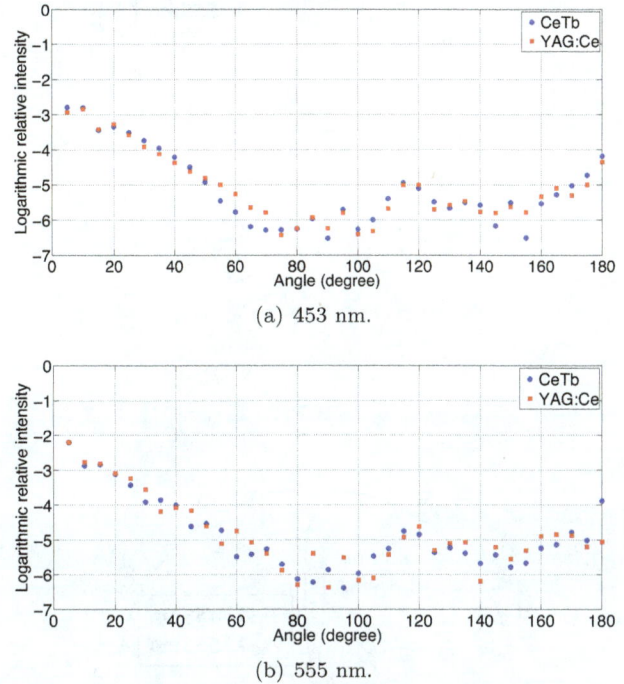

(a) 453 nm.

(b) 555 nm.

Fig. 3: Emission spectra of multiple phosphors.

3. Simulation Results and Discussions

In this research, it is necessary to keep the MCW-LED work at high CCTs of 7000 K and 8500 K for meeting the LEDs product specification. If the green CeTb phosphor concentration grows, its yellow YAG:Ce phosphor concentration needs to be decreased in order to fix the CCTs of 7000 K and 8500 K. The weight percentage of the LED phosphor layer could be calculated as:

$$\sum W_{\text{phosphor layer}} = W_{\text{yellow phosphor}} + W_{\text{silicone}} + W_{\text{green phosphor}} = 100\,(\%), \quad (3)$$

where the $W_{\text{phosphor layer}}$, $W_{\text{yellow phosphor}}$ and $W_{\text{green phosphor}}$ are the weight percentage of the silicone glue, the yellow YAG:Ce phosphor and the green CeTb phosphor, respectively.

Figure 5 presented the angular color deviation of MCW-LEDs with CeT band without CeTb in the phosphor compounding for CPP (Fig. 5(a)) and IPP (Fig. 5(b)). From the simulation results, the CCT

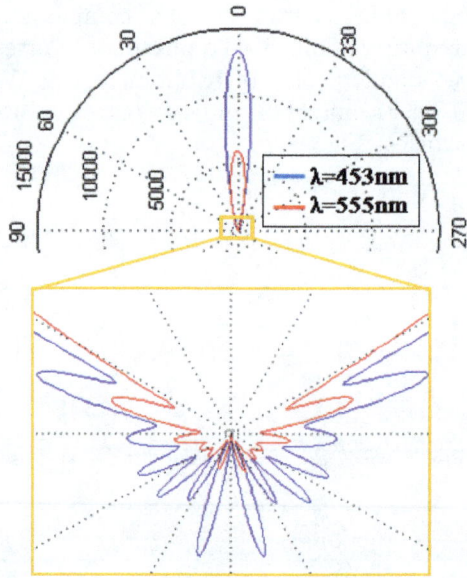

(a) CPP.

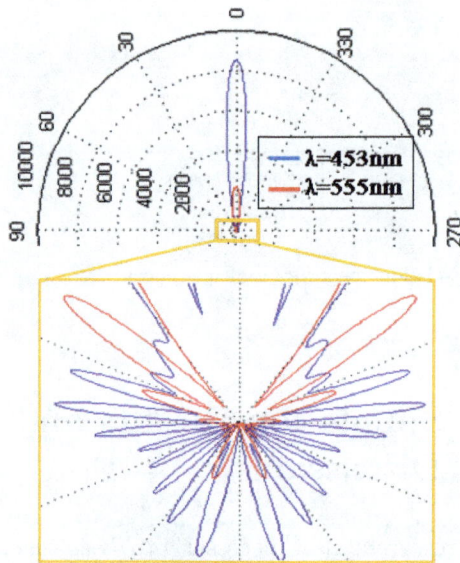

(b) IPP.

Fig. 4: The angular scattering amplitudes of the diffusional particles.

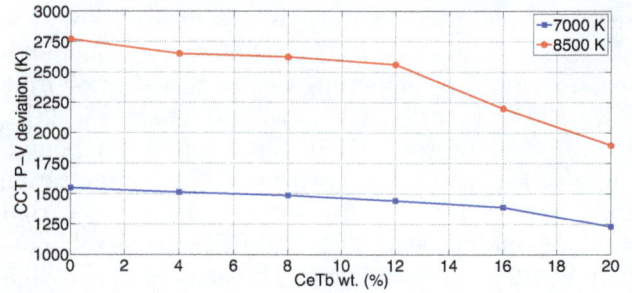

(a) CPP.

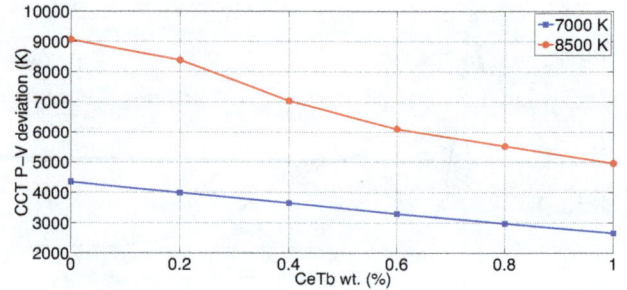

(b) IPP.

Fig. 5: The angular CCT peak-valley (P-V) deviation with various CeTb concentration.

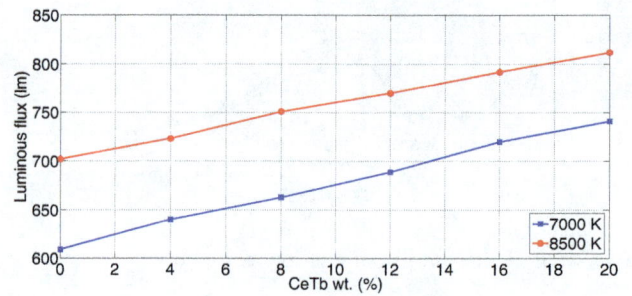

(a) CPP.

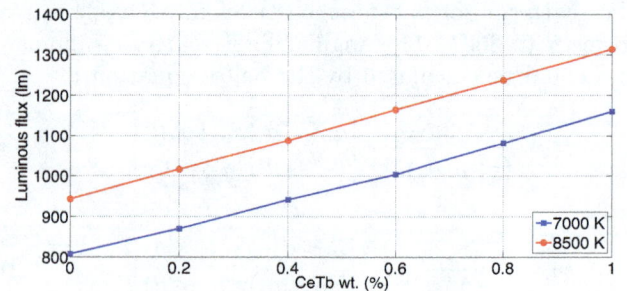

(b) IPP.

Fig. 6: The lumen output at average CCTs 7000 K and 8500 K with different CeTb concentration.

peak-valley Deviation (DCCT) reduces significantly after involving CeTb. It means that the spatial color distribution of MCW-LEDs with CeTb is much flatter than the non-CeTb case. In addition to that, the growing of the weight of CeTb increases continually from 0 % to 20 % for the Conformal Phosphor Packaging (CPP) and 0 % to 1.0 % for the Phosphor In-cup Packaging (IPP). Referring to the simulation results in Fig. 6, it is clear that the lumen output grows remarkably with increasing of the concentration of CeTb.

In summary, the MCW-LED packages with better-Correlated Color Temperature (CCT) uniformity could

be accomplished. However, referring to the accomplished simulation results in Fig. 6 and Fig. 7, it can be found that the higher CeTb concentration is involved, the higher the luminous flux, but less CRI can be obtained for both CPP and IPP MCW-LEDs.

(a) CPP.

(b) IPP.

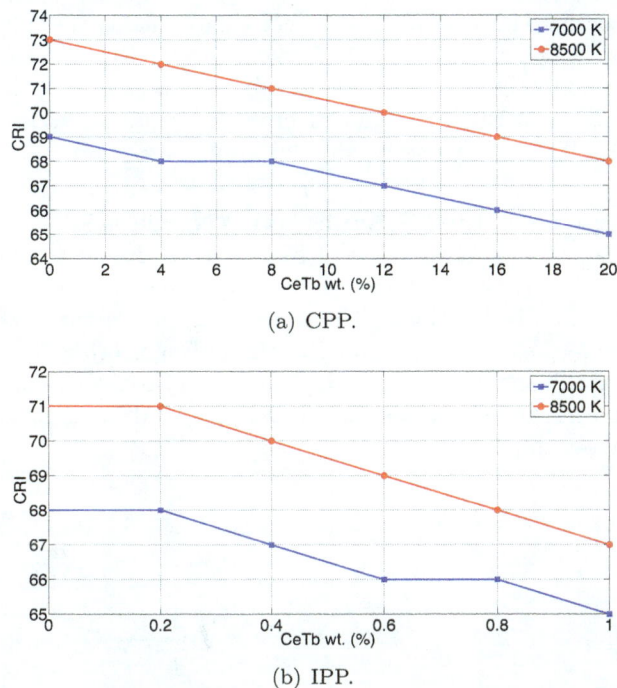

Fig. 7: The color rendering index at average CCTs 7000 K and 8500 K with different CeTb concentration.

4.   Conclusion

In this paper, the green CeTb phosphor is proposed for enhancing the spatial color uniformity and the luminous efficiency of both CPP and IPP MCW-LED lamps. This study provided that with growing concentration of green CeTb phosphor the ACU and the lumen output of MCW-LED lamps increased significantly in comparison with the related works. Although, decreasing CRI with increasing concentration of the green CeTb phosphor is a limitation of this research. This study is the first step towards enhancing our understanding of green CeTb phosphor effect on the lighting performance of MCW-LED lamps. In the future work, the influence of green CeTb size on lighting performance and how to increase CRI of MCW-LED lamps with using green CeTb phosphor must be investigated.

Acknowledgment

This paper was supported by Professor Hsiao-Yi Lee, Department of Electrical Engineering National Kaohsiung University of Applied Sciences, Kaohsiung, Taiwan.

References

[1] QUOC, A. N.-D., M.-F. LAI, H.-Y. MA and H.-Y. LEE. Enhancing of correlated color temperature uniformity for multi-chip white-light LEDs by adding SiO2 in phosphor layer. *Journal of the Chinese Institute of Engineers*. 2015, vol. 38, iss. 3, pp. 297–303. ISSN 0253-3839. DOI: 10.1080/02533839.2014.981214.

[2] LIU, Z., S. LIU, K. WANG and X. LUO. Analysis of factors affecting color distribution of white LEDs. In: *International Conference on Electronic Packaging Technology & High Density Packaging (ICEPT-HDP)*. Shanghai: IEEE, 2008, pp. 1–8. ISBN 978-1-4244-2739-0. DOI: 10.1109/ICEPT.2008.4607013.

[3] LAI, M.-F., A. N.-D. QUOC, H.-Y. MA and H.-Y. LEE. Scattering effect of SiO2 particles on correlated color temperature uniformity of multi-chip white light LEDs. *Journal of the Chinese Institute of Engineers*. 2016, vol. 39, iss. 4, pp. 468–472. ISSN 0253-3839. DOI: 10.1080/02533839.2015.1117950.

[4] LIU, S. and X. LUO. *LED Packaging for Lighting Applications: Design, Manufacturing and Testing*. 1st ed. Singapore: Wiley, 2011. ISBN 978-0-470-82783-3.

[5] WON, Y.-H., H. S. JANG, K. W. CHO, Y. S. SONG, D. Y. JEON and H. K. KWON. Effect of phosphor geometry on the luminous efficiency of high-power white light-emitting diodes with excellent color rendering property. *Optics Letters*. 2009, vol. 34, iss. 1, pp. 1–3. ISSN 0146-9592. DOI: 10.1364/OL.34.000001.

[6] OH, J. H, Y. J. EO, S. J. YANG and Y. R. DO. High-color-quality multipackage phosphor-converted LEDs for yellow photolithography room lamp. *IEEE Photonics Journal*. 2015, vol. 7, iss. 2, pp. 1–9. ISSN 1943-0655. DOI: 10.1109/JPHOT.2015.2415674.

[7] ZHENG, M., W. DING, F. YUN, D. XIA, Y. HUANG, Y. ZHAO, W. ZHANG, M. ZHANG, M. GUO and Y. ZHANG. Study of High CRI White Light-emitting Diode Devices with Multichromatic Phosphor. In: *64th Electronic Components and Technology Conference (ECTC)*. Orlando: IEEE, 2008, pp. 2236–2240. ISBN 978-147992407-3. DOI: 10.1109/ECTC.2014.6897614.

About Authors

Nguyen Doan Quoc ANH was born in Khanh Hoa province, Vietnam. He has been working at the Faculty of Electrical and Electronics Engineering, Ton Duc Thang University, Ho Chi Minh City, Vietnam. He received his Ph.D. degree from National Kaohsiung University of Applied Sciences, Taiwan. His research interest is optoelectronics (such as Multi-chip white light LEDs, free-form lens, and optical material) and Power system.

Tran Hoang Quang MINH defended his Ph.D. thesis at Tomsk Polytechnic University, Tomsk City, Russian Federation. The author's major fields of study are High-voltage Power System, Relay Protections, and Optoelectronics. He is working as Lecturer in Faculty of Electrical and Electronics Engineering, Ton Duc Thang University, Ho Chi Minh City, Vietnam.

Nguyen Huu Khanh NHAN defended his Ph.D. thesis at Institute of Research and Experiments for Electrical and Electronic Equipment, Moscow, Russian Federation. He is working as Lecturer in Faculty of Electrical and Electronic Engineering, Ton Duc Thang University, Ho Chi Minh City, Vietnam. His research interests include VLSI, MEMS and LED driver chips.

10

NON-DESTRUCTIVE FIBER-OPTIC SENSOR SYSTEM FOR THE MEASUREMENT OF SPEED IN ROAD TRAFFIC

Jan NEDOMA, Marcel FAJKUS, Lukas BEDNAREK, Vladimir VASINEK

Department of Telecommunications, Faculty of Electrical Engineering and Computer Science, VSB–Technical University of Ostrava, 17. listopadu 15, 70833 Ostrava-Poruba, Czech Republic

jan.nedoma@vsb.cz, marcel.fajkus@vsb.cz, lukas.bednarek@vsb.cz, vladimir.vasinek@vsb.cz

Abstract. *Fiber-optic sensors offer an attractive option to existing sensors for the measurement of the vehicle speed in road traffic. This article describes the measuring scheme of two interferometric sensor units including input-output components for the measurement of the vehicle speed. The interferometric sensors operate on a principle of Mach-Zehnder interferometer. The sensors are constructed to detect a vibration caused by vehicles moving on roads. The sensor system processes the vibrational response, and the vehicle speed is calculated in a time domain. DFB laser was used with a wavelength of 1550 nm and output power of 1– 5 mW. The solution provides very high sensitivity. The performance of the proposed system was verified by a series of experimental measurements of the speed. The vehicle speed was monitored by GPS. The highest relative difference of the evaluated speed against GPS data was 7.7 %, the smallest was 1.36 %. When recalculated on kph, the absolute error ranged in the tolerance of ±3 kph, which denotes segmental measuring systems in CZE.*

Keywords

Interferometer, measurement of speed, non-destructive sensor, speed.

1. Introduction

Traffic detectors are devices that scan input data and information for further transport telematics systems such as ITS (Intelligent Transport Systems). Detectors can use different physical principles, the data are usually collected without limiting the traffic flow while vehicles move on roads. The obtained data are used for the subsequent processing of crucial traffic-engineering parameters, including the current vehicle speed. The occupancy of the detector and the time of the detector's occupancy are the basic entries for the evaluation of traffic data (i.e. the pass-by of a vehicle or the stopping of a vehicle in a certain lane). Systems, which solve the issue of the measurement of vehicle speed, can be divided into stationary or mobile according to their basic operation. Stationary systems are devices directly connected to the road (e.g. induction loops, pneumatic detectors, etc.) or they can be installed as a part of other devices (e.g. camera system of toll gates). Mobile systems are intended to be used in specific situations. The used principle of the detection of vehicles is another criterion of the classification in which touch detectors, ultrasonic, electromagnetic with stationary field or light field are the most widely used. According to the installation procedure, the detectors were divided into destructive and non-destructive. Destructive detectors interfere with their construction elements into the road or its surface, and they consequently disrupt the integrity of the road. Until recently, destructive detectors were used as majority detectors.

As an interesting alternative to existing conventional sensors, there is a possibility of using fiber-optic sensors. The subject of this article is the verification of this fact using fiber-optic interferometry. These compiled sensors can both replace the current detectors and also open up a new application potential. In many cases, due to their properties, sensors can find applications in areas, where their use is still not possible or expensive (a passive mode from the viewpoint of power supply towards conventional electrical sensors). Substantial advantage of optical fibers is the insensitivity to electromagnetic interference, the material does not rust, and it can operate over a wide temperature range. The flexibility and size of optical fibers allow their simplified installation. The massive expansion of fiber-optic cables offers the possibility of connecting the existing telecommunications fiber-optic networks along roads. This fact confirms that the direction of the development field was correctly chosen, and it also underlines the considerable list of positive characteristics of fiber-optic sensors.

Strictly defined distance between two fiber-optic interferometers can be used to determine the desired speed of the object [1]. Speed is calculated using a time interval between interferometers. Interferometers can be also used for the detection of vehicle axles. This fact is proved further in this article in the section focused on the methods of the speed measurement.

This article [2] describes a distributed fiber-optic sensor with Fabry-Perot interferometer which is used for collecting fundamental information about road vehicles or about traffic in general. The measurement was verified in real traffic. We verified the capability of detection of parameters like speed, vehicle classification, a weight of vehicle or traffic flow. The optical fiber was used as a sensor medium. Fabry-Perot interferometer was used for the evaluation. The sensing optical fiber was stored in a special metal protection casing, and it was installed on the road surface. The results of experimental measurements have proved that it is not necessary to install the fiber into the road, but it is required to implement it to the road surface.

The utility model [3] considers a complex information system with the focus on sensing the speed and weight of vehicles. Fabry-Perot interferometer is used for the evaluation. Interferometer and evaluation unit can be placed out of the roadway. The optical fiber needs to be built into the roadway. When a car is passing through the optical fiber embedded into the road, it causes a phase change in the transmitted light waves, and these interference patterns are subsequently evaluated. The use of two optical fibers is necessary for the calculation of velocity. Vehicle speed can be determined based on the time interval of passing between the fibers.

The patent [4] relates to the detection of vehicles passing by on roads. Optical interferometer and optical fiber embedded in the roadway were used for the detection. Monitored transport system comprises at least one sensor placed in the roadway for a vehicle detection. The integration of two fibers within a single sensor or a combination of more sensor units (at least two) are required for the measurement of speed. Here, we can use strictly define the distance. When the vehicle is passing by the fiber, it changes the phase of light waves by the pressure of the vehicle. The resulting phase shift is further evaluated.

Utility model [5] relates to the detection and measurement of the speed of vehicles using two fiber-optic Mach-Zehnder interferometers. Measuring and reference branches have a length of 5.5 m. These branches have a shape of the ring which is built in the roadway in a protective metal enclosure with plastic filling diameter of 80 mm. The pass-by of the vehicle changes the phase of the light wave by the pressure of the tires on fiber embedded in the roadway. The detection system evaluates changes in the interference patterns.

The patent [6] reveals new facts about the fiber-optic interferometric sensor for the monitoring of traffic. The authors tested the possibility of speed detection for cars without the need to implement the optical fiber on or into the roadway. The sensor uses a well-known link with Mach-Zehnder interferometer. The authors tested the ability to analyze the frequencies, which are used to detect the type of a vehicle which is passing by, for example, car vs. truck. The authors specify the possible extension of the application possibilities of both measuring the speed and measuring the weight of vehicles.

The patent [7] relates to the detection of vehicle speed using the two interferometers of the Michelson type. The pass-by of the vehicle around a first interferometric box will detect the reflected beam from the vehicle and switch the timer. The pass-by of the vehicle around the second interferometer box again detects the reflected beam from the vehicle, and the timer is turned off. Then, we can calculate speed from a strictly defined distance of interferometers and from the time interval of the pass-by of the vehicle between interferometers. Therefore, it is an extrinsic non-destructive sensor.

Existing sensing systems for the detection of vehicle speed, which can be placed on a roadway or in a railway yard, possibly inside a roadway, are formed especially by inductive loops, microwave detectors and camera systems. Based on the literature review from the field of interferometric measurements, the main contribution and advances of this work is in the creation of a new fiber-optic intrinsic sensor system which can measure the speed of vehicles, trams or trains. This system is based on the evaluation of time-shifted signals. The system is non-destructive to a roadway or railyard, and its output could be directly connected to the existing telecommunication optical networks by a suitable design of the interface.

Pilot measurements were carried out for the road traffic, further measurements are currently being prepared for the railway traffic. The measurement system can improve the traffic safety on roads, it can be also used in the railway traffic. Reasons are obvious – one of the characteristic features of optical technologies is the maximum resistance to electric and electromagnetic interferences when especially electric and electromagnetic systems have problems with the functional reliability due to the introduction of new tractive technologies into power engines. The reason is a considerable increase in electromagnetic interference appearing in the vicinity of modern power engines, and interference which spreads in rail tracks that are superimposed by reverse traction currents. Next problem of electric detection systems is a small resistance to effects or a damage caused by atmospheric discharge, or more precisely, by lightning strike into or near the railway installations. Installations, which have metallic couplers, can be affected by the appearance of undesirable inductive loops passing through their electrically conductive circuits which causes that the protection against such an undesirable influence or damage is very difficult. The proposed detection system using optical fibres should eliminate the above mentioned problems.

2. Operating Principles

Interferometry is an optical method that can monitor the phase difference between two optical beams which pass through similar (if possible identical) optical paths. Phase shift arises in the interferometer. Interferometry is able to detect three parameters. These parameters affect the optical beam propagating along the optical path:

- change of the propagation speed,

- change of the wavelength,

- change of the route length.

If the change occurs in any of these parameters, then a change also occurs in a wave phase. This change depends on the length of the path L, the refractive index n, and the wavelength λ according to the equation:

$$\phi = 2\pi L \frac{n}{\lambda} = kLn, \qquad (1)$$

where L is the length of used fiber, n is the refractive index of the core, λ is the wavelength of the radiation source and k is the size of the wave vector.

An interference maximum is a place where two waves with the same phase are joined, and it is given by:

$$\Delta s = 2k \frac{\lambda}{2}. \qquad (2)$$

An interference minimum is a place where two waves with the opposite phase are joined, and it is given by:

$$\Delta s = (2k + 1) \frac{\lambda}{2}, \qquad (3)$$

where Δs is the path difference, k is the size of the wave vector ($\frac{2\pi}{\lambda}$).

The output intensity of the interferometer can be expressed by the relation:

$$I = \frac{I_0 \alpha}{2} (1 + \cos \Delta \phi), \qquad (4)$$

where α expresses the optical loss of the interferometer, I0 is the light intensity on the input of coupler and $\Delta \phi = \phi_r - \phi_s$ is the phase difference between both arms of the interferometer.

Intensity on the output of detector creates electrical current of:

$$i = \epsilon \cdot I_0 \cdot \alpha \cdot \cos(\phi_d + \phi_p \cdot \sin \omega t), \qquad (5)$$

where ϵ is the responsivity of the photodetector, and phase difference $\Delta \phi$ may be separated into the signal term of amplitude ϕ_p, frequency ω and slowly varying phase shift ϕ_d.

This resulting electric signal is further processed and converted into the amplitude domain. The proposed system for the measurement of the traffic speed works with two interferometric units. These units use the modified fiber-optic Mach-Zehnder interferometer (MZI) as their structural basis. In Fig. 1, we can see a simplified diagram of the measuring unit with a light source in the form of a laser and a photodetector which converts the resultant beam of light into a measurable electric current. MZI has two couplers. First coupler splits the optical beam (power) into two optical parts (the reference labeled L_2 and measuring labeled L_1) in a defined ratio of 1:1. The second coupler merge again optical beam.

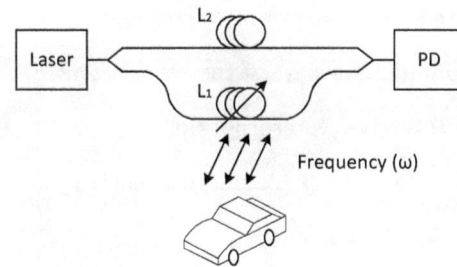

Fig. 1: Simplified scheme of a fiber-optic interferometer MZI and the influence of vibrations when passing cars.

The reference part must be designed in such a way so as to maximize the elimination of unwanted signals. Above Eq. (1) the wavelength λ is not changed due to using a stable light source (laser). The isolation of the reference part must be made in a way that even the remaining two parameters such as refractive index n and the length L do not change. The measured variable, acting on the measuring fiber, then causes a change in the optical length of the arm (the product of refractive index n and geometric length).

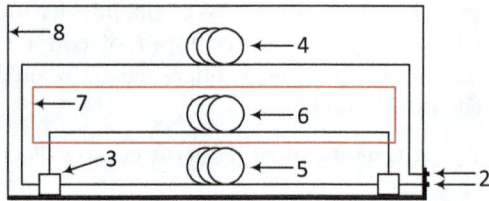

Fig. 2: Cross-section of the modified measurement sensor unit.

3.   Experimental Setup

To carry out the experiment, we increased the compactness, reduced the size and weight, and implemented a new idea of the reference channel that was positioned in such way to suppress undesirable signals. Furthermore, I/O interface sensory unit was modified including storage in the water resistant casing. The interface consists of two FC adapters. Longitudinal section of the modified prototype is shown in Fig. 2.

The list of referential marks (applies to Fig. 2 and Fig. 5):

- 1 - Laser radiation source,

- 2 - I/O interface,

- 3 - Coupler,

- 4 - Conventional optical fiber G.652D,

- 5 - Measuring part of the interferometer,

- 6 - Reference part of interferometer,

- 7 - Dampening part of the reference arm,

- 8 - Protective waterproof box,

- 9 - Photodetector (or photodetector system),

- 10 - Coaxial cable,

- 11 - Part of signal processing,

- 12 - Conventional optical fiber for connection of the sensor units.

Red color denotes isolated reference arm designed to be most immune to variation in the parameter L and the refractive index n. The referential arm of the interferometer is covered by a polystyrene layer. This material was chosen because of its good insulation characteristics. The measuring interferometric arm is mounted on a resonant surface. Vibrations caused by vehicles are that of low frequencies thereby resonant pad is formed from sufficiently massive glass sheet. Due to an elasticity of resonant pad it well transfers vibrations from the road to the attached optical fiber. Figure 3

Fig. 3: Functional prototype of measuring unit.

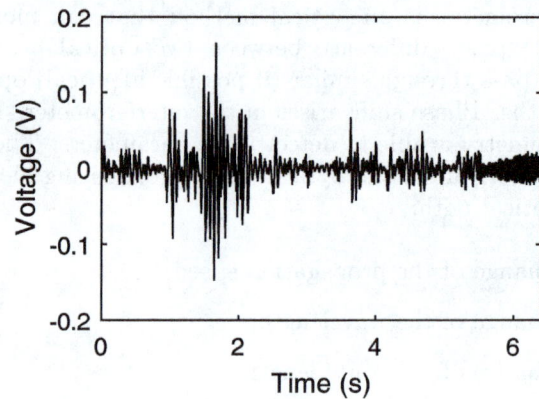

(a)

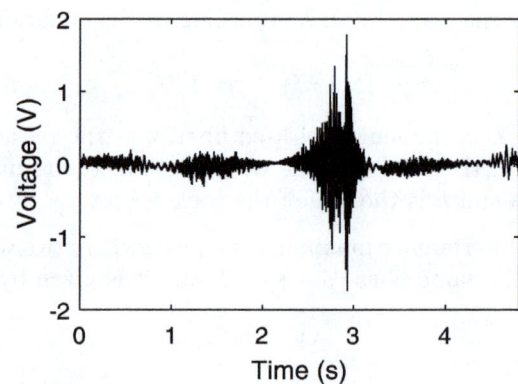

(b)

Fig. 4: Time response of the original prototype (a), time response of the new prototype (b) on the passage of the same vehicle and speed.

shows the functional prototype with the resonance pad (glass pad).

The measurement results (Fig. 4) show that there was a significant increase in the voltage response due to the pass-by of the same vehicle. For this reason we can say that the development of design modifications was chosen correctly. The value of SNR (signal-to-noise ratio) of the same type of the car was increased tenfold in the current prototype comparing it with the original.

Apart from analyzing the vehicle types [8], the assembled interferometer units can be also used for the measurement of other parameters in traffic. One of the preferred parameters is the vehicle speed. The assembled arrangement (Fig. 5) is based on two identical units placed in strictly defined distance L apart. Based on the measured time span between measuring units ΔT and the known distance L, we can evaluate the speed of passing vehicles Eq. (6):

$$v_{car} = \frac{L}{\Delta T}. \tag{6}$$

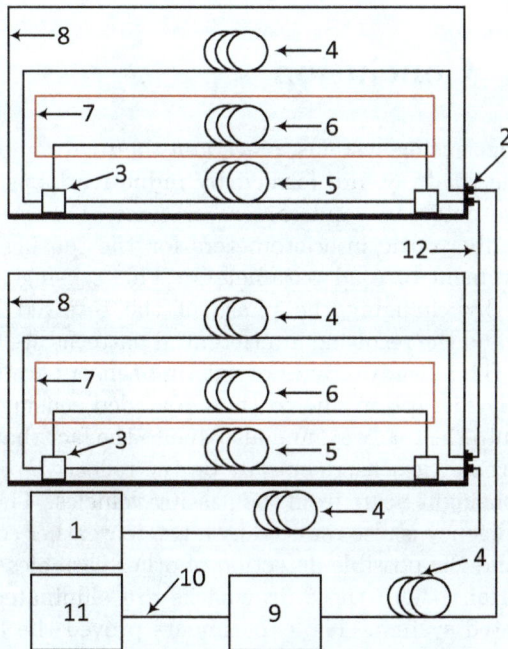

Fig. 5: Scheme for measurement of the speed of vehicles.

We tested DFB laser source with the output power in a range of 1 to 10 mW. The conventional SM (single mode) fiber G.652.D is used in patch cords of sensor units. The testing cable length was selected in the range from 1 to 250 m. The length of the patch cord for connecting sensor units was tested depending on the distance spacing between units in the range of 1 to 50 m. However, the minimum distance is 1 m due to the resolution of the evaluation software. The couplers have the split ratio of 50:50 with a tolerance of ±5 %. The interferometer is connected and mounted in a protective waterproof box to obtain greatest sensitivity to low frequency and best detection of car vibration response.

The photodetector detects a signal due to the interference of optical beams from the reference and measuring arms and converts it into a measurable electric current. The signal processing unit uses a high-pass filter with a cutoff frequency of 8 Hz for ensuring zero offset voltage. The amplifier is 16-bit analog-to-digital converter with the sampling rate of 250 kS·s^{-1}. Evaluation software handles signals in the time domain. The application displays the progress of the signal as a voltage versus time.

The application makes the reference measurement of the background noise. It is necessary to set the parameter L (in the range of 1–50 m) which determines the distance spacing between interferometric units. The trigger value (1 V) was determined five times greater than the noise value (0.2 V). The timer switches on if the first unit detects the pass-by of the vehicle (if the signal has a sufficient level of SNR - trigger level). The timer switches off if the second unit detects the pass-by of the vehicle (if the signal has a sufficient level of SNR - trigger level). The application calculates speed from a fixed defined distance L of interferometers and from the time interval ΔT of the pass-by of the vehicle between interferometers Eq. (6).

The more accurate detected signal occurs due to the involvement of two units. Further research should be directed to the development of the condition when application compare the two signals and determines whether it is the same car using frequency analysis. Typical symptoms characterizing a vehicle are given by maximum values, which are almost the same for the identical type of vehicle. We can state that identical measuring units and the same source of radiation are used (Fig. 6).

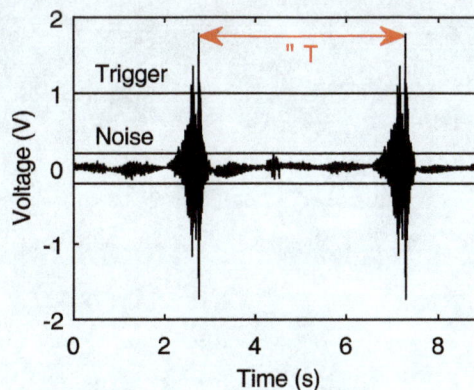

Fig. 6: Detection of maximum amplitude (the evaluation of time interval).

Table 1 shows the first experimental measurement of speed up to 55 kph. The measured data are compared with the GPS data which have an accuracy of 1.08 kph. Ten measurements were accomplished for each speed value. The table gives the average speed

Tab. 1: Experimental measurement of speed.

Reference speed (kph) [GPS]	Measurement speed (kph)	Relative difference (%)	Absolute difference (kph)
10	10.77	7.7	0.77
15	15.43	2.86	0.43
21	21.87	4.14	1.87
26	26.74	2.84	0.74
31	31.82	2.64	0.82
35	35.81	2.31	0.81
41	42.53	3.73	1.53
45	46.12	2.48	1.12
50	51.28	2.56	1.28
55	55.75	1.36	0.75

values. The maximum absolute deviation was 1.53 kph with ±1 kph the tolerance of GPS. We can say that the system demonstrates reliable values (including tolerance GPS) considering the tolerance of $pm3$ kph indicated for segmental measuring systems in the CZE for measurement of speed to 100 kph. The system for the measurement of the vehicle speed can be used in a wide variety of application areas. For example, in railway traffic or other areas, where there is the same problem with the installation and use of electronic devices.

Figure 7 shows the location of two sensor units in strictly defined distance L in which units are placed on roadsides. We verified even more suitable locations (outside the roadside). The main advantage is the non-destructive performance towards the road due

to the use of conventional elements and standard fibers G.652.D. Other advantages are low cost and the possibility of connection to existing telecommunication networks. This connection is possible without the use of converters, for example, when using PM (Polarization Maintaining) fibers.

4. Conclusion

Existing sensing systems, placed on the roadway or inside the roadway, are formed by inductive loops, microwave detectors, and camera systems. Our sensors utilize fiber-optic interferometers for the sensing and measurement of road parameters. This sensing is reflected by changing the phase of the received light beam, and the resulting interferential patterns are evaluated. The disadvantages of existing sensors are both roadway disruption due to the embedded sensing element into the roadway, and also due to the fact that the interferometric sensors operate on the reflective principle of the light beam from the passing vehicles. The device efficiency is also reduced by adverse weather conditions and the possible detection of other variables (e.g. pedestrians). All these drawbacks are eliminated by our tested system. The experiments proved the maximum absolute error of 1.53 kph with ±1 kph of the tolerance of GPS. Indicated tolerance of ±3 kph, which corresponds to the segmental measuring systems in the CZE for the speed to 100 kph, rated the testing system as a reliable speed measuring device in traffic. The results of the experiment also proved that due to a high sensitivity of the unit it is not necessary to implement it only into the road traffic. There are a variety of sectors where this speed measuring system can be used. Testing is currently focused on the railway traffic.

Acknowledgment

This article was supported by the projects of the Technology Agency of the Czech Republic TA03020439 and TA04021263, and by the Ministry of Education, Youth

Fig. 7: Experimental measurement of vehicle speed.

and Sports of the Czech Republic within the project no. SP2016/149 and project no. CZ.1.07/2.3.00/20.0217 within the frame of the operation programme Education for competitiveness financed by the European Structural Funds and from the state budget of the Czech Republic. This article was supported too by project VI20152020008.

References

[1] KRAUTER, K. G., G. F. JACOBSON, J. R. PATTERSON, J. H. NGUYEN and W. P. AMBROSE. Single-mode Fiber, Velocity Interferometry. *Review of Scientific Instruments*. 2011, vol. 82, no. 4, pp. 45–110. ISSN 0034-6748. DOI: 10.1063/1.3574797.

[2] FENG, L. L., Y. T. WANG, C. RUAN and S. TAO. Road Vehicle Information Collection System Based on Distributed Fiber Optics Sensor. *Advanced Materials Research*. 2014, vol. 1030–1032, iss. 8, pp. 2105–2109. ISSN 1662-8985. DOI: 10.4028/www.scientific.net/AMR.1030-1032.2105.

[3] LIU, Z. and R. CHI. *Distributed Optical Fiber Vehicle Comprehensive Information Detecting System*. Utility model CN201498105. 2nd June 2010.

[4] YU, Z., Z. MINGSHENG, Z. XINGCHUN, P. WU and Y. FENGLEI. *Road traffic monitoring system*. Utility model CN104504916. 8th April 2015.

[5] GE, Z. P. *Novel optical fiber vehicle detector*. Utility model CN200962255. 17th October 2007.

[6] VASINEK, V., S. KEPAK, J. CUBIK, and T. KAJNAR. *Optical-fiber interferometric sensor for monitoring traffic operations*. Patent 305889. 9th March 2016.

[7] HE, Y., M. HUI and G. JIHUA. Road vehicle running speed detecting method and device. Utility model CN1641359. 20th July 2005.

[8] NEDOMA, J., O. ZBORIL, M. FAJKUS, P. ZAVODNY, S. KEPAK, L. BEDNAREK, R. MARTINEK and V. VASINEK. Fiber optic system design for vehicle detection and analysis. In: *Proceedings of SPIE 9889: Optical Modelling and Design IV*. Brussel: SPIE, 2016, pp. 1–7. ISBN 978-151060134-5. DOI: 10.1117/12.2239549.

About Authors

Jan NEDOMA was born in 1988 in Prostejov. In 2012 he received a Bachelor's degree from VSB–Technical University of Ostrava, Faculty of Electrical Engineering and Computer Science, Department of Telecommunications. Two years later, he received his Master's degree in the field of Telecommunications in the same workplace. Currently he is an employee and Ph.D. student of Department of Telecommunications. He works in the field of biomedical engineering and fiber-optic sensor systems.

Marcel FAJKUS was born in 1987 in Ostrava. In 2009 he received a Bachelor's degree from VSB–Technical University of Ostrava, Faculty of Electrical Engineering and Computer Science, Department of Telecommunications. Two years later, he received his Master's degree in the field of Telecommunications in the same workplace. Currently he is an employee and Ph.D. student of Department of Telecommunications. He works in the field of biomedical engineering and fiber-optic sensor systems.

Lukas BEDNAREK was born in 1988 in Frydek-Mistek. In 2011 he received a Bachelor's degree from VSB–Technical University of Ostrava, Faculty of Electrical Engineering and Computer Science, Department of Telecommunications. Three years later, he received his Master's degree in the field of Telecommunications in the same workplace. Currently he is Ph.D. student of Department of Telecommunications. He works in the field of optical communications and aging of the optical components.

Vladimir VASINEK was born in Ostrava. In 1980 he graduated in Physics, specialization in Optoelectronics, from the Science Faculty of Palacky University. He was awarded the title of RNDr. at the Science Faculty of Palacky University in the field of Applied Electronics. The scientific degree of Ph.D. was conferred upon him in the branch of Quantum Electronics and Optics in 1989. He became an associate professor in 1994 in the branch of Applied Physics. He has been a professor of Electronics and Communication Science since 2007. He pursues this branch at the Department of Telecommunications at VSB–Technical University of Ostrava. His research work is dedicated to optical communications, optical fibers, optoelectronics, optical measurements, optical networks projecting, fiber optic sensors, MW access networks. He is a member of many societies - OSA, SPIE, EOS, Czech Photonics Society; he is a chairman of the Ph.D. board at the VSB–Technical University of Ostrava. He is also a member of habitation boards and the boards appointing to professorship.

11

Influence of Scattering Enhancement Particles CaCO$_3$, CaF$_2$, SiO$_2$ and TiO$_2$ on Color Uniformity of White LEDs

Nguyen Huu Khanh NHAN, Tran Hoang Quang MINH, Nguyen Doan Quoc ANH

Faculty of Electrical and Electronics Engineering, Ton Duc Thang University, 19 Nguyen Huu Tho Street, Tan Phong Ward, District 7, Ho Chi Minh City, Vietnam

nguyenhuukhanhnhan@tdt.edu.vn, tranhoangquangminh@tdt.edu.vn, nguyendoanquocanh@tdt.edu.vn

Abstract. *In this paper, the influence of scattering enhancement particles CaCO$_3$, CaF$_2$, SiO$_2$ and TiO$_2$, adding to YAG:Ce phosphor compounding, on color uniformity of white LEDs (W-LEDs) was presented. Firstly, the physical model of multi-chip W-LEDs is simulated and demonstrated by using commercial Light-Tools 8.1.0 program. After that, the influence of scattering enhancement particles on color uniformity is calculated and analyzed. With using the Monte Carlo simulation and the Mie-scattering theory, the color uniformity improvement of an 8500 K W-LEDs is demonstrated convincingly. From the researched results, the best color uniformity can be accomplished with TiO$_2$ particles. The results and discussions provided a practical approach for higher-quality manufacturing W-LEDs.*

Keywords

CaCO$_3$, CaF$_2$, color uniformity, SiO$_2$, TiO$_2$, white LEDs.

1. Introduction

Nowadays, W-LEDs are becoming increasingly important light sources for illumination applications, because they are long-life, compact, mercury-free and energy-efficient. Color uniformity is the main optical properties of W-LEDs and it could be improved in many previous papers [1], [2] and [3]. All these studies started from the scattering enhancement in phosphor-converted white-LEDs (PC-LEDs). In fact, the structure of PC-LEDs is the combination of YAG:Ce phosphor and silicone glue. The YAG:Ce phosphor absorbs the exciting blue light from the chips to stimulate the yellow light and thus result in white light with the desired color temperature [4]. In other words, in these studies, the color uniformity of LEDs was improved by optimizing the state of the phosphor or the optical structure of PC-LEDs. In conclusions, the spatial color uniformity of PC-LEDs can be controlled by the thickness and the concentration of the phosphor [9]. Moreover, the location of phosphor material in the silicone layer significantly effects on the color performance. The color temperature of PC-LEDs has demonstrated the strong influence of the refractive indexes of the silicone matrix and the phosphor materials and the size of phosphor particles [10].

In this study, we concentrated on finding one particle from scattering enhancement particles CaCO$_3$, CaF$_2$, SiO$_2$ and TiO$_2$, which is employed for manufacturing higher-quality W-LEDs. The target of study is an improvement the color uniformity of W-LEDs. This research paper can be divided into three main sections: In Section 2. , the physical model of 8500 K W-LEDs is simulated and demonstrated by using commercial LightTools 8.1.0 program. In Section 3. , by adding one of scattering enhancement particles CaCO$_3$, CaF$_2$, SiO$_2$ and TiO$_2$ to YAG:Ce phosphor compounding, the color uniformity is simulated, calculated and analyzed: In Section 4. , the simulation can be convinced by using the Monte Carlo simulation and the Mie-scattering theory. In this study, the results demonstrated that the best color uniformity of 8500 K W-LEDs could be accomplished with TiO$_2$ particles. This results can consider the prospective solution for higher-quality manufacturing W-LEDs in the near future.

2. Physical Model

In this work, an 8500 K W-LEDs with the conformal phosphor structure is simulated by using the commercial LightTools software based on the Monte Carlo ray-tracing method. To perform optical simulations, we built 3-D models (Fig. 1). In this research, W-LEDs has commonly configured:

- The reflector has a bottom length of 8 mm, a height of 2.07 mm and a length of 9.85 mm at its top surface.

- The conformal phosphor layer with a fixed thickness of 0.08 mm covers the 9 LED chips.

- Each LED chip with a square base of 1.14 mm and a height of 0.15 mm is bound in the cavity of the reflector (Fig. 1(b)). The radiant flux of each blue chip is 1.16 W at wavelength 455 nm.

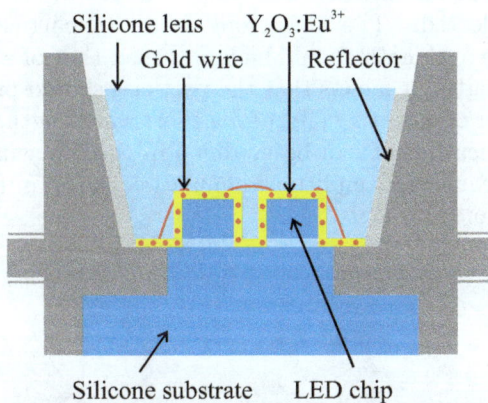

(a) The conformal phosphor structure.

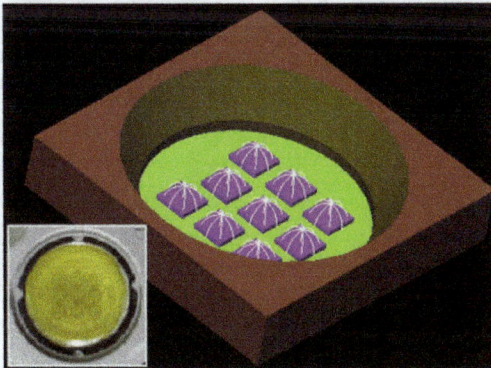

(b) The original lamps and physical model.

Fig. 1: W-LEDs structure.

To maintain the average Correlated Color Temperature (CCT) of 8500 K, the YAG:Ce concentration changes to the concentration of $CaCO_3$, CaF_2, SiO_2 and TiO_2. The refractive index of the diffusors such as $CaCO_3$, CaF_2, SiO_2 and TiO_2 are chosen as 1.66, 1.44, 1.47 and 2.87, respectively. The diffusers are assumed

to be spherical and have radius 0.5 µm. The average radius of the phosphor particles are 7.25 µm and have a refractive index of 1.83 at all wavelengths of light. The refractive index of the silicone glue is 1.5. The diffusional particle density is varied for optimizing illumination CCT uniformity and output efficiency by the expression:

$$W_{phosphor} + W_{silicone} + W_{diffusor} = 100\ \%, \quad (1)$$

where $W_{silicone}$, $W_{phosphor}$ and $W_{diffusor}$ are the weight percentages of the silicone, phosphor and diffuser of the W-LEDs, respectively. To maintain the mean CCT value of 8500 K, the weight of YAG:Ce phosphor should be decreased when the weight percentage of the diffuser is increased.

3. Results and Discussion

For improving the light quality of the W-LEDs, the difference of angular CCT Deviation (D-CCT) between the normal and large angle is an important standard to evaluate in the solid-state lighting application [9]. The larger D-CCT can cause the yellow ring phenomenon and generate the non-uniform white color at the different angle [14]. In this study, the D-CCT is expressed as D-CCT = CCT (Max) – CCT (Min). Here CCT (Max) and CCT (Min) are the maximal CCT at the zero degree of viewing angle and minimal CCT at the 70 degree of viewing angle, respectively. The scattered light of each particle in PC-LEDs is different, resulting in varying the optical properties of W-LEDs. If the scattered blue light is enhanced enough, the D-CCT can be reduced significantly. Conversely, the D-CCT should be increased with lack or redundancy of the scattered blue light in W-LEDs. The scattered blue light not only combines with the converted yellow but also combine the yellow ring for emitting white light, resulting in a reduction of yellow ring phenomenon of W-LEDs. It can be seen in Fig. 2, where the D-CCT of $CaCO_3$ and TiO_2 cases have a downward trend. Meanwhile, the D-CCT of CaF_2 and SiO_2 cases grow with their concentration.

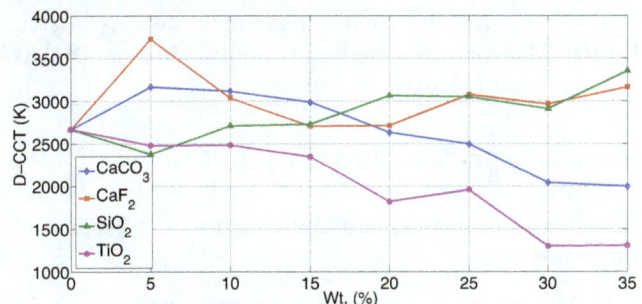

Fig. 2: The impact of the diffusive particles concentration on CCT deviations.

4.   Scattering Description

Simulation results can be investigated and demonstrated by Matlab software using Mie-scattering theory [11]. The scattering coefficient $\mu_{sca}(\lambda)$, anisotropy factor $g(\lambda)$ and reduced scattering coefficient $\delta_{sca}(\lambda)$ are calculated by expression Eq. (2), Eq. (3) and Eq. (4):

$$\mu_{sca}(\lambda) = \int N(r) C_{sca}(\lambda, r) dr, \qquad (2)$$

$$g(\lambda) = \int \int_{-1}^{1} p(\theta, \lambda, r) f(r) \cos\theta d\cos\theta dr, \qquad (3)$$

$$\delta_{sca} = \mu_{sca}(1 - g), \qquad (4)$$

where $N(r)$ is the number density distribution of diffusional particles (per cubic millimeter), C_{sca} is the scattering cross sections (per square millimeter), $p(\theta, \lambda, r)$ is the phase function, λ is the wavelength of the incident light (nanometers), r is the radius of particles (micrometers), θ is the scattering angle (degree) and $f(r)$ is the size distribution function of the diffusers in the phosphor layer.

$$f(r) = f_{dif}(r) + f_{phos}(r), \qquad (5)$$

$$\begin{aligned} N(r) &= N_{dif}(r) + N_{phos}(r) = \\ &= K_N \cdot [f_{dif}(r) + f_{phos}(r)], \end{aligned} \qquad (6)$$

where $N(r)$ is composed of the diffusive particle number density $N_{dif}(r)$ and the phosphor particle number density $N_{phos}(r)$. $f_{dif}(r)$ and $f_{phos}(r)$ are the size distribution function data of the diffusor and phosphor particle. If the phosphor concentration c (milligrams per cubic millimeter) of the mixture is known, K_N denotes the number of the unit diffusor for one diffuser concentration and K_N can be obtained by:

$$c = K_N \int M(r) dr. \qquad (7)$$

To obtain K_N, we should first know the mass distribution $M(r)$ (milligrams) of the unit diffusor. Below equation can calculate $M(r)$:

$$M(r) = \frac{4}{3}\pi r^3 \left[\rho_{dif} f_{dif}(r) + \rho_{phos} f_{phos}(r)\right], \qquad (8)$$

where ρ_{dif} and ρ_{phos} are the density of diffusor and phosphor crystal.

In Mie theory, C_{sca} is normally presented:

$$C_{sca} = \frac{2\pi}{k^2} \sum_{0}^{\infty} (2n - 1)(|a_n|^2 + |b_n|^2), \qquad (9)$$

where k is the wavenumber $(2\pi/\lambda)$ and a_n and b_n are the expansion coefficients with even symmetry and odd symmetry, respectively. These coefficients can be calculated by equations below:

$$a_n(x, m) = \frac{\Psi'_n(mx)\Psi_n(x) - m\Psi_n(mx)\Psi'_n(x)}{\Psi'_n(mx)\xi_n(x) - m\Psi_n(mx)\xi'_n(x)}, \qquad (10)$$

$$a_n(x, m) = \frac{m\Psi'_n(mx)\Psi_n(x) - \Psi_n(mx)\Psi'_n(x)}{m\Psi'_n(mx)\xi_n(x) - \Psi_n(mx)\xi'_n(x)}, \qquad (11)$$

where x is the size parameter $(= k \cdot r)$, m is the refractive index of the scattering diffusive particles. $\Psi_n(x)$ and $\xi_n(x)$ are the Riccati - Bessel function.

According to Eq. (3), the theoretical results of $g(\lambda)$ are calculated and shown in Fig. 3, Fig. 4 and Fig. 3. Results show that the variation of the diffuser concentration has a slight impact on the anisotropy factor $g(\lambda)$ and the increase of $g(\lambda)$ by the diffusional particle density is so small that the increase can be neglected. The anisotropy factor of particles for a long wavelength should be larger than that of a short wavelength. It means that the particles should present stronger a scattering effect for a short wavelength. This theoretical result can be modified in the following angular scattering amplitudes simulation shown in Fig. 3, Fig. 4 and Fig. 5.

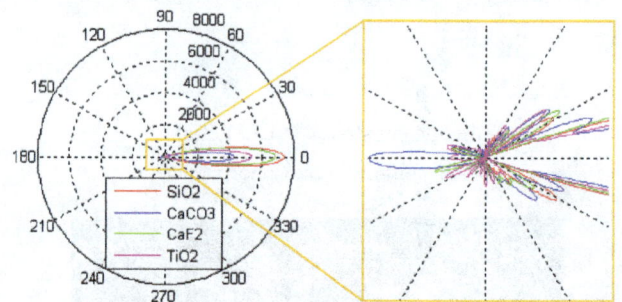

Fig. 3: The angular scattering amplitudes of the various diffusional particles with sphere diameter = 1 μm for blue light = 455 nm.

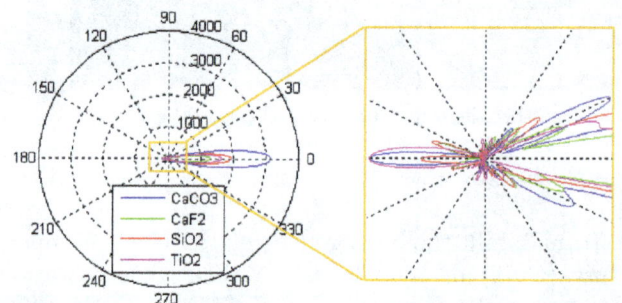

Fig. 4: The angular scattering amplitudes of the various diffusional particles with sphere diameter = 1 μm for yellow light = 595 nm.

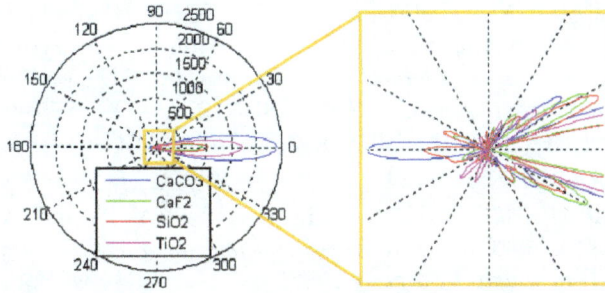

Fig. 5: The angular scattering amplitudes of the various diffusional particles with sphere diameter = 1 μm for red light = 680 nm.

In the mixture of phosphor, diffusor and silicone, the refractive index of embedded silicone (n_{sil}) is 1.53 and the refractive index of diffusor (n_{dif}) are 1.66, 1.44, 1.47 and 2.87 respectively. Silicone and diffusors are considered to be transparent for the blue light and the yellow light. The refractive index of the phosphor particle (n_{phos}) has a complex form. Therefore, the relative refractive indices of diffusor (m_{dif}) and phosphor (m_{phos}) in the silicone are $m_{dif} = n_{dif} \cdot n_{sil}^{-1}$ and $m_{phos} = n_{phos} \cdot n_{sil}^{-1}$. For small spheres, the phase function $p(\theta, \lambda, r)$ can be calculated according to the following equation [12] and [13]:

$$p(\theta, \lambda, r) = \frac{4\pi \beta(\theta, \lambda, r)}{k^2 C_{sca}(\lambda, r)}, \qquad (12)$$

where $\beta(\theta, \lambda, r)$ is the dimensionless scattering function, which is obtained by the scattering amplitude functions $S_1(\theta)$ and $S_2(\theta)$:

$$\beta(\theta, \lambda, r) = \frac{1}{2}\left[|S_1(\theta)|^2 + |S_2(\theta)|^2\right]. \qquad (13)$$

$$S_1 = \sum_{n=1}^{\infty} \frac{2n+1}{n(n+1)}\left[\begin{array}{c} a_n(x,m)\pi_n(\cos\theta) \\ +b_n(x,m)\tau_n(\cos\theta) \end{array}\right]. \qquad (14)$$

$$S_2 = \sum_{n=1}^{\infty} \frac{2n+1}{n(n+1)}\left[\begin{array}{c} a_n(x,m)\tau_n(\cos\theta) \\ +b_n(x,m)\pi_n(\cos\theta) \end{array}\right]. \qquad (15)$$

In equations Eq. (14) and Eq. (15), the angular dependent functions and are expressed in the angular scattering patterns of the spherical harmonics.

5.   Conclusion

In this research, the influence of $CaCO_3$, CaF_2, SiO_2 and TiO_2 on color uniformity of 8500 K MCW-LEDs was presented, calculated, analyzed and demonstrated. From the researched results, some conclusions are proposed:

- The CCT deviation has a decreasing tendency when the concentration of $CaCO_3$ and TiO_2 increases.

- Meanwhile the CCT deviation of CaF_2 and SiO_2 cases grow with their concentration.

- The best color uniformity of W-LEDs can be obtained in TiO_2 case. In summary, TiO_2 particles should be chosen for improving the color uniformity of W-LEDs. This research provided an important technical implication for the selection of phosphors in WLED manufacturing and development of phosphor materials for WLED applications. In further research, color rending index and luminous efficiency of MCW-LEDs by adding $CaCO_3$, CaF_2, SiO_2 and TiO_2 particle into the phosphor compounding is necessary to analyze and demonstrate.

Acknowledgment

This paper was supported by Professor Hsiao-Yi Lee, Department of Electrical Engineering, National Kaohsiung University of Applied Sciences, Kaohsiung, Taiwan.

References

[1] LIU, Z., S. LIU, K. WANG and X. LUO. Optical Analysis of Color Distribution in White LEDs With Various Packaging Methods. *IEEE Photonics Society.* 2008, vol. 20, iss. 24, pp. 2027–2029. ISSN 1941-0174. DOI: 10.1109/LPT.2008.2005998.

[2] HU, R., X. LUO and S. LIU. Effect of the amount of phosphor silicone gel on optical property of white light-emitting diodes packaging. In: *12th International Conference on Electronic Packaging Technology and High Density Packaging (ICEPT-HDP).* Shanghai: IEEE, 2011, pp. 1–4. ISBN 978-1-4577-1770-3. DOI: 10.1109/ICEPT.2011.6067015.

[3] ZHENG, H., X. LUO, R. HU, B. CAO, X. FU, Y. WANG and S. LIU. Conformal phosphor coating using capillary microchannel for controlling color deviation of phosphor-converted white light-emitting diodes. *Optics Express.* 2012, vol. 20, iss. 5, pp. 5092–5098. ISSN 1094-4087. DOI: 10.1364/OE.20.005092.

[4] ANH, N. D. Q., M.-F. LAI, H.-Y. MA and H.-Y. LEE. Enhancing of correlated color temperature uniformity for multi-chip white-light

LEDs by adding SiO_2 in phosphor layer. *Journal of the Chinese Institute of Engineers*. 2015, vol. 38, iss. 3, pp. 297–303. ISSN 0253-3839. DOI: 10.1080/02533839.2014.981214.

[5] CHEN, H.-C., K.-J. CHEN, C.-C. LIN, C.-H. WANG, H.-V. HAN, H.-H. TSAI, H.-T. KUO, S.-H. CHIEN, M.-H. SHIH and H.-C. KUO. Improvement in uniformity of emission by ZrO_2 nano-particles for white LEDs. *Nanotechnology*. 2012, vol. 23, no. 26, pp. 1–5. ISSN 1361-6528. DOI: 10.1088/0957-4484/23/26/265201.

[6] MONT, F. W., J. K. KIM, M. F. SCHUBERT, E. F. SCHUBERT and R. W. SIEGEL. High-refractive-index TiO_2-nanoparticle-loaded encapsulants for light-emitting diodes. *Journal of Applied Physics*. 2008, vol. 103, iss. 8, pp. 1–6. ISSN 0021-8979. DOI: 10.1063/1.2903484.

[7] LAI, M.-F., N. D. Q. ANH, H.-Y. MA and H.-Y. LEE. Scattering effect of SiO_2 particles on correlated color temperature uniformity of multi-chip white light LEDs. *Journal of the Chinese Institute of Engineers*. 2016, vol. 39, iss. 4, pp. 468–472. ISSN 0253-3839. DOI: 10.1080/02533839.2015.1117950.

[8] LIU, S. and X. B. LUO. *LED Packaging for Lighting Applications: Design, Manufacturing and Testing*. 1st ed. Singapore: John Wiley & Sons, 2011. ISBN 978-0-470-82785-7. DOI: 10.1002/9780470827857.fmatter.

[9] SHUAI, Y., Y. HE, N. T. TRAN and F. G. SHI. Angular CCT Uniformity of Phosphor Converted White LEDs: Effects of Phosphor Materials and Packaging Structures. *IEEE Photonics Technology Letters*. 2010, vol. 23, iss. 3, pp. 137–139. ISSN 1941-0174. DOI: 10.1109/LPT.2010.2092759.

[10] SOMMER, C., F. REIL, J. R. KRENN, P. HARTMANN, P. PACHLER, H. HOSCHOPF and F. P. WENZL. The Impact of Light Scattering on the Radiant Flux of Phosphor-Converted High Power White Light-Emitting Diode. *Journal of Lightwave Technology*. 2011, vol. 29, iss. 15, pp. 2285–2291. ISSN 0733-8724. DOI: 10.1109/JLT.2011.2158987.

[11] ZHONG, J., M. XIE, Z. OU, R. ZHANG, M. HUANG and F. ZHAO. Mie Theory Simulation of the Effect on Light Extraction by 2-D Nanostructure Fabrication. In: *2011 Symposium on Photonics and Optoelectronics (SOPO)*. Wuhan: IEEE, 2011, pp. 1–4. ISBN 978-1-4244-6554-5. DOI: 10.1109/SOPO.2011.5780566.

[12] JONASZ, M. and G. R. FOURNIER *Light Scattering by Particles in Water: Theoretical and Experimental Foundations*. 1st ed. London: Academic Press, 2007. ISBN 978-0-12-388751-1. DOI: 10.1016/B978-0-12-388751-1.50011-9

[13] MISHCHENKO, M. I., L. D. TRAVIS and A. A. LACIS. *Scattering, Absorption and Emission of Light by Small Particles*. 1st ed. New York: Cambridge University Press, 2002. ISBN 978-0-52-178252-4.

[14] HUANG K.-C., T.-H. LAI and C.-Y. CHEN. Improved CCT uniformity of white LED using remote phosphor with patterned sapphire substrate. *Applied Optics*. 2013, vol. 52, iss. 30, pp. 7376–7381. ISSN 1559-128X. DOI: 10.1364/AO.52.007376.

[15] OH, J. H., Y. J. EO, S. J. YANG and Y. R. DO. High-Color-Quality Multipackage Phosphor-Converted LEDs for Yellow Photolithography Room Lamp. *IEEE Photonics Journal*. 2015, vol. 7, iss. 2, pp. 1–8. ISSN 1943-0655. DOI: 0.1109/JPHOT.2015.2415674.

[16] PENG, H. Y., H. S. HWANG and M. DEVARAJAN. High-Color-Quality Multipackage Phosphor-Converted LEDs for Yellow Photolithography Room Lamp. In: *2014 IEEE Region 10 Symposium*. Kuala Lumpur: IEEE, 2014, pp. 293–296. ISBN 978-1-4799-2027-3. DOI: 10.1109/TENCONSpring.2014.6863044.

[17] YU, H. J., W. CHUNG and S. H. KIM. White Light Emission from Blue InGaN LED with Hybrid Phosphor. In: *10th IEEE Conference on Nanotechnology (IEEE-NANO)*. Seoul: IEEE, 2010, pp. 958–961. ISBN 978-1-4244-7031-0. DOI: 10.1109/NANO.2010.5697998.

[18] LI, Z.-T., Y. TANG, Z.-Y. LIU, Y.-E. TAN and B.-M. ZHU. Detailed Study on Pulse-Sprayed Conformal Phosphor Configurations for LEDs. *Journal of Display Technology*. 2013, vol. 9, iss. 6, pp. 433–440. ISSN 1558-9323. DOI: 10.1109/JDT.2012.2225019.

[19] SCHRATZ, M., C. GUPTA, T. J. STRUHS and K. GRAY. Reducing energy and maintenance costs while improving light quality and reliability with led lighting technology. In: *Pulp and Paper Industry Technical Conference (PPIC)*. Charlotte: IEEE, 2013, pp. 43–49. ISBN 978-1-4673-5100-3. DOI: 10.1109/PPIC.2013.6656043.

About Authors

Nguyen Huu Khanh NHAN defended his Ph.D. thesis at Institute of Research and Experiments for Electrical and Electronic Equipment, Moscow, Russian Federation. He is working as Lecturer in Faculty of Electrical and Electronic Engineering, Ton Duc Thang University, Ho Chi Minh City, Vietnam. His research interests include VLSI, MEMS and LED driver chips.

Tran Hoang Quang MINH defended his Ph.D. thesis at Tomsk Polytechnic University, Tomsk City, Russian Federation. The author's major fields of study are High-voltage Power System, Relay Protections and Optoelectronics. He is working as Lecturer in Faculty of Electrical and Electronics Engineering, Ton Duc Thang University, Ho Chi Minh City, Vietnam.

Nguyen Doan Quoc ANH was born in Khanh Hoa province, Vietnam. He has been working at the Faculty of Electrical and Electronics Engineering, Ton Duc Thang University, Ho Chi Minh City, Vietnam. He received his Ph.D. degree from National Kaohsiung University of Applied Sciences, Taiwan in 2014. His research interest is optoelectronics (such as Multi-chip white light LEDs, free-form lens and optical material).

12

Non-Invasive Fiber Optic Probe Encapsulated Into PolyDiMethylSiloxane for Measuring Respiratory and Heart Rate of the Human Body

Jan NEDOMA[1], Marcel FAJKUS[1], Petr SISKA[1],
Radek MARTINEK[2], Vladimir VASINEK[1]

[1]Department of Telecommunications, Faculty of Electrical Engineering and Computer Science,
VSB–Technical university of Ostrava, 17. listopadu 15, 708 33 Ostrava, Czech Republic
[2]Department of Cybernetics and Biomedical Engineering, Faculty of Electrical Engineering and Computer
Science, VSB–Technical university of Ostrava, 17. listopadu 15, 708 33 Ostrava, Czech Republic

jan.nedoma@vsb.cz, marcel.fajkus@vsb.cz, petr.siska@vsb.cz, radek.martinek@vsb.cz, vladimir.vasinek@vsb.cz

Abstract. *This article describes the design and the functional verification of fiber optic system with an innovative non-invasive measuring probe for monitoring respiratory and heart rate. The measuring probe is based on Fiber Bragg Grating (FBG), and it is encapsulated in the PolyDiMethylSiloxane polymer (PDMS). PDMS offers a unique combination of suitable properties for the use in biomedical applications. The main advantages include inert to human skin and immunity to electromagnetic interference. The measuring probe is a part of contact strip which is placed on the chest of the patient. The measurement is based on sensing the movements of the thoracic cavity of the patient during breathing. Movement (mechanical stress) is transferred to FBG using the contact strip. Respiratory and heart rate are analyzed using the spectral evaluation of the measured signals. This monitoring method is fully dielectric; thus the absolute safety of the patient is ensured. The main contributions of the article are a design of non-invasive probe encapsulated into a PDMS polymer and implementation of the probe for humans using a contact strip. This combination forms an essential element of the measuring system. The set of experimental measurements verified functionality with respect to the position of the patient. Performed experiments proved the functionality of the presented solution so it can be utilized for further research in biomedical applications.*

Keywords

FBG, Fiber Bragg Gratings, fiber optic sensor, Heart Rate (HR), Non-invasive, PolyDiMethyl-Siloxane (PDMS), Respiration Rate (RR).

1. Introduction

The current trend of developments clearly indicates that future monitoring essential vital functions of the human body tend to sophisticated diagnostic equipment and methods in biomedical applications. The aim is to merge more diagnostic parameters to a single universal probe or a universal measuring system. Recently, the utilization of Optical Sensors has been growing in variety of emerging biomedical applications [1] and [2]. Based on the study of the literature review, authors of this article introduce an innovative combination of non-invasive measuring probe encapsulated into PDMS and implementation into the clamping contact strip. Monitoring respiratory and heart rate can be performed using one universal probe. The measuring probe based on Bragg grating technology and the probe encapsulated in a particular shape into a PDMS polymer. Fiber-optical sensors have been increasingly utilized in the biomedical applications, e.g. for measuring respiratory and cardiac activity [3], for specific heart rate monitoring [4], for early detection of health deterioration through a network of FBG [5], or for sensing respiratory parameters and the chemical reactions of human skin to external influences by using a sensor based on an enlarged-taper tailored Fiber Bragg Grating [6]. This fact is mainly due to characteristic features of the optical sensor-power supply independence and electromagnetic immunity. The optical fiber sensors can be utilized without electrical interference in the presence of other electrical equipment. Thus, the safety of patient monitoring is not affected, or more precisely, it makes the measurement more safety. The combination with PDMS material further enhances comfort during monitoring parameters. Biocompatibility is one of the primary factors for the patient comfort. PDMS is inert to the human skin, and it does not affect the patient's body. Authors focus on both description of the innovative non-invasive measuring probe and on implementation of the probe for measurement on human. The aim is not a comparison with existing diagnostic tools and methods.

2.　　Gait Motion Capture

Siloxanes are the compounds which contain a Si-O-Si bond in the molecule. This chemical group is specific for its stability; therefore, one can prepare an endless chain with composed of -(O-Si-O-Si-O)-. The last two free bonds of the silicon atom can be occupied by various -HO groups or organic ligands such as -CH_3. The most commonly used compound is PolydDiMethylSiloxane which is defined by the chemical formula in the form: $(CH_3)_3SiO[SiO(CH_3)_2]nSi(CH_3)_3$. The siloxanes are completely stable in standard conditions and are not subject to degradation in the presence of water or oxygen. The resulting products are solid or liquid substances according to the number of ligands and siloxane groups. Their other properties are both hydrophobicity and almost complete inertness to living organisms.

The production of PDMS involves three elements, namely technical silicon, hydrochloric acid, and methanol. This combination creates the so-called chloromethane. The manufacturing process contains four chemical phases: synthesis, rectification, hydrolysis, and polycondensation. Figure 1 describes chemical composition of PDMS. Methyl (CH_3) [7] and [8] represents the organic substituent in most cases.

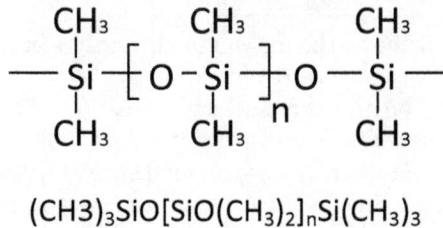

$$(CH3)_3SiO[SiO(CH_3)_2]_nSi(CH_3)_3$$

Fig. 1: Chemical composition of PDMS.

Encapsulation of measuring probe was made by PDMS with the designation of Sylgard 184. Sylgard 184 is a two-component casting compound; wherein A component creates own pre-polymer and B component is a curing agent. Both components are mixed according to datasheet in a weight ratio of 10:1 (A:B). Bubbles and microbubbles which result from the combination of the pre-polymer and the curing agent

can be removed using an ultrasonic bath. Homogeneity of connection is provided using a laboratory shaker.

Sylgard 184 belongs to the moderately viscous liquid elastomers. The primary characteristic of PDMS is its temperature stability. PDMS is formed by the bonds Si-CH_3 and Si-O having a high binding energy (452 kJ · mole^{-1}). PDMS withstands temperatures ranging from −60 °C through 200 °C, in short-term processes to 350 °C. At temperatures around 100 °C, curing of Sylgard 184 can be performed within several minutes.

Fiber-optical Bragg gratings are most commonly used fiber-optical sensors for their spectral characteristics [9] and [10]. Repeated changes in refractive index of the core of an optical fiber creates the grating. Spectral reflection of a selected wavelength, known as Bragg wavelength, occurs on periodic interfaces. All the other wavelengths pass through the Bragg grating without damping. Figure 2 shows the structure of the FBG.

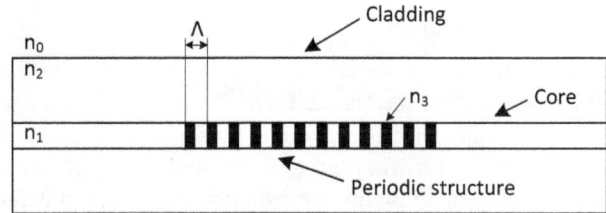

Fig. 2: Example of structure of Fiber Bragg Grating.

The Bragg wavelength λ_B is given by following equation:

$$\lambda_B = 2n_{eff}\Lambda, \tag{1}$$

where n_{eff} is the effective refractive index of the optical fiber with Bragg grating and Λ is the period of changes in the refractive index pitch for the fiber's core. The effective refractive index in a single-mode optical fiber can be approximated using the formula:

$$n_{eff} \cong \sqrt{n_2 + \frac{\lambda^2}{4\pi^2 r^2} \cdot (1.1428V - 0.996)^2}, \tag{2}$$

where n_2 is the refractive index of the cladding, λ is the wavelength of transmitted light, r is the core diameter and V is the normalized frequency.

The primary use of FBG is based on the deformational and temperature sensitivities. Based on the temperature evolution of mechanical stress, the Bragg wavelength shift can be defined as:

$$\frac{\Delta\lambda}{\lambda_0} = k\varepsilon + (\alpha_\Delta + \alpha_n)\Delta T, \tag{3}$$

where k is the deformational coefficient, α_n is the optical temperature coefficient, α_Δ is the coefficient of thermal expansion, ΔT is the temperature change and ε is the applied deformation. Deformational dependence and temperature dependence are determined both by the parameter values and the central Bragg wavelength. Normalized deformational and temperature coefficients [11] are introduced for determination of the individual sensitivities. Normalized deformational coefficient is given by:

$$\frac{1}{\lambda_B}\frac{\Delta\lambda_B}{\Delta\varepsilon} = 0.78 \cdot 10^{-6} \quad (\mu\varepsilon^{-1}), \tag{4}$$

and normalized temperature coefficient by:

$$\frac{1}{\lambda_B}\frac{\Delta\lambda_B}{\Delta\varepsilon} = 6.678 \cdot 10^{-6} \quad (°C^{-1}). \tag{5}$$

FBGs are single-point sensors, with multiplexing techniques we can join them together and obtain a multipoint measuring probe [12] and [13].

3. Results

Implemented manufacturing technology of FBG encapsulation is based on the casting of the liquid PDMS into the desired form (dimensions of the measuring probe). We take into account the size, weight, and shape within the design of the form. We create the rectangular shape with dimensions of 80×40×5 mm. The casting insert is created by a substrate into which is stored the bare FBG. Process of encapsulation is divided into three independent steps: the integration of FBG into liquid PDMS, the curing at a temperature of 100 °C $\pm$ 3 % in a temperature box for 60 minutes and 24 hour relaxation time. Results in the article [14] show that this type of encapsulation does not affect the structure of the FBG.

The measuring probe consists of the uniform FBG with polyamide protection with Bragg wavelength of 1554.1207 nm. The width of the reflecting spectrum is 2.3241 nm and reflectivity is 95.7 %. Monitoring of the basic parameters of FBG was performed during curing and after 24 hour relaxation time. Broad-spectrum LED (Light Emitting Diode) light source with a central wavelength of 1550 nm and an optical spectrum analyzer with a sampling frequency of 300 Hz were used for monitoring the parameters. The final probe is shown in Fig. 3.

Three different techniques for attaching the measuring probe were tested within the implementation of the probe on the human body. Based on the post-analysis, we defined the most efficient suitable method of attachment seems to be utilizing the contact strip placed around the chest of the patient. The position of the probe is in an area of the heart. The measurement is

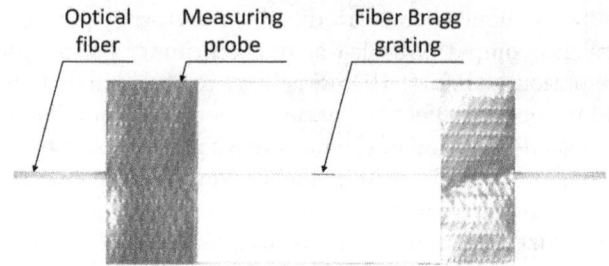

Fig. 3: The measuring probe for monitoring respiratory and heart rate.

based on sensing the widening of the ribcage of the patient during breathing. Movement (mechanical stress) is transferred to FBG using the contact strip. Respiratory and heart rate are analyzed by the spectral evaluation of the measured signals. Broad-spectrum LED with a central wavelength of 1550 nm ensures a source of light radiation. The methodology of the test was based on sensing a minute measurement of breath and heart rate with a sampling frequency of 300 Hz at five tested people. Three positions of patients were tested: static position in standing, static position in sitting, and static position on the back. Figure 4 shows measuring diagram for analyzing both heart and respiratory rate.

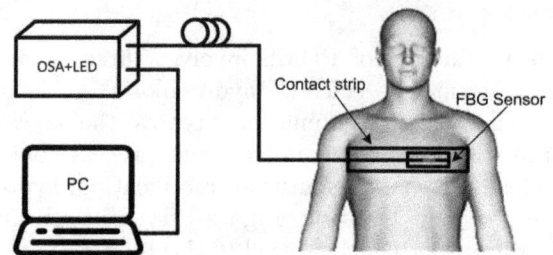

Fig. 4: Measuring diagram of respiratory and heart rate.

Figure 5 shows the measured changes of Bragg wavelength of the respiratory rate for the tested person in depending on the three selected positions. The respiratory rate was derived using the dominant frequency found via Fourier transform of the given waveform. A 20 second time is always shown for better clarity.

Based on the post-analysis, digital IIR (Infinite Impulse Response) filter utilized the measured waveforms of the respiratory rate. This IIR filter is bandpass of Butterworth type with a lower cut-off frequency of 1 Hz and upper cut-off frequency of 5 Hz. The magnitude response of the used digital IIR filter is shown in Fig. 7.

Figure 6 shows resulting superimposed courses of pulse activity over the courses of the breath for the tested persons in depending on the three defined positions. Heart rate was calculated based on the Fourier

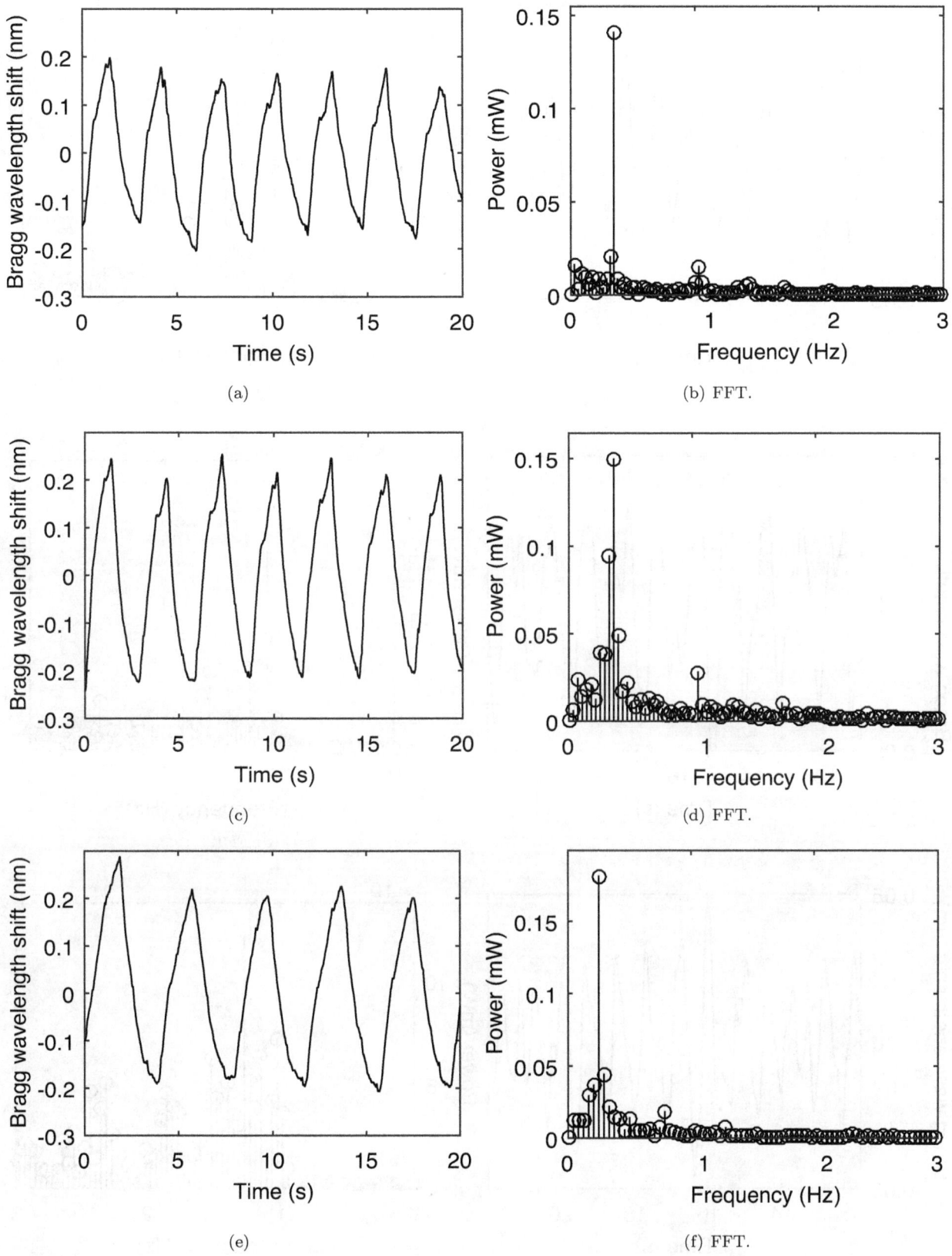

Fig. 5: Courses of respiratory rate and her frequency spectra: (a)–(b) static position in standing, (c)–(d) static position in sitting, (e)–(f) static position on the back.

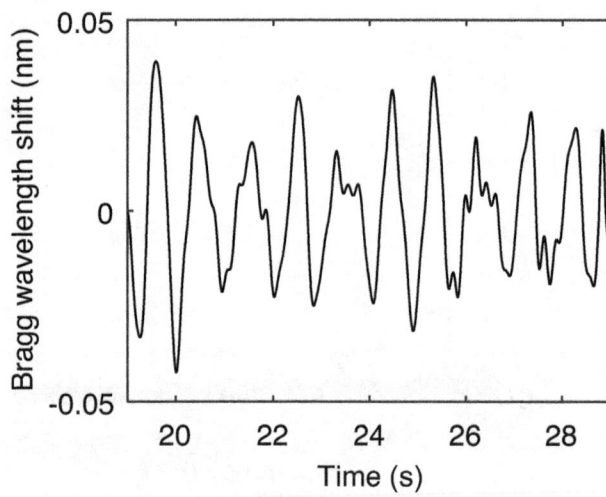

(a)

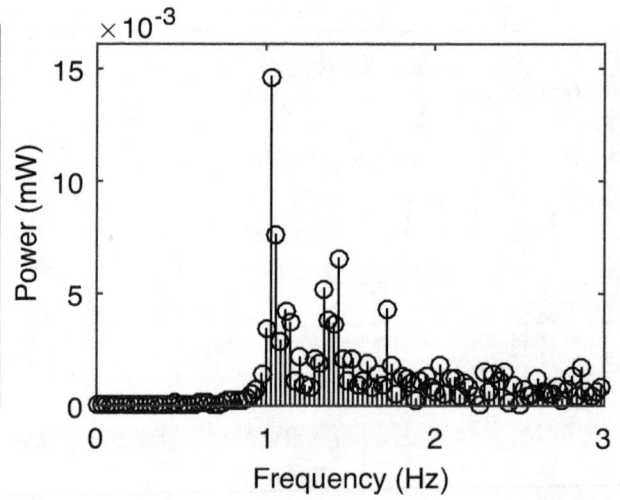

(b) FFT.

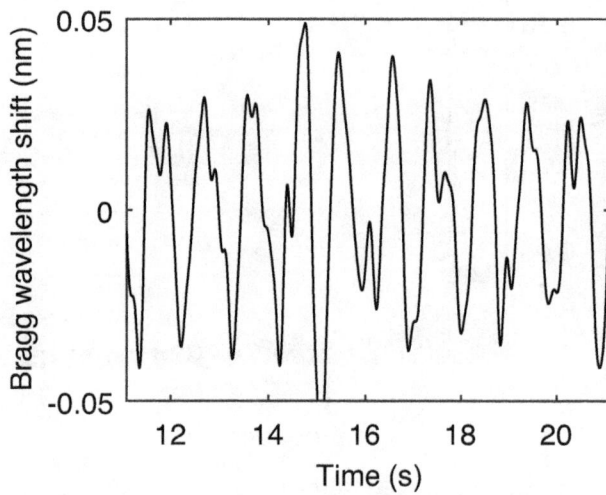

(c)

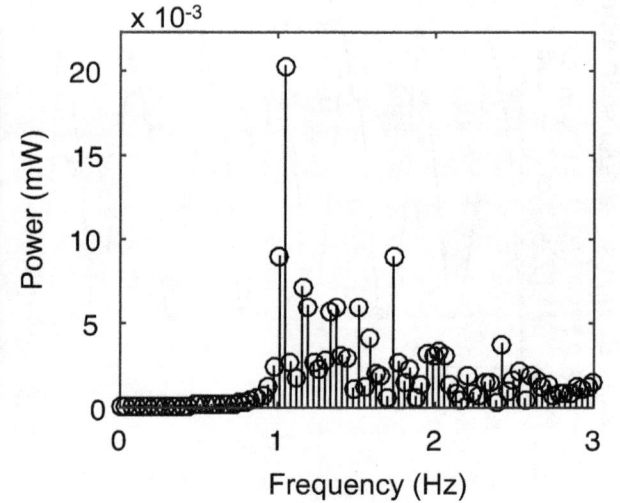

(d) FFT.

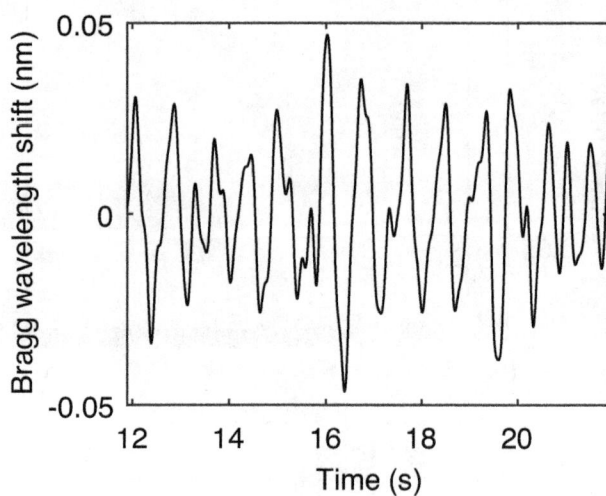

(e)

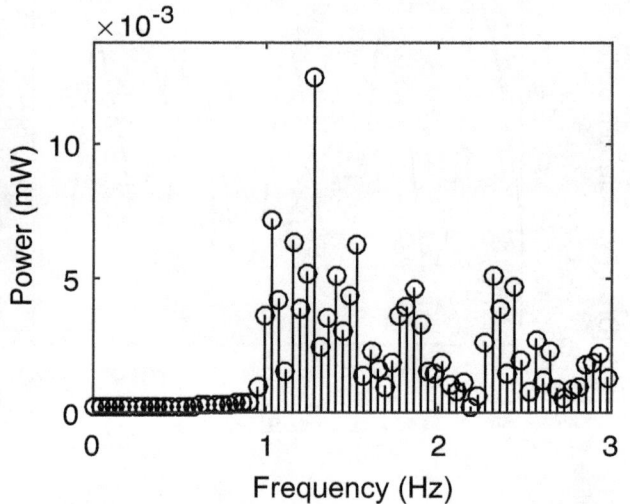

(f) FFT.

Fig. 6: Courses of heart rate and her frequency spectra: (a)–(b) static position in standing, (c)–(d) static position in sitting, (e)–(f) static position on the back.

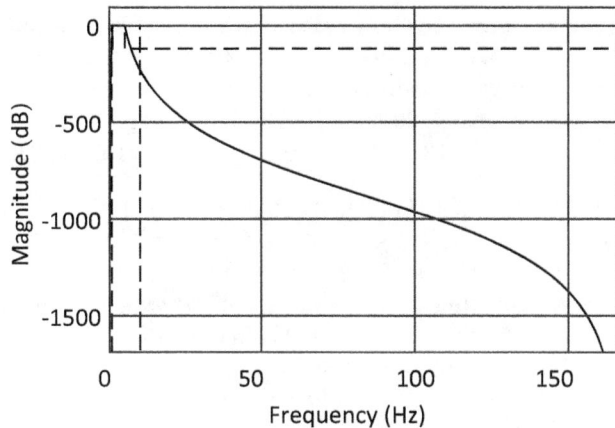

Fig. 7: The magnitude response of the used digital IIR filter.

transform, or more precisely, from the dominant frequency of the given waveforms. For clarity, only 10 seconds of the signal is plotted

Table 1 shows the dominant frequency and respiratory rate for the displayed courses in Fig. 5. Table 2 shows the dominant frequency and heart rate for the displayed courses in Fig. 6.

Tab. 1: Statistical data for courses of respiratory rate in Fig. 5.

Defined position	Dominant frequency (Hz)	Respiratory Rate (min^{-1})
Static position in standing	0.3611	21.6676
Static position in sitting	0.3439	20.6386
Static position on the back	0.2479	14.8741

Tab. 2: Statistical data for courses of heart rate in Fig. 6.

Defined position	Dominant frequency (Hz)	Heart Rate (min^{-1})
Static position in standing	1.0473	62.8363
Static position in sitting	1.0319	61.9159
Static position on the back	1.2808	76.8499

4. Conclusion

The authors described the design, implementation and verification (by experimental measurement) of the innovative prototype of the non-invasive measuring probe to monitoring respiratory and heart rate of the human body. PDMS has a unique combination of properties, and is suitable for use in biomedical applications. The main advantages include inert to human skin, immunity to electromagnetic interference, mechanical durability, and temperature stability. The main contribution of this paper are the design, implementation of non-invasive measuring probe encapsulated into PDMS, and implementation of the probe to the human using clamping contact strip. This combination creates a fundamental element of the measuring system. The repeated test of assembled prototype confirmed the functionality. Experimental results, which were carried out by the measuring probe in the laboratory, demonstrated the functionality of the proposed solution. Three general positions for measurements of mentioned vital parameters were analyzed. Based on the post-analysis, we can say that positions do not affect the functionality of the measuring probe. Respiratory and heart rate were derived based on the Fourier transform, or more precisely, from the dominant frequency component of the given waveform. The article does not include all aspects which can arise or affect monitoring of the mentioned fundamental life functions within the innovative measuring probe. It is the initial step to the new little-described medical field of non-invasive monitoring of the human body. The comprehensive characterization of our novel sensor and its comparison to other existing sensors, and the further signal processing will be the aim of our future articles.

Acknowledgment

This article was supported by the project of the Technology Agency of the Czech Republic TA04021263 and by Ministry of Education of the Czech Republic within the projects Nos. SP2017/128 and SP2017/79. The research has been partially supported by the Ministry of Education, Youth and Sports of the Czech Republic through the grant project no. CZ.1.07/2.3.00/20.0217 within the frame of the operation programme Education for competitiveness financed by the European Structural Funds and from the state budget of the Czech Republic. This article was also supported by the Ministry of the Interior of the Czech Republic within the projects Nos. VI20152020008 and VI2VS/444. The research has been partially supported by the COST action MP1401. This article was also supported by the Ministry of Industry and Trade of the Czech Republic within the project No. FV 10396.

References

[1] FAJKUS, M., J. NEDOMA, P. SISKA and V. VASINEK. FBG sensor of breathing encapsulated into polydimethylsiloxane. In: *Optical Materials and Biomaterials in Security and Defence Systems Technology XIII*. Edinburg: SPIE, 2016, pp. 1–5. ISBN 978-151060392-9. DOI: 10.1117/12.2241663.

[2] FAJKUS, M., J. NEDOMA, R. MARTINEK, V. VASINEK, H. NAZERAN and P. SISKA. A Non-Invasive Multichannel Hybrid Fiber-Optic Sensor System for Vital Sign Monitoring. *Sensors*. 2016, vol. 17, iss. 1, pp. 1–17. ISSN 1424-8220. DOI: 10.3390/s17010111.

[3] CHETHANA, K., G. A. S. PRASAD, S. N. OMKAR and S. ASOKAN. Fiber bragg grating sensor based device for simultaneous measurement of respiratory and cardiac activities. *Journal of Biophotonics*. 2017, vol. 10, iss. 2, pp. 278–285. ISSN 1864-0648. DOI: 10.1002/jbio.201500268.

[4] ZHU, Y., V. F. S. FOOK, E. H. JIANZHONG, J. MANIYERI, C. GUAN, H. ZHANG, E. P. JILIANG and J. BISWAS. Heart rate estimation from FBG sensors using cepstrum analysis and sensor fusion. In: *36th Annual International Conference of the IEEE Engineering in Medicine and Biology Society (EMBC)*. Chicago: IEEE, 2016, pp. 5365–5368. ISBN 978-1-4244-7929-0. DOI: 10.1109/EMBC.2014.6944838.

[5] TAN, C. H., Y. G. SHEE, B. K. YAP and F. R. M. ADIKAN. Fiber Bragg grating based sensing system: Early corrosion detection for structural health monitoring. *Sensors and Actuators A: Physical*. 2016, vol. 246, iss. 1, pp. 123–128. ISSN 0924-4247. DOI: 10.1016/j.sna.2016.04.028.

[6] LIANG, Y., G. YAN and S. HE. Enlarged-taper tailored Fiber Bragg grating with polyvinyl alcohol coating for humidity sensing. In: *International Conference on Optical Instruments and Technology: Optical Sensors and Applications*. Beijing: SPIE, 2015, pp. 1–6. ISBN 978-1-4244-7929-0. DOI: 10.1117/12.2193414.

[7] FENDINGER, N. J. *Organosilicon Chemistry Set: From Molecules to Materials*. 1st ed. New York: Wiley, 2008. ISBN 978-3-5276-2077-7. DOI: 10.1002/9783527620777.ch103c.

[8] NEDOMA, J., M. FAJKUS and V. VASINEK. Influence of PDMS encapsulation on the sensitivity and frequency range of fiber-optic interferometer. In: *Optical Materials and Biomaterials in Security and Defence Systems Technology XIII*. Edinburgh: SPIE, 2016, pp. 1–7. ISBN 978-151060392-9. DOI: 10.1117/12.2243170.

[9] LIU, Z. and H.-Y. TAM. Industrial and medical applications of fiber Bragg gratings. *Chinese Optics Letters*. 2016, vol. 14, iss. 12, pp. 1–5. ISSN 1671-7694. DOI: 10.3788/COL201614.120007.

[10] FAJKUS, M., J. NEDOMA, P. SISKA, L. BEDNAREK, S. ZABKA and V. VASINEK. Perimeter System Based on a Combination of a Mach-Zehnder Interferometer and the Bragg Gratings. *Advances in Electrical and Electronic Engineering*. 2016, vol. 14, no. 3, pp. 318–324. ISSN 1336-1376. DOI: 10.15598/aeee.v14i3.1752.

[11] KERSEY, A. D., M. A. DAVIS, H. J. PATRICK, M. LEBLANC, K. P. KOO, C. G. ASKINS, M. A. PUTNAM and E. J. FRIEBELE. Fiber grating sensors. *Journal of Lightwave Technology*. 1997, vol. 15, no. 8, pp. 1442–1463. ISSN 1558-2213. DOI: 10.1109/50.618377.

[12] FAJKUS, M., I. NAVRUZ, S. KEPAK, A. DAVIDSON, P. SISKA, J. CUBIK and V. VASINEK. Capacity of Wavelength and Time Division Multiplexing for Quasi-Distributed Measurement Using Fiber Bragg Gratings. *Advances in Electrical and Electronic Engineering*. 2015, vol. 13, no. 5, pp. 575–582. ISSN 1336-1376. DOI: 10.15598/aeee.v13i5.1508.

[13] FAJKUS, M., J. NEDOMA, S. KEPAK, L. RAPANT, R. MARTINEK, L. BEDNAREK, M. NOVAK and V. VASINEK. Mathematical model of optimized design of multi-point sensoric measurement with Bragg gratings using wavelength divison multiplex. In: *Optical Modelling and Design IV*. Brussel: SPIE, 2016, pp. 1–7. ISBN 978-151060134-5. DOI: 10.1117/12.2239551.

[14] NEDOMA, J., M. FAJKUS, L. BEDNAREK, J. FRNDA, J. ZAVADIL and V. VASINEK. Encapsulation of FBG Sensor into the PDMS and its Effect on Spectral and Temperature Characteristics. *Advances in Electrical and Electronic Engineering*. 2016, vol. 14, no. 4, pp. 460–466. ISSN 1336-1376. DOI: 10.15598/aeee.v14i4.1786.

About Authors

Jan NEDOMA was born in 1988 in Prostejov. In 2012 he received a Bachelor's degree from VSB–Technical University of Ostrava, Faculty of Electrical Engineering and Computer Science, Department of Telecommunications. Two years later, he received his Master's degree in the field of Telecommunications in the same workplace. He is currently employee and a Ph.D. student of Department of Telecommunications at VSB–Technical University of Ostrava. He works in the field of optical communications and fiber optic sensor systems.

Marcel FAJKUS was born in 1987 in Ostrava. In 2009 he received a Bachelor's degree from VSB–Technical University of Ostrava, Faculty of Electrical Engineering and Computer Science, Department of

Telecommunications. Two years later, he received a Master's degree in the field of Telecommunications in the same workplace. He is currently employee and a Ph.D. student of Department of Telecommunications at VSB–Technical University of Ostrava. He works in the field of optical communications and fiber optic sensor systems.

Petr SISKA was born in 1979 in Kromeriz. In 2005 he finished M.Sc. study at VSB–Technical University of Ostrava, Faculty of Electrical Engineering and Computer Science, Department of Electronic and Telecommunications. Three years later, he finished Ph.D. study in Telecommunication technologies. Currently he is employee of Department of Telecommunications. He is interested in Optical communications, Fiber optic sensors and Distributed Temperature Sensing systems.

Radek MARTINEK was born in 1984 in Czech Republic. In 2009 he received Master's degree in Information and Communication Technology from VSB–TU of Ostrava. Since 2012 he worked here as a research fellow. In 2014 he successfully defended his dissertation thesis titled „The use of complex adaptive methods of signal processing for refining the diagnostic quality of the abdominal fetal electrocardiogram".

He works as an assistant professor at VSB–TUO: Technical University of Ostrava since 2014. In 2016 he became the Layout Editor in journal Advances in Electrical and Electronic Engineering (ISSN 1336-1376).

Vladimir VASINEK was born in Ostrava. In 1980 he graduated in Physics, specialization in Optoelectronics, from the Science Faculty of Palacky University. He was awarded the title of RNDr. at the Science Faculty of Palacky University in the field of Applied Electronics. The scientific degree of Ph.D. was conferred upon him in the branch of Quantum Electronics and Optics in 1989. He became an associate professor in 1994 in the branch of Applied Physics. He has been a professor of Electronics and Communication Science since 2007. He pursues this branch at the Department of Telecommunications at VSB–Technical University of Ostrava. His research work is dedicated to optical communications, optical fibers, optoelectronics, optical measurements, optical networks projecting, fiber optic sensors, MW access networks. He is a member of many societies - OSA, SPIE, EOS, Czech Photonics Society; he is a chairman of the Ph.D. board at the VSB–Technical University of Ostrava. He is also a member of habitation boards and the boards appointing to professorship.

13

Germania and Alumina Dopant Diffusion and Viscous Flow Effects at Preparation of Doped Optical Fibers

Jens KOBELKE, Kay SCHUSTER, Joerg BIERLICH, Sonja UNGER,
Anka SCHWUCHOW, Tino ELSMAN, Jan DELLITH, Claudia AICHELE,
Ron FATOBENE ANDO, Hartmut BARTELT

Leibniz Institute of Photonic Technology, Albert-Einstein-Strasse 9, 07745 Jena, Germany

jens.kobelke@leibniz-ipht.de, kay.schuster@leibniz-ipht.de, joerg.bierlich@leibniz-ipht.de,
sonja.unger@leibniz-ipht.de, anka.schwuchow@leibniz-ipht.de, tino.elsmann@leibniz-ipht.de,
jan.dellith@leibniz-ipht.de, claudia.aichele@leibniz-ipht.de, ron.spittel@leibniz-ipht.de,
hartmut.bartelt@leibniz-ipht.de

Abstract. *We report on germania and alumina dopant profile shift effects at preparation of compact optical fibers using packaging methods (Stack-and-Draw method, Rod-in-Tube (RiT) technique). The sintering of package hollow volume by viscous flow results in a shift of the core-pitch ratio in all-solid microstructured fibers. The ratio is increased by about 5 % in the case of a hexagonal package. The shift by diffusion effects of both dopants is simulated for typical slow speed drawing parameters. Thermodynamic approximations of surface dissociation of germania doped silica suggest the need of an adequate undoped silica barrier layer to prevent an undesired bubble formation at fiber drawing. In contrast, alumina doping does not estimate critical dissociation effects with vaporous aluminium oxide components. We report guide values of diffusion length of germania and alumina for the drawing process by kinetic approximation. The germania diffusion involves a small core enlargement, typically in the submicrometer scale. Though, the alumina diffusion enlarges it by a few micrometers. A drawn pure alumina preform core rod transforms to an amorphous aluminosilicate core with a molar alumina concentration of only about 50 % and a non-gaussian concentration profile.*

Keywords

Alumina, diffusion, doped silica, germania, optical fiber, sensor fiber.

1. Introduction

For about two decades, fibers with extreme refractive index contrast, for instance holey and all-solid microstructured fibers, have been intensively investigated due to their unique optical properties, e.g. unusual dispersion, endlessly single mode transmission, photonic band gap propagation, etc. [1], [2] and [3]. Such fibers are suitable for various sensor applications, e.g. in chemical analytics by absorption, Raman or Brillouin spectroscopy. Fibers with extreme material contrast between core and cladding are also interesting for different applications, e.g. nonlinear devices [4], or sensors for temperature, strain or pressure monitoring with core-inscribed Bragg gratings [5].

The typical fabrication approach for both fiber types follows packaging procedures. The Rod-in-Tube process combines a core glass rod with a different cladding glass hollow cylinder, whereas the Stack-and-Draw method arranges different doped rods or hollow cylinders surrounded by a jacketing tube. The interstitial volume between the mostly hexagonal packaged circular elements is removed at least in the final drawing procedure by evacuation. Alternatively, by drawing at atmospheric pressure the interspace collapse can also be affected by surface tension. The profiles of the doped fibers are smoothed by diffusion, compared to the preform. On the diffusion of GeO_2 in silica matrix by doped layer annealing is reported in [6]. Diffusion and dissociation effects of germania at all-solid fiber fabrication are shown in [7], about the alumina diffusion with low Al_2O_3 concentration reports [8].

2. Simulation of Dopant Dissociation and Diffusion

The germania dissociation is the cause for the development of the central dip in typical gradient and step index fibers. The dissociation can be described by the following simplified reaction:

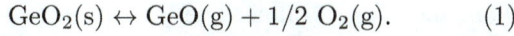

$$GeO_2(s) \leftrightarrow GeO(g) + 1/2\ O_2(g). \qquad (1)$$

Figure 1 shows an exemplary equilibrium partial pressure of the gaseous reaction product GeO starting with a glass composition 10 mol% GeO_2 - 90 mol% SiO_2. The chemical equilibrium was calculated with the program HSC [9]. It demonstrates the high evaporation tendency at fiber drawing temperature of about 1900 °C. The partial pressure of GeO is about 6 % of atmospheric pressure at this temperature. In contrast, for alumina doping no significant vaporous components at this temperature were found. All considered gaseous species, e.g. AlO(g), AlO_2(g), Al_2O(g), Al_2O_2(g), Al_2O_3(g) [10], show an equilibrium partial pressure $p << 10^{-6}$ Pa in the simulation.

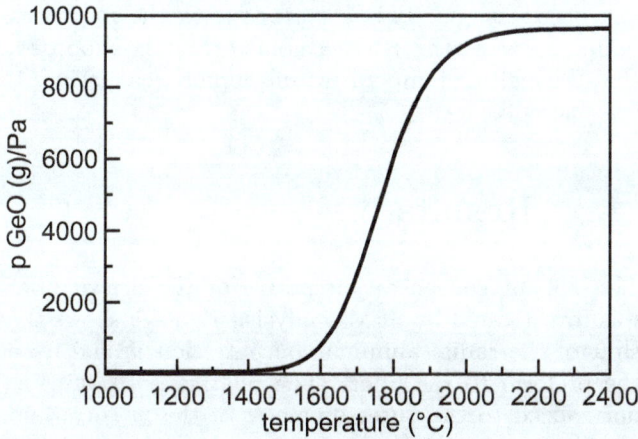

Fig. 1: Calculation of partial pressure vs. temperature for evaporation reaction of germania-silica glass (10 mol% GeO_2), total pressure: 1 bar.

However, in variation to chemical equilibrium consideration the experimental dopant distribution is limited by diffusion effects, depending on temperature and dopant concentration. Figure 2 shows the expected diffusion coefficients at drawing temperature dependent on the dopant concentration. The variation of germania concentration from 1 mol% to 20 mol% at 1900 °C increases the diffusion coefficient by a factor of about 3, from $10^{-10.5}$ cm^2·s^{-1} to 10^{-10} cm^2·s^{-1} [6].

The investigation of annealed aluminosilicate layers by [8] indicates a significantly higher diffusion coefficient with a strong nonlinearity compared to germania doping. In the drawing temperature range (1900–2200 °C) it is expected to be from about

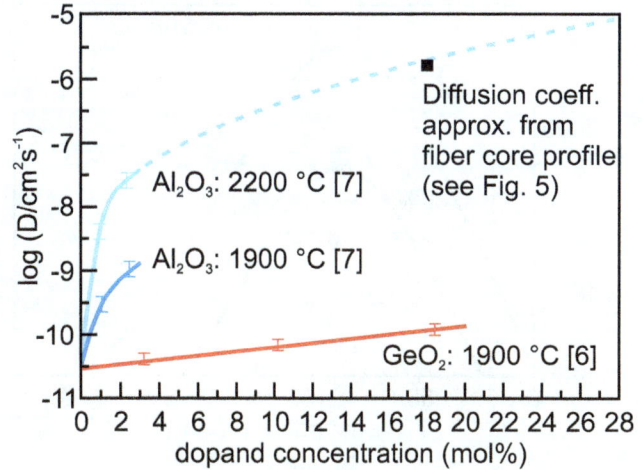

Fig. 2: Approximated concentration dependence of diffusion coefficients for germania (corr. to [6]) and alumina (corr. to [8]) in a silica preform at 1900 °C and 2200 °C, respectively.

10^{-8} cm^2·s^{-1} up to 10^{-9} cm^2·s^{-1} even for low alumina concentration. Though the investigated maximum Al_2O_3 concentration was about 3 mol% in [8].

The effective diffusion time τ_{DF} during fiber drawing was calculated from the axial diffusion zone length z_D and the fiber drawing speed v_F:

$$\tau_{DF} = \frac{z_D}{v_F}. \qquad (2)$$

The axial diffusion zone length was approximated from the axial temperature profile, measured along the central axis of the drawing furnace. Based on the equivalent temperature T_{eq} (K), the diffusion zone length z_D can be approximated from the temperature profile (see Fig. 3).

T_{eq} is associated with the activation energy of diffusion E_a given by Eq. (3):

$$\frac{1}{T_{eq}} = \frac{1}{T_{\max}} + \frac{R}{E_a}, \qquad (3)$$

where $T_{\max}$ (K) is the maximum temperature of the axial temperature profile and R is the universal gas constant. E_a is given in [6] and [8] presuming an Arrhenius dependence of the diffusion coefficient.

Figure 3 shows the graphically determined diffusion zone lengths with $T_{\max} = 1900$ °C for germania doped fiber drawing and $T_{\max} = 2200$ °C for aluminosilicate fiber drawing. The value of z_D is about 38 mm for a germania doping level between 1 mol% and 20 mol% [7]. For alumina diffusion z_D is approximated to be 52 mm for an Al_2O_3 concentration of 2 mol% and extrapolated to be 60 mm for 20 mol% Al_2O_3.

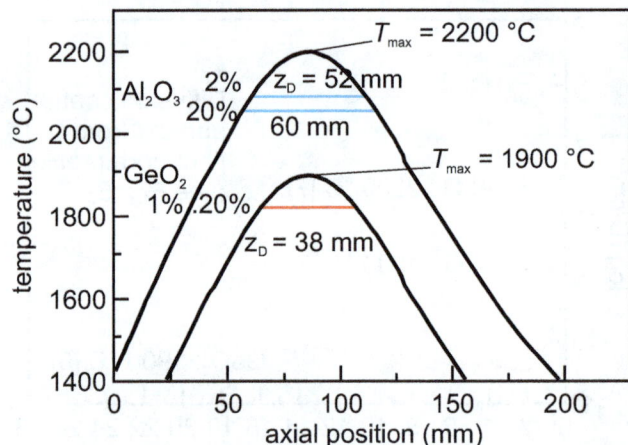

Fig. 3: Approximated effective diffusion zone lengths for alumina and germania doped silica preforms in the drawing furnace.

3. Drawing Experiments

An all-solid microstructured fiber with 19 germanium doped cores (core diameter: $d = 5.0$ µm, pitch: $\Lambda = 10.0$ µm) was manufactured by stack-and-draw technique for a setup in a Coherent Antiresonant Raman Spectroscopic Probe [11]. The preform package arrangement and the drawn fiber are shown in Fig. 4.

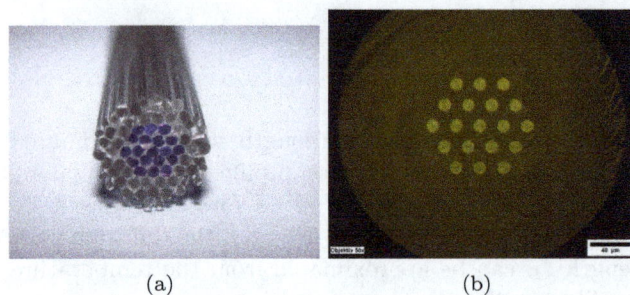

Fig. 4: (a) preform package arrangement of the 19 core fiber. The germania doped rods are marked blue. (b) micrograph of the fiber after drawing.

The shift of the ratio d/Λ was found to be about 5 %. It is caused by the evacuation of the interstitial volume during drawing. The core diameter of the GeO_2 doped preform rods was 504 µm, the outer diameter of 1000 µm, which corresponds to the pitch of the preform arrangement. The GeO_2 concentration was 6 mol%. The core was deposited by Modified Chemical Vapor Deposition (MCVD) with 25 doped layers. After the drawing the core-pitch-ratio has changed from 0.504 (preform) to ratio d/Λ: 0.532–0.533 in the fiber. This corresponds to the expectation by sintering of the interspace volume. The diameter varies over all core elements 5.31–5.57 µm, the pitch: 9.98–10.45 µm, respectively. Due to the relatively low diffusivity of germanium, no significant additional broadening of the core

diameter is experimentally observed. The pure silica cladding layer thickness of about 2.35 µm is sufficient for avoidance of bubble formation according Eq. (1).

An aluminosilicate core fiber with extreme high alumina concentration was prepared by RiT technique. A crystalline alumina rod (sapphire) with 2.8 mm diameter was inserted in a silica tube (Heraeus F300) with outer diameter $OD = 30$ mm and central hole diameter $ID = 3$ mm. The preform was drawn with a drawing speed of $v_F = 40$ m·min^{-1} to an UV-acrylate coated fiber with an outer diameter of 125 µm. The drawing furnace temperature, measured with an IR-thermometer, was 2200 °C. The drawing force F was between 0.1 N and 0.17 N. The effective drawing temperature was calculated from the effective drawing viscosity given by Eq. (4) proposed in [12]:

$$\eta(T) = \frac{4F\Delta z_b}{3v_P\left(OD^2 - ID^2\right)\ln\left(\dfrac{v_F}{v_P}\right)}, \qquad (4)$$

where Δz_b is the axial width of the heating zone for viscous flow in the neck down region of the preform. It is about 35 mm for the used drawing furnace. The approximated viscosity was calculated as about $\log(\eta/Pas) \approx 3$. This corresponds to the measured temperature of the IR-thermometer. It is supposed, that the melting temperature of sapphire (ca. 2050 °C) was exceeded.

4. Results

The dopant concentration profile of the drawn fiber was investigated by SEM analysis. Figure 5 shows the shift of the radial alumina concentration profile from the preform to the fiber. The relative positioning is normalized to the outer diameter of the preform and the fiber, respectively.

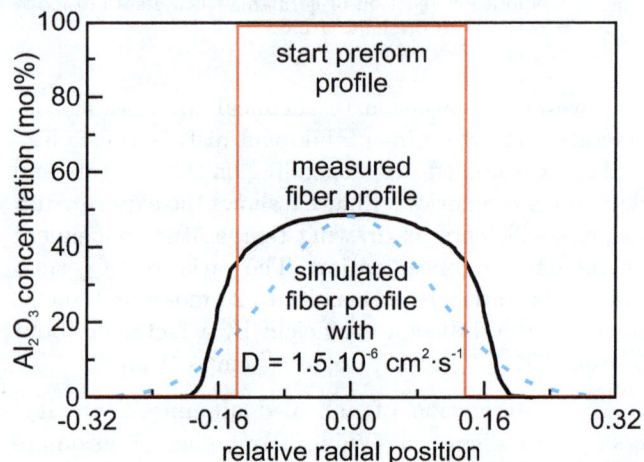

Fig. 5: Alumina concentration profile of the drawn fiber compared to the starting profile of the preform.

The composition of the core changes from pure alumina in the preform to an aluminosilicate glass composition of 50.6 % SiO_2 and 49.4 % Al_2O_3 in the center of the fiber core. The mean diffusion coefficient calculates to be $1.5 \cdot 10^{-6}$ cm^2·s^{-1} by Fick's law. The approximation is based on an axial diffusion zone length of 60 mm in the drawing furnace and a drawing speed of 40 m·min^{-1}. By extrapolating the specified concentration dependence in [8], the found diffusivity corresponds to an aluminosilicate glass with a mean concentration of 18 mol% Al_2O_3 (see Fig. 2). Nevertheless, the experimentally found alumina concentration profile in the fiber cannot be explained only by a simple thermal activated alumina dopant concentration flow balance.

We found structural crystalline changes in the thermal induced transition region from preform core to the fiber core. The investigation of the preform neck-down region by X-ray diffraction analysis shows beside a strong amorphous phase intense peaks of mullite (see Fig. 6). Similar recrystallization effects were also described in [13]. Obviously, the formation of mullite is caused by the long dwell time (approx. 50 min) during passing the neck-down region. After removing the preform out of the drawing furnace it was quenched with natural air convection. The cooling rate is here strong limited by the low surface-volume ratio of the upper neck-down region of the preform. Compared to this low cooling rate under static air quenching conditions, the drawn fiber suffers an extremely high cooling rate by a drawing speed of 40 m·min^{-1}. The duration for passing a length of the neck-down region in this case is only 0.05 s.

The investigation of the fiber core (shown in Fig. 7) by electron backscattering shows an amorphous core cross section.

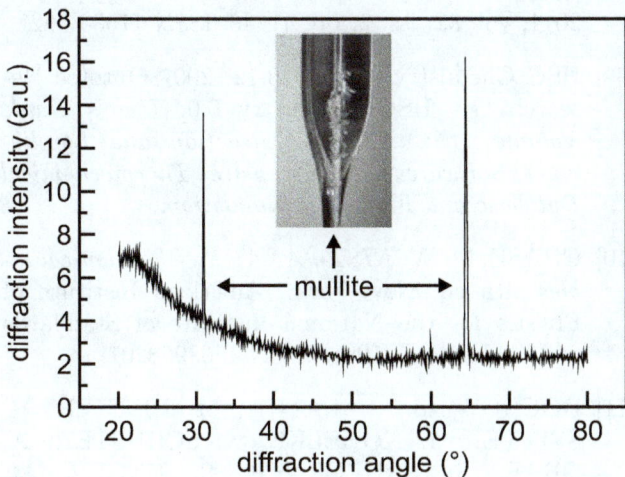

Fig. 6: X-ray diffraction pattern of the core material in the preform neck-down region (inset: photograph of the preform neck-down region, arrowhead shows to the measuring point).

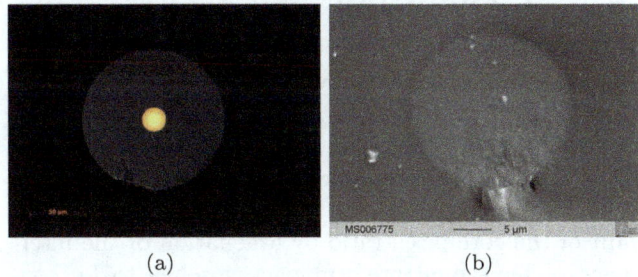

Fig. 7: (a) white light micrograph of the fiber end face, (b) SEM picture of the fiber core.

We have not found the appearance of any separate crystalline phase in the fiber core. This correspondes with the high transmission and low scattering level of the aluminosilicate core. Figure 7(a) shows the micrograph of the cleaved fiber with white light transmission illumination. The brightness contrast between core and cladding is caused by the high refractive index difference. The SEM picture in Fig. 7(b) demonstrates the excellent homogeneity of the core region. No separate crystals or heterophase separations were observed. The fracture surface at the core cladding interface is caused by the strong tension mismatch between the silica clad and the aluminosilicate core, which has a much higher thermal expansion then silica.

Figure 8 shows the spectral loss of the multimode fiber with a core diameter of 21 µm and a numerical aperture of NA = 0.54.

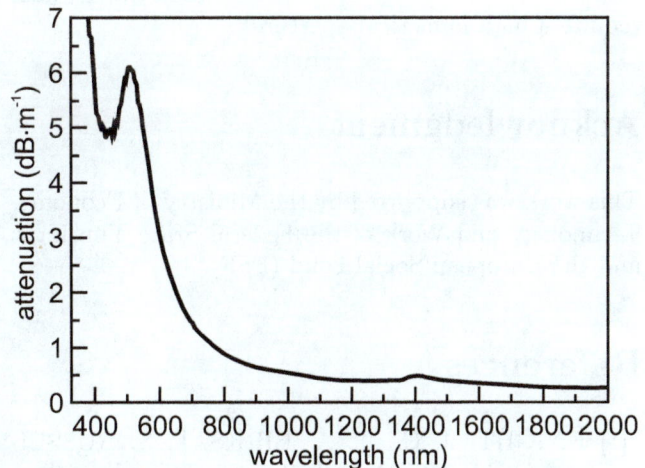

Fig. 8: Attenuation spectrum of the aluminosilicate core fiber.

The minimum loss was found at a wavelength of 1850 nm with 0.27 dB·m^{-1}. For a typical application, like Bragg grating inscription at a wavelength of 1550 nm [5], the loss is 0.37 dB·m^{-1}. The peak at 550 nm is attributed to impurities of the sapphire starting material.

5.   Conclusion

Dopant diffusion combined with thermal induced dissociation can cause variations and disturbances in dopant profiles of all-solid microstructured fibers. For germania doping, a sufficient silica barrier avoids the undesired bubble formation by thermal dissociation. The shift of the core-pitch-ratio by evacuation of the interspace of hexagonal preform packages has to be considered for the design of the final compact fiber. The drawing of a sapphire core in silica cladding does not result in a fiber with a pure alumina core. We found a maximum alumina concentration of about 50 mol% in the fiber core. Obviously, the sufficient softening of the complete preform at fiber drawing presumes the melting of the sapphire rod. The necessarily applied high drawing temperature $T > 2050$ °C causes an intense reciprocal diffusion of alumina and silica. We found an intermediate recrystallization to mullite in the neckdown region of the preform. The drawn fiber shows an aluminosilicate core with a non-gaussian alumina concentration profile and a maximum Al_2O_3 concentration of about 50 mol%. The diameter of the core is enlarged about by factor 2 compared to the preform cross section. Although the high alumina glasses show devitrification tendency, we found no crystalline phase in the final fiber core. Obviously, the large cooling rate during fiberization of the preform allows overcome the high recrystallization tendency. The fiber core shows a low spectral loss minimum < 1 dB·m^{-1} due to its low scattering and good axial homogeneity. Future applications of this fiber type are to be seen in high temperature sensing and complex all-solid fiber devices, which require a high numerical aperture.

Acknowledgment

This work was supported by the Ministry of Economy, Technology and Work of the Federal State Thuringia and the European Social Fond (ESF).

References

[1] KNIGHT, J. C., T. A. BIRKS, P. S. RUSSEL and D. M. ATKIN. All-silica single-mode optical fiber with photonic crystal cladding. *Optics Letters*. 1996, vol. 21, iss. 19, pp. 1547–1549. ISSN 0146-9592.

[2] RUSSELL, P. Photonic crystal fibers. *Science*. 2003, vol. 299, iss. 5605, pp. 358–362. ISSN 1095-9203. DOI: 10.1126/science.1079280.

[3] LÆGSGAARD, J. and A. BJARKLEV. Microstructured Optical Fibers–Fundamentals and Applications. *Journal of the American Ceramic Society*. 2006, vol. 89, iss. 1, pp. 2–12. ISSN 1551-2916. DOI: 10.1111/j.1551-2916.2005.00798.x.

[4] KARRAS, C., W. PAA, D. LITZKENDORF, S. GRIMM, K. SCHUSTER and H. STAFAST. SiO_2-Al_2O_3-La_2O_3 glass - a superior medium for optical Kerr gating at moderate pump intensity. *Optical Material Express*. 2016, vol. 6, iss. 1, pp. 125–130. ISSN 2159-3930. DOI: 10.1364/OME.6.000125.

[5] ELSMANN, T., A. LORENZ, N. S. YAZD, T. HABISREUTHER, J. DELLITH, A. SCHWUCHOW, J. BIERLICH, K. SCHUSTER, M. ROTHHARDT, L. KIDO and H. BARTELT. High temperature sensing with fiber Bragg gratings in sapphire-derived all-glass optical fibers. *Optics Express*. 2014, vol. 22, iss. 22, pp. 26831–26839. ISSN 1094-4087. DOI: 10.1364/OE.22.026825.

[6] KIRCHHOF, J., S. UNGER, B. KNAPPE and J. DELLITH. Diffusion in binary GeO_2-SiO_2 glasses. *Physics and Chemistry of Glasses: European Journal of Glass Science Technology part B*. 2007, vol. 48, iss. 3, pp. 129–133. ISSN 1753-3562.

[7] KOBELKE, J., J. BIERLICH, K. WONDRACZEK, C. AICHELE, Z. PAN, S. UNGER, K. SCHUSTER and H. BARTELT. Diffusion and Interface Effects during Preparation of All-Solid Microstructured Fibers. *Materials*. 2014, vol. 7, iss. 9, pp. 6879–6892. ISSN 1996-1944. DOI: 10.3390/ma7096879.

[8] UNGER, S., J. DELLITH, A. SCHEFFEL and J. KIRCHHOF. Diffusion in Yb_2O_3-Al_2O_3-SiO_2 glass. *Physics and Chemistry of Glasses: European Journal of Glass Science Technology part B*. 2011, vol. 52, iss. 2, pp. 41–46. ISSN 1753-3562.

[9] HSC Chemistry, version 6.12, 2007 Outotec Research Oy, HSC Chemistry 6.0. *User's Guide Volume 1/2: Chemical Reaction and Equilibrium Software with Extensive Thermochemical Database and Flowsheet Simulation*.

[10] CHASE, M. W. *NIST-JANAF thermochemical tables*. 4th ed. New York: American Institute of Physics for the National Institute of Standards and Technology, 1998. ISBN 1563968207.

[11] DOCHOW, S., I. LATKA, M. BECKER, R. SPITTEL, J. KOBELKE, K. SCHUSTER, A. GRAF, S. BRUECKNER, S. UNGER, M. ROTHHARDT, B. DIETZEK, C. KRAFFT and J. POPP. Multicore fiber with integrated fiber Bragg gratings for background-free Raman sensing. *Optics Express*. 2012, vol. 20,

iss. 18, pp. 20156–20169. ISSN 1094-4087. DOI: 10.1364/OE.20.020156.

[12] KIRCHHOF, J., K. GERTH, J. KOBELKE and K. SCHUSTER. Photonic Crystal Fibers - Viscous Behaviour of Silica Tubes During Collapsing and Hollow Fiber Drawing Poc. In: *6th ESG Conference*. Montpellier: Glass Odyssey, 2002, pp. 196–201.

[13] PAN, Z. *Consolidation of high melting SiO_2-Al_2O_3-La_2O_3 glass powders with gas pressure vacuum viscous sintering technology*. Jena, 2016. Dissertation thesis. Friedrich-Schiller-Universitaet Jena.

About Authors

Jens KOBELKE was born 1959 in Merseburg, Germany. He studied chemistry received his Ph.D. from Technische Hochschule Merseburg in 1986. Since this time he works at Institute of Photonic Technology (IPHT) Jena on the development and preparation of special optical fibers.

Kay SCHUSTER was born 1965 in Beetzendorf, Germany. He studied chemistry at Martin-Luther-University Halle and received his Ph.D. 1995 from the University Karlsruhe. Since 1996 he is with the IPHT Jena. He is the leader of the fiber technological group at IPHT and engaged in research on optical speciality fibers.

Joerg BIERLICH was born 1973 in Jena, Germany. He studied medical engineering at University of Applied Science Jena and received his Ph.D. 2008 from the Bergakademie Freiberg. Since 2001 he works at IPHT Jena on preparation of specialty optical fibers.

Sonja UNGER was born 1951 in Rathen, Germany. She received the Ph.D. degree in inorganic chemistry from the College of Education, Erfurt, Germany, in 1977. She works at the IPHT in the Optical Fiber Technology Group as a research collaborator in the field of MCVD technology. She is engaged on research in the development of materials and high silica specialty optical fibers (active fibers for laser and amplifier, photosensitive and nonlinear fibers).

Anka SCHWUCHOW was born in 1970 in Jena, Germany. From 1988 to 1991 she studied microelectronics at Technical University of Sofia (VMEI) in Sofia (Bulgaria) and from 1991 telecommunications at Technische Universitaet Dresden. There she obtained her first german degree. Since 1994 she has been a member of IPHT Jena and has specialized in characterisation of optical fibers and preforms and rare earths in glasses.

Tino ELSMANN was born 1986 in Zeulenroda, Germany. He studied physics at University Jena and received the diploma degree 2010. Actually he is a Ph.D. student at the IPHT Jena. He is interested in fiber optical devices, e.g. fiber Bragg gratings for special applications.

Jan DELLITH was born 1973 in Jena, Germany. He studied material engineering at University of Applied Science Jena and received his Ph.D. 2008 from the Bergakademie Freiberg. Since 2000 he works at IPHT Jena in particular in the field of solid state analytics.

Claudia AICHELE was born 1969 in Jena, Germany. She studied material science at the Bergakademie Freiberg. There she received the diploma engineer degree in 1993. Since 1995 she is with the IPHT Jena. Her interests are material investigation and preparation by MCVD method and thermodynamical simulation.

Ron FATOBENE ANDO was born 1981 in Jena, Germany. He studied physics at the Friedrich Schiller University Jena and received his Ph.D. in 2016. His interests comprise the design and simulations of specialty optical fibers.

Hartmut BARTELT was born 1951. He studied physics at University of Karlsruhe and Nuernberg-Erlangen. After receiving Ph.D. degree 1980 and habilitation 1985 he worked at Siemens AG Erlangen. Since 1994 he is Professor at the University of Jena and head of the Fiber Optics Division of the IPHT. His interests are in theory and application of fiber optics.

14

Comparision of Splitting Properties of Various 1×16 Splitters

Catalina BUTSCHER [1, 2], *Dana SEYRINGER* [1], *Michal LUCKI* [2]

[1]Research Centre for Microtechnology, Vorarlberg University of Applied Sciences,
Hochschulstrasse 1, 6820 Dornbirn, Austria
[2]Department of Telecommunication Engineering, Faculty of Electrical Engineering,
Czech Technical University in Prague, Technicka 2, 16627 Prague 6, Czech Republic

catalina.burtscher@fhv.at, dana.seyringer@fhv.at, luckimic@fel.cvut.cz

Abstract. *Optical Access Networks (OAN) mostly use optical splitters to distribute the services from Optical Line Terminal (OLT) on the provider's side to the subscribers in Optical Network Unit (ONU). Optical splitters are the key components in such access networks as for example GPON and XG-PON by ITU-T. In this paper we investigate the optical properties of 1×16 Y-branch splitter and 1×16 MMI splitters based on different widths of multimode interference section and different lengths of the output ports. These two splitters were designed, simulated and the obtained results of both were studied and compared with each other. Additionally, we show that the used standard waveguide core size (usually 6×6 μm^2 to match the diameter of the single mode input/output fibers, i.e. to keep the coupling loses as low as possible) supports not only propagation of the single mode but of the first mode too, leading to an asymmetric splitting ratio (increasing non-uniformity of split power over all the output waveguides). Decreasing waveguide core size, it is possible to suppress presence of the first mode and this way to reduce non-uniformity.*

Keywords

GPON, MMI splitter, multimode interference splitter, Optical Access Networks (OAN), Optical Network Unit (ONU), optical splitting, XG-PON, Y-branch splitter.

1. Introduction

Optical splitters play an important role in the integrated optics allowing several customers to share the same connection, bringing high-speed networking, digital television and telephone services to residences using fiber-optic cables [1].

There are two main approaches used to split one optical signal into N output signals. One of these approaches is to use a MultiMode Interference (MMI) coupler, where the splitting of the optical signal is based on a self-imaging effect [1] and [2]. MMI splitters feature a large splitting number and a stable splitting ratio, ensuring good uniformity over all output signals [3]. They exhibit good fabrication tolerance since the splitting is performed in a large multimode section. However, their main disadvantage results from the fact that the length of the MMI section is wavelength-dependent, i.e. MMI splitters are designed solely for one wavelength and can only operate in a narrow wavelength band. They are also polarization-dependent, but it was shown that for strong guidance waveguide structures this dependence is negligible [4].

Another possibility to split an optical signal is to make it as a cascade of one-by-two waveguide branches called Y-branch splitting. Y-branch splitters are the key components in Fiber-To-The-x (FTTx) networks because they are polarization and wavelength-independent, i.e. one device can be used in the whole operating wavelength window. However, they have the disadvantage that the processing of branching points, where two waveguides start to separate, is technologically very difficult, leading to an asymmetric splitting ratio of the split power over all the output waveguides. Furthermore, Y-branch splitters, especially high channel optical splitters, are much larger in comparison to MMI splitters.

In the MMI approach, optical properties of a splitter depend on the width of a multimode coupler. Therefore, in this paper we studied the splitting properties of

(a) Geometry of the MMI splitter.

(b) Simulation of MMI splitter.

(c) Field distribution at the end of the splitter.

(d) Detailed view on the field distribution showing the non-uniformity, ILu and the insertion loss, IL.

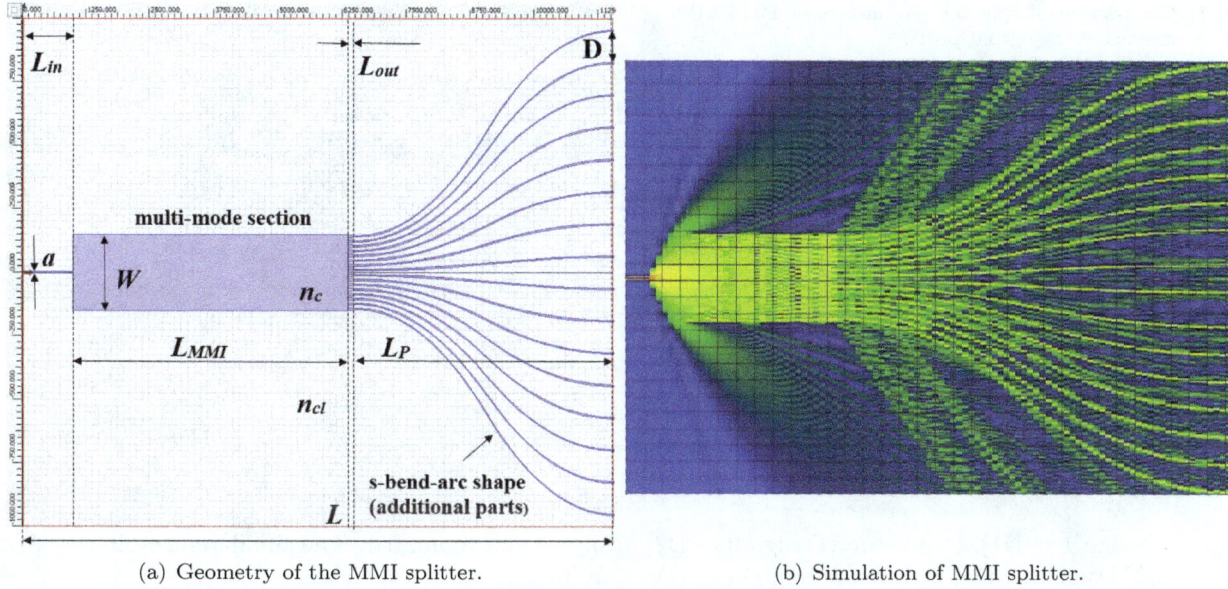

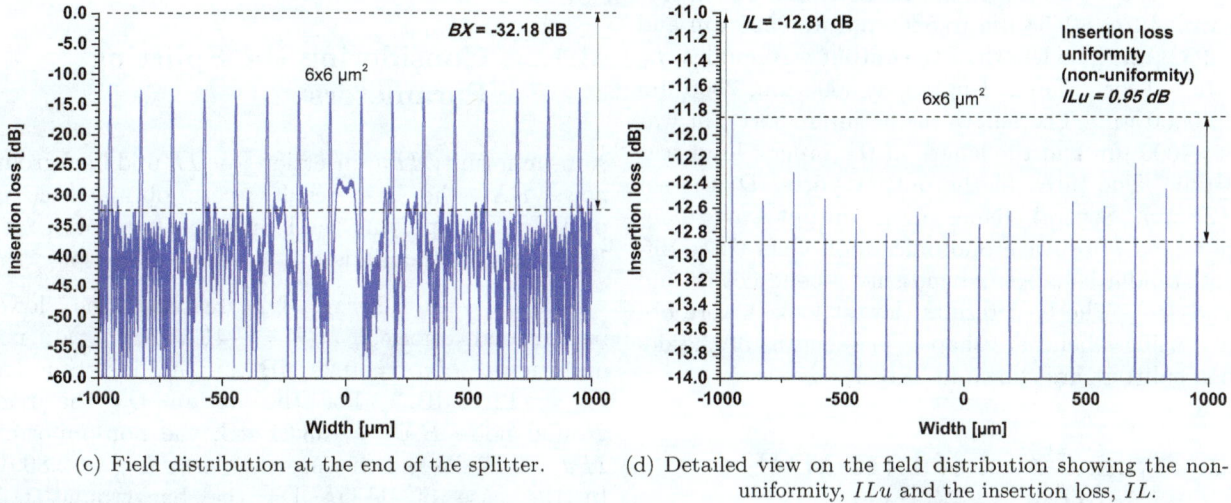

Fig. 1: 1×16 MMI splitter (Design 1).

1×16 MMI splitter in terms of different widths of the multimode interference section. The best 1×16 MMI splitter design was compared with 1×16 Y-branch splitter to show their advantages and disadvantages.

In Y-branch splitters, additionally, we show that not only technology but also the waveguide core size has strong influence on the non-uniformity of the split power.

2.   Design and Simulation of 1×16 MMI Splitter

The 1×16 MMI splitter operating at wavelength $\lambda = 1550$ nm was designed and simulated using Optiwave tool (OptiBPM Designer using Beam-Propagation Method).

2.1.   Design of 1x16 MMI Splitter

In Fig. 1(a) the geometry of 1×16 MMI splitter is shown together with its design parameters: n_{cl} - refractive index of the cladding, n_c - refractive index of the core, W - width of the MMI section, $LMMI$ - length of the MMI section, a - width of the input/output waveguides, Lin - length of the input waveguide, $Lout$ - length of the tapered part of output waveguide, Lp - length of the output waveguides and D is the port pitch.

The design of the splitter was focused on weakly guiding glass waveguides with the refractive index of the cladding $n_{cl} = 1.445$ and of the core $n_c = 1.456$. The core size of the input/output waveguides was set to 6×6 μm^2 to support the single mode propagation only. To study the properties of the 1×16 MMI splitter the width of the multimode section, W was set to

Tab. 1: Comparison of splitting parameters of 1×16 MMI splitter for the different widths of multimode section and different lengths of the output ports.

Design	W (μm)	LMMI (μm)	Lp (μm)	Non-uniformity ILu (dB)	Insertion loss, IL (dB)	Background noise, BX (dB)	Length of, splitter, L (μm)
D1	200	2458	5000	0.95	-12.81	-32.18	8638
D1	200	2458	7500	1.04	-12.83	-36.85	11138
D1	200	2458	10000	1.28	-12.80	-37.72	13638
D2	**300**	**5368**	**5000**	**0.51**	**-12.54**	**-36.8**	**11548**
D2	300	5368	7500	0.39	-12.56	-37.71	14048
D2	300	5368	10000	0.26	-12.44	-41.35	16548
D3	400	9529	5000	0.92	-12.93	-35.06	15709
D3	400	9529	7500	0.61	-13.27	-32.76	18209
D3	400	9529	10000	0.57	-12.89	-33.21	20709
D4	500	14932	5000	1.72	-14.89	-37.69	21112
D4	500	14932	7500	1.91	-14.44	-30.17	23612
D4	500	14932	10000	0.83	-13.85	-33.63	26612

200 μm (Design 1 = D1), to 300 μm (Design 2 = D2), to 400 μm (Design 3 = D3) and finally to 500 μm (Design 4 = D4). The length of the MMI section, LMMI has varied from 2458 μm to 5368 μm, to 9529 μm and to 14932 μm. The length of the output waveguides, L_p was tested for different lengths, as 5000 μm, 7500 μm and 10000 μm. The length of the input port Lin was set to 1000 μm and the length of the tapers, Lout was 180 μm. The pitch of the output ports, D was set to 127 μm. Several shapes of the output waveguides were tested (Optiwave photonics tool offers three different standard shapes: s-bend-sine, s-bend-cosine and s-bend-arc). The best results (lowest losses) were obtained using s-bend-arc shape. The lengths of the designed splitters are shown in Tab. 1.

2.2. Simulation of 1×16 MMI Splitter

Figure 1(b) shows the top view of the simulated 1×16 MMI splitter (design D1 with the length of the output waveguides, L_p = 5000 μm) performed in Optiwave tool. Figure 1(c) presents the simulation results of 1×16 MMI splitter, i.e. field distribution at the end of the simulated structure together with background noise parameter, BX = −32.18 dB. Figure 1(d) shows the detailed view of the field distribution with the non-uniformity parameter (difference between the highest and the lowest peak, also called insertion loss non-uniformity), ILu = 0.95 dB and insertion loss (the worst peak), IL = −12.81 dB.

3. Different Widhts of MMI Section

1×16 MMI splitters with different widths, W of the MMI section and lengths, L_P of output ports were de-

signed and simulated. The simulated results are shown in Tab. 1.

3.1. Considering the Splitting Parameters

Non-uniformity ILu, insertion loss IL and background noise BX, the best results were obtained as expected when applying the longest output ports, Lp = 10000 μm in all designs.

Namely, the design D2 reached the lowest background noise, BX = −41.35 dB, a non-uniformity, ILu = 0.26 dB and an insertion loss, IL = −12.44 dB. For the design D3 the background noise BX = −33.21 dB, the non-uniformity ILu = 0.57 dB and the insertion loss IL = −12.89 dB. In the case of design D4 the background noise BX = −33.63 dB, the non-uniformity ILu = 0.83 dB and the insertion loss IL = −13.85 dB. From the simulation results it can be concluded that the best splitting characteristics were obtained for the design D2 with the width of the MMI section, W = 300 μm.

3.2. Considering the Whole Length of the MMI Splitter

The simulation results, in the case of the shortest splitter, where the paramount design parameter, Lp = 5000 μm in each design, are presented in Fig. 2. In the design D2 the whole length of MMI splitter reached L = 11548 μm (see Tab. 1). The background noise, BX is −36.8 dB (see Fig. 2(a)), the non-uniformity, ILu = 0.51 dB and the insertion loss, IL is −12.54 dB (see Fig. 2(b)). For the design D3, the whole length of the splitter, L = 15709 μm. The background noise, BX = −35.06 dB (see Fig. 2(c)), the non-uniformity, ILu = 0.92 dB and the insertion loss, IL = −12.93 dB (Fig. 2(d)). D4 design reached the whole length of the

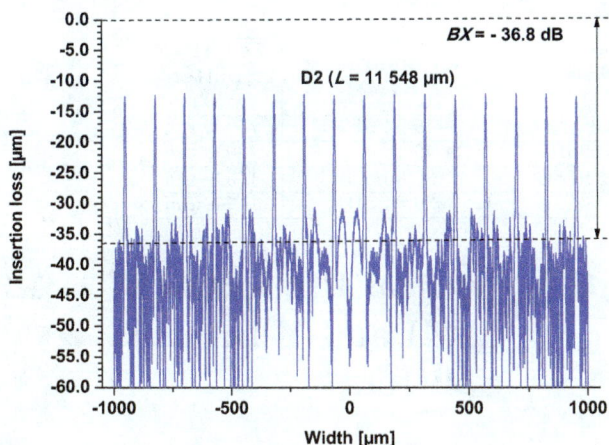

(a) Field distribution at the end of the splitter structure.

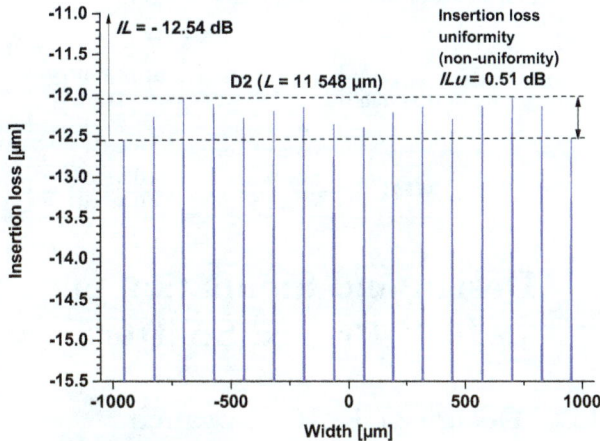

(b) The non-uniformity, ILu and the insertion loss, IL.

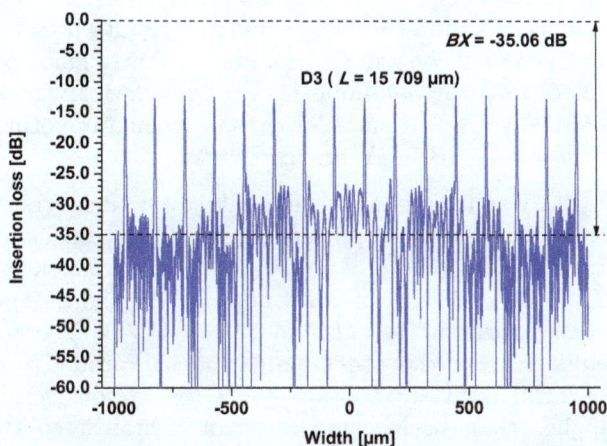

(c) Field distribution at the end of the splitter structure.

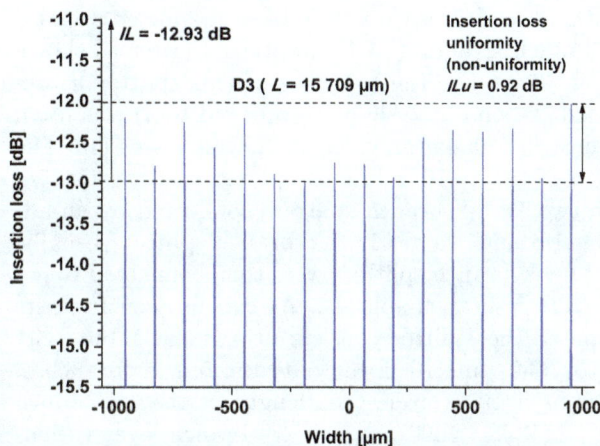

(d) The non-uniformity, ILu and the insertion loss, IL.

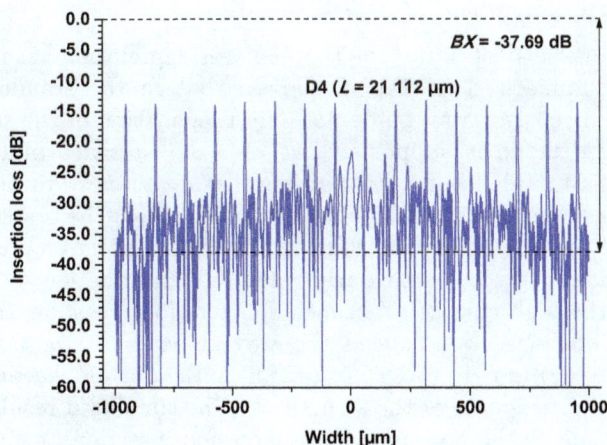

(e) Field distribution at the end of the splitter structure.

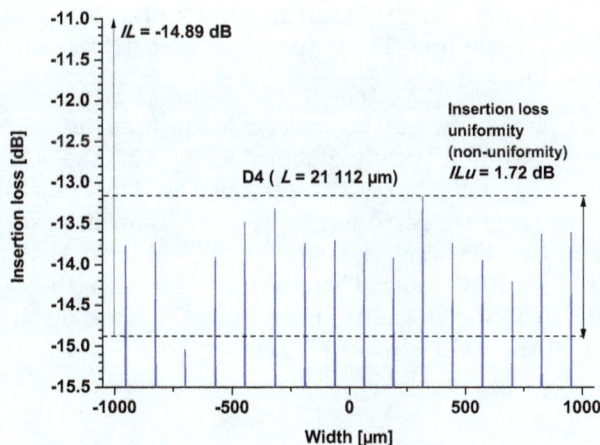

(f) The non-uniformity, ILu and the insertion loss, IL.

Fig. 2: Simulation results for different widths, W of MMI multimode section: field distribution at the end of the splitter structure (a), (c), (e) and detailed view of the field distribution showing the non-uniformity, ILu and the insertion loss, IL (b), (d), (f).

MMI splitter, $L = 21112$ μm. This splitter features the background noise, $BX = -37.69$ dB (see Fig. 2(e)), the non-uniformity, $ILu = 1.72$ dB and an insertion loss, $IL = -14.89$ dB (see Fig. 2(f)). It is obvious from the simulation results that the design D2 reached the best optical splitting parameters.

Tab. 2: Comparison of splitting parameters of 1×16 Y-branch and 1×16 MMI splitters.

Y-branch		Non-uniformity ILu	Insertion loss IL	Background noise BX	Length of splitter L
	6×6 µm²	1.77 dB	13.09 dB	−49.04 dB	78000 µm
	5.5×5.5 µm²	0.89 dB	13.09 dB	−49.18 dB	78000 µm
MMI	6×6 µm²	0.51 dB	12.548 dB	−36.80 dB	11548 µm
	5.5×5.5 µm²	0.46 dB	12.432 dB	−36.84 dB	11548 µm

4. Design and Simulation of 1×16 Y-Branch Splitter

4.1. Design of 1×16 Y-branch Splitter

For the design of the 1×16 Y-branch splitter we used a Y-branch structure of 1×4 optical splitter as shown in Fig. 3(a). For the branches of this splitter a pre-defined "s-bend-arc" shape (OptiBPM tool) was used, because this shape provides the lowest losses [4]. The design of the 1×16 Y-branch splitter was constructed from two 1×4 Y-branch splitters connected by an additional branch to get 1×8 Y-branch splitter. Finally, two 1×8 Y-branch splitters were then connected to get the 1×16 Y-branch splitter. As can be seen from its geometry the splitter consists of a linear input port set to 1000 µm, 16 linear outputs and 15 branches, distributed on 4 layers (the length of the 1st branch layer, $L(1^{st}) = 5000$ µm, the 2nd branch layer is doubled, $L(2^{nd}) = 10000$ µm). To keep further the constant bending shape, the 3rd branch layer was also doubled i.e. $L(3^{rd}) = 20000$ µm and the 4th branch layer, $L(4^{th}) = 40000$ µm. The output port's length was set to 2000 µm.

The pitch between the waveguides in each branch layer was automatically doubled, i.e. in the 1st branch layer $W(1^{st}) = 127$ µm, in the 2nd branch layer $W(2^{nd}) = 254$ µm, in the 3rd branch layer $W(3^{rd}) = 508$ µm and in the 4th branch layer $W(4^{th}) = 1016$ µm. Thereby the whole length of the 1×16 Y-branch splitter reached 78000 µm and the width of the splitter was 1905 µm (= 15×127 µm).

4.2. Simulation of 1×16 Y-branch Splitter

Figure 3(b) presents the top view of the simulated 1×16 Y-branch splitter using Optiwave tool. Figure 3(c) shows the corresponding field distribution at the end of the structures. The background noise of Y-branch splitter, $BX = -49.04$ dB (Fig. 3(c)). The uniformity of the split power over all the output channels, $ILu = 1.77$ dB and the insertion loss $IL = -13.09$ dB for Y-branch approach (Fig. 3(d)).

5. Optical Properties of 1×16 MMI and Y-Branch Splitters

MMI splitters have some advantages over Y-branch splitters. The main advantage is their low non-uniformity and size. The simulated results presented in the Tab. 2 confirm that for standard waveguide core (6×6) µm² the non-uniformity, $ILu = 0.51$ dB in case of MMI splitter is much lower than the non-uniformity, $ILu = 1.77$ dB for Y-branch splitter.

MMI splitter is approximately seven times shorter than the Y-branch splitter, namely 11548 µm in contrast to 78000 µm. Furthermore, the insertion loss, $IL = -12.548$ dB for MMI approach is slightly lower than insertion loss, $IL = -13.017$ dB for Y-branch splitter. On the other hand, the background noise of MMI splitter, $BX = -36.8$ dB is considerably higher than background noise of Y-branch splitter, $BX = -49.04$ dB. Insertion loss, $IL = -13.017$ dB for Y-branch splitter is slightly higher than for MMI splitter, where $IL = -12.548$ dB.

The deep study of the achieved simulation results summarized in Tab. 2 showed that in the standard (6×6) µm² waveguide not only propagation of the single mode is supported but also the presence of the first mode is already so strong that it causes additional asymmetric splitting of the optical signal at the branching points in Y-branch splitters. This becomes a dominant factor, particularly when reducing the length of the high channel Y-branch splitters [6]. To show this influence, we reduced the waveguide core size from (6×6) µm² to (5.5×5.5) µm² in both splitters, keeping the same size of the structures. The simulated results, namely the non-uniformity, ILu and insertion loss, IL of both splitters are shown in Fig. 4.

As can be seen in Fig. 4(b), the splitting parameters of the MMI splitter are only slightly improved since the splitting appears in the large coupler.

Y-branch splitter, consisting of many waveguides, features strong improvement of its optical properties, particularly the non-uniformity is strongly reduced from $ILu = 1.77$ dB (for (6×6) µm²) to $ILu = 0.89$ dB (for (5.5×5.5) µm²), that is less than one half of its original value (see Fig. 4(a)).

(a) Geometry of Y-branch splitter.

(b) Simulation of Y-branch splitter.

(c) Field distribution at the end of Y-branch splitter.

(d) Detailed view on the field distribution showing the non-uniformity, ILu and the insertion loss, IL.

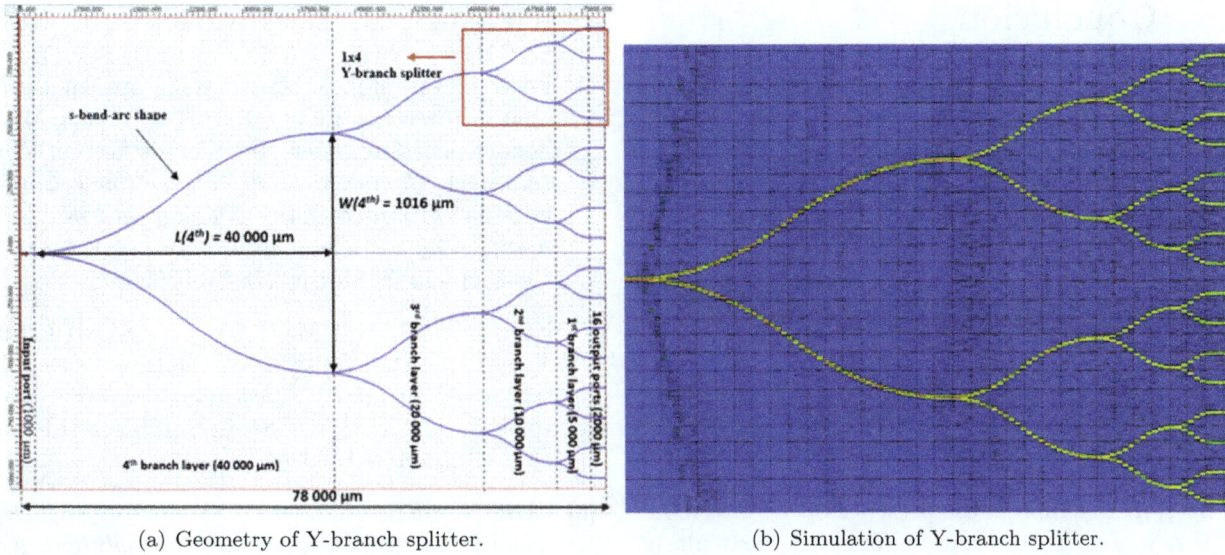

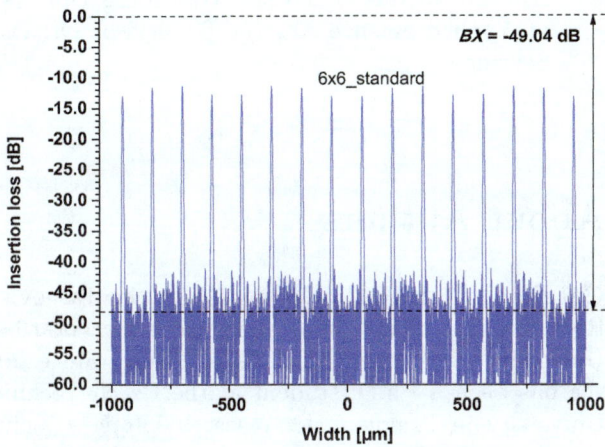

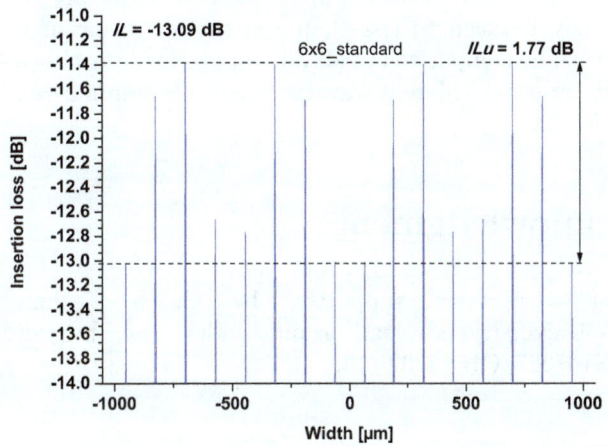

Fig. 3: 1×16 Y-branch splitter.

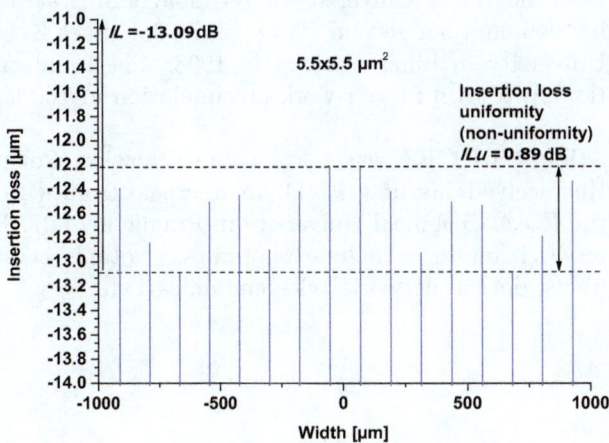

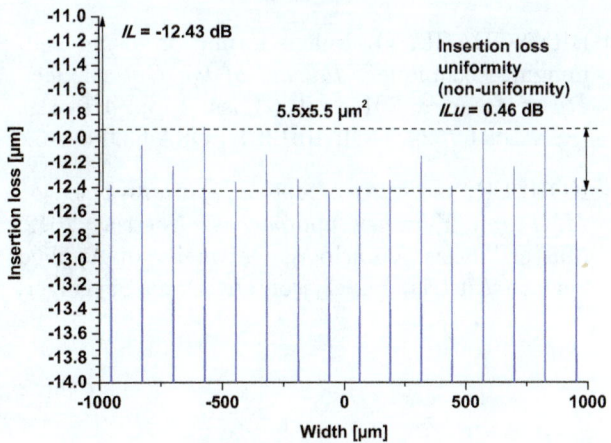

(a) 1×16 Y-branch splitter.

(b) 1×16 MMI splitter.

Fig. 4: Non-uniformity and insertion loss of 1×16 Y-branch splitter (a) and 1×16 MMI (b) splitters with the waveguide core size of (5.5×5.5) μm^2 .

6. Conclusion

In this paper we studied the optical properties of 1×16 MMI splitter based on different widths of MMI sections and 1×16 Y-branch splitter. It is evident from the simulation results that the width, W of the multimode interference section together with the lengths of the output ports are important design parameters. The simulation results showed that the optimal width of the MMI coupler, for which we were able to get the best performance of the MMI splitter, was obtained in D2 design. It was also showed that non-uniformity, ILu and insertion loss, IL parameters can be improved when adjusting the length of the output waveguides, Lp. Additionally, by decreasing the waveguide core size it is possible to suppress the presence of the first mode and this way to reduce non-uniformity. Particularly, in the case of Y-branch splitter the non-uniformity is strongly reduced to less than one half of its original value. The comparison of all optical properties of both splitters using different waveguide cores is summarized in Tab. 2.

Acknowledgment

This work was supported by Czech Technical University student grant under the project SGS16/227/OHK3/3T/13.

References

[1] KEISER, G. *FTTX Concepts and Applications.* 1st ed. New Jersey: Willey-IEEE Press, 2006. ISBN 978-0-471-70420-1.

[2] BRYNGDAHL, O. Image formation using self-imaging techniques. *Journal of the Optical Society of America.* 2015, vol. 63, iss. 4, pp. 416–419. ISSN 1084-7529. DOI: 10.1364/JOSA.63.000416.

[3] KOHLER, L. *Study of optical splitters based on MMI and Y-branch approaches.* Austria, 2012. Master Thesis. Vorarlberg University of Applied Science Austria. Supervisor Dr. Dana Seyringer.

[4] BACHMAN, M., M. K. SMITH, P. A. BESE, E. GINI, H. MELCHIOR and L. B. SOLDANO. Polarization-intensive low voltage optical waveguide swith using InGaAsP/InP four port Mach-Zender interferometer. In: *Conference on Optical Fiber Communication/International Conference on Integrated Optics and Optical Fiber Communication.* San Jose: Optical Society of America, 1993, pp. 32–33. ISBN 978-0780309982.

[5] NOURSHARGH, M., E. M. STARR and T. M. ONG. Integrated Optic 1×4 splitter in SiO_2/GeO_2. *Electronics Letters.* 1989, vol. 25, iss. 15, pp. 981–982. ISSN 0013-5194. DOI: 10.1049/el:19890656.

[6] BURTSCHER, C. *Design, Simulation and Optimization of High Channel Optical Splitters.* Austria, 2014. Master Thesis. Vorarlberg University of Applied Science Austria. Supervisor Dr. Dana Seyringer.

About Authors

Catalina BURTSCHER was born in Targoviste, Romania. She received her M.Sc. from the Vorarlberg University of Applied Sciences in 2014. Since 2014 she has been a Ph.D. student at the Czech Technical University in Prague. Her research interests include integrated optic, passive optical components for telecommunication.

Dana SEYRINGER was born in Martin, Slovakia. She received her first Ph.D. in microelectronics from the Slovak University of Technology in Bratislava in 1996, and her second Ph.D. at the Johannes Kepler University in Linz, Austria, in 1998. She is internationally known for her work on simulation methods.

Michal LUCKI was born in Starachowice, Poland. He received his first Ph.D. in microelectronics from the Czech Technical University in Prague in 2007. His research interests include photonics, photonic crystal fibers, optical network, telecommunication.

15

Fluorescence Properties of Chlorella sp. Algae

Tibor TEPLICKY[1,2], *Miroslava DANISOVA*[1,2], *Martin VALICA*[2],
Dusan CHORVAT Jr.[1], *Alzbeta MARCEK CHORVATOVA*[1,2]

[1]Department of Biophotonics, International Laser Center, Ilkovicova 3, 841 01 Bratislava, Slovak Republic
[2]Department of Ecochemistry and Radioecology, Faculty of Natural Sciences, University of SS.
Cyril and Methodius, Namesti J. Herdu 2, 917 01 Trnava, Slovak Republic

t.teplicky@gmail.com, mirka.danisova@gmail.com, martin.valica.mv@gmail.com,
dusan.chorvat@ilc.sk, alzbeta.chorvatova@ilc.sk

Abstract. *Water quality and its fast and reliable monitoring is the challenge of the future. Design of appropriate biosensors that would be capable of non-invasive identification of water pollution is an important prerequisite for such challenge. Chlorophylls are pigments, naturally presented in all plants that absorb light. The main forms of chlorophyll in algae are chlorophyll a and chlorophyll b, other pigments include xantophylls and beta-carotenes. Our aim was to characterize endogenous fluorescence of the Chlorella sp. algae, present naturally in drinking water. We recorded spatial, spectral and lifetime fluorescence distribution in the native algae. We noted that the fluorescence was evenly distributed in the algae cytosol, but lacked in the nucleus and reached maximum at 680–690 nm. Fluorescence decay of chlorella sp. was double-exponential, and clearly shorter than that of its isolated pigments. For the first time, fluorescence lifetime image of the algae is presented. Study of the fluorescence properties of algae is aimed at the improvement of water supply contamination detection and cleaning.*

Keywords

Chlorella sp., chlorophylls, confocal microscopy, FLIM fluorescence lifetime spectroscopy.

1. Introduction

Chlorophylls are pigments naturally present in all plants that absorb light. As highly conjugated compounds, they absorb in wide-range, from ultraviolet, via visible to infrared light [1]. Two to three percent of the absorbed sun energy is then re-emitted from the pigment system as the fluorescence. Light energy, captured in the form of radiation, is subsequently - in a series of transfers to other molecules and complexes - converted into chemical energy in the form of ATP. The main forms of chlorophyll in plants, including algae, are chlorophyll a and chlorophyll b, derived from protoporphyrins [2]. In the chemical sense, chlorophylls are tetrapyrol rings of porphyrin, chlorine, or bacteriochlorine, characterized by the fifth isocyclic ring that is biosynthetically derived from the C-13 propionic acidic side chain of protoporphyrin [3], characterized by the presence of a central atom of magnesium Mg^{2+}. Algae exhibit strong autofluorescence from photosynthetic pigments, namely chlorophyll a-d, phycobilins and carotenoids, the emission properties of which vary dependently on metabolic activities and physiological state of algae [4].

Absorption and emission properties of plant natural pigments are well described [1], [2] and [3]. Lately, time-resolved techniques proved to be valuable for evaluation of endogenous fluorophores and their sensitivity to the environment, as fluorescence lifetimes are independent on the fluorophore concentration, but react to changes in local chemical environmental conditions, namely oxygen saturation, or binding [5]. Evaluation of the fluorescence lifetime properties in algae can therefore serve as non-invasive sensor of their physiological state.

Chlorella, as most algae, is famous for removing heavy metal and other synthetic toxins from the body, and/or from its natural source (lake, pond, or swamp). Chlorella sp. can therefore serve for identification of water pollution and its cleaning. Consequently, it can be employed for designing optical biosensors used for monitoring of water pollution, e.g. presence of toxins, herbicides, etc. [6].

The aim of this study is to characterize properties of the endogenous fluorescence in the Chlorella sp. green algae, naturally present in drinking water and compare them to isolated pigments. Advanced microscopy and spectroscopy methods are employed to obtain spatial, spectral and/or lifetime distribution of the endogenous fluorescence in the algae.

2. Material and Methods

2.1. Preparation of the Chlorella sp.

Chlorella sp. was obtained from the Faculty of Natural Sciences, University of SS. Cyril and Methodius in Trnava collection of the green algae. The green algae of genus Chlorella sp. were previously isolated from the main drinking water supply. Green algae were cultivated in Hoagland cultivation medium [7].

2.2. Isolation of Pigments from Chlorella sp.

Pigments were extracted from Chlorella sp. algae. After a centrifugation at 45000 rpm for 15 minutes, pellets were dehydrated, and then crushed with the sea sand. Pigments were extracted with n-hexane, and resulting components were divided using silica gel chromatography in the solution of n-hexane: acetone (7:3) [2]. Resulting bands were separated according to colours (Fig. 1). A Blue-Green (BG), a Yellow-Green (YG) and a Yellow (Y) band, representing the chlorophyll a, the chlorophyll b and the carotenoids respectively [2], were dissolved in the DMSO and used for comparison.

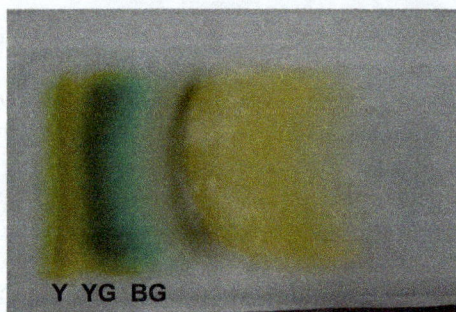

Fig. 1: Extracted Yellow (Y), Yellow-Green (YG) and Blue-Green (BG) bands separated by gel chromatography.

2.3. Confocal Microscopy Imaging and Spectroscopy

Confocal images of the algae autofluorescence were gathered with a laser scanning confocal microscope, equipped with LSM 510 META detector coupled to Axiovert 200 inverted microscope, employing C-Apochromat 40×, 1.2 NA objective (all Carl Zeiss, Germany). Algae were excited with 632 nm laser (Lasos Lasertechnik), using a 16 channel META detector. For spectrally-resolved microscopy measurements, data were recorded by META detector in the range of 650 nm to 740 nm with an 11 nm step.

2.4. Time-Correlated Single Photon Counting (TCSPC) Measurements

Time-Correlated Single Photon Counting (TCSPC) method was applied to measure fluorescence decays of Chlorella sp. and isolated pigments. Fluorescence decays were detected at room temperature in a cuvette after excitation by 635 nm picosecond laser diode BHL-635 (output power < 0.5 mW, pulse width 50 ps, pulse frequency 50 MHz) using SPC-130 TCSPC card (both Becker & Hickl, Germany). Fluorescence was detected by photon counting, using a PMC-100 detector (Becker& Hickl, Germany) after passing through a spectrograph (monochromator PRA B102 Photochemical Research Associates, Canada). To achieve spectral resolution, fluorescence decays were recorded at individual wavelengths from 640 to 740 nm with a 10 nm step.

2.5. Fluorescence Lifetime Imaging Microscopy (FLIM)

FLIM images were recorded using TCSPC technique coupled to the confocal microscope. In these experiments, a 475 nm picoseconds laser diode (BDL-475, Becker& Hickl, Germany) was used. The laser beam was reflected to the sample through the epifluorescence path of the LSM 510 META microscope (Zeiss, Germany) with C-Apochromat 40×, 1.2 NA lens. The emitted fluorescence was separated from laser excitation using LP 500 nm filter and detected by HPM 100-40 photomultiplier array (Becker&Hickl, Germany) employing SPC-830 TCSPC imaging board.

2.6. Data Analysis

Confocal data were visualized by ZEN 2011 software (Zeiss, Germany). FLIM images were processed using proprietary software packages SPCImage (Becker & Hickl, Germany), fitted by up to a three-exponential decay to gain $\chi^2 \leq 1.3$. Results were visualized as a map and as a distribution of calculated fluorescence lifetimes.

3.   Results

In this work, we aimed to identify fluorescence properties of endogenous fluorescence in Chlorella sp. Although capable of absorbing light from visible up to infrared regions [1], for identification of the algae' endogenous fluorescence we have primarily chosen the 632–635 nm wavelength for excitation, which is harmless for work with living organisms.

3.1.   Confocal Microscopy Imaging and Spectroscopy of Endogenous Fluorescence in Chlorella sp.

Our first aim was to identify spatial and spectral distribution of endogenous fluorescence after excitation at 632 nm using confocal microscopy imaging. As expected, we observed that algae had round shape with diameter of around 10–15 μm (Fig. 2), exhibiting bright fluorescence in the red spectral region.

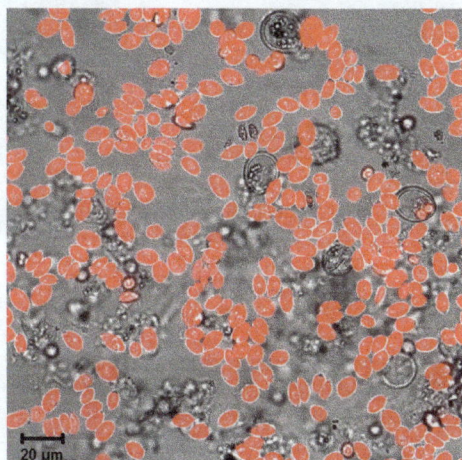

Fig. 2: Transmission and fluorescence image of Chlorella sp., exc. 632 (nm), LP 650 (nm), Scale: 10 (μm).

Spectrally-resolved images were recorded at individual wavelengths (Fig. 3(a)) and uncovered maximum fluorescence at around 680 nm (Fig. 3(b)).

3.2.   Fluorescence Lifetimes of Chlorella sp.

Our second aim was to evaluate fluorescence spectra together with fluorescence lifetimes of endogenous fluorescence in Chlorella sp. TCSPC method was applied to measure fluorescence decays of Chlorella sp. Fluorescence signal was recorded at individual wavelengths from 640 to 740 nm. We noted that Chlorella sp. had maximum emission at 690 nm (Fig. 4(a)). Gathered fluorescence decay (Fig. 4(b)) was best fitted with

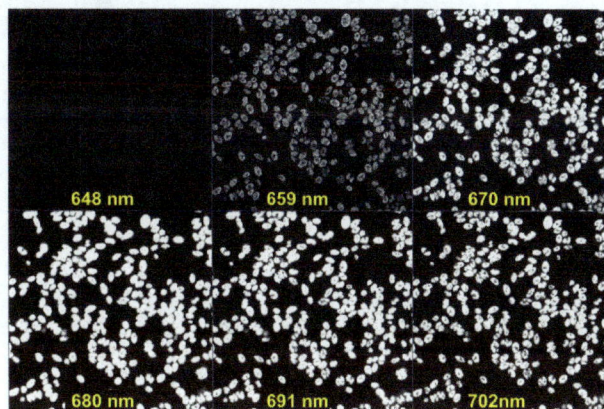

(a)

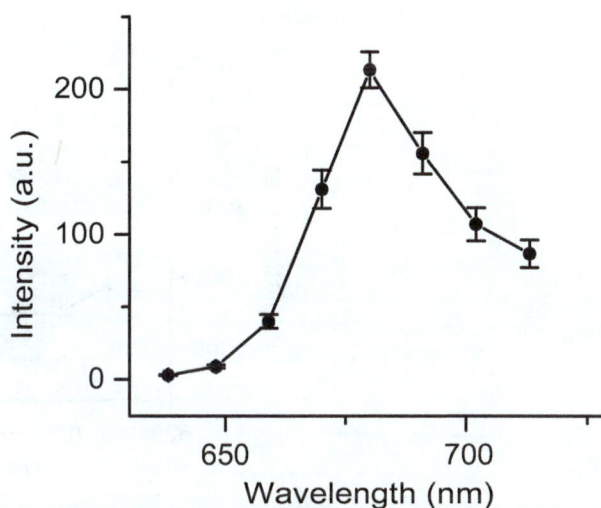

(b)

Fig. 3: Confocal microscopy spectra of Chlorella sp., exc. 632 (nm), emission 648–713 (nm), step 11 (nm). Fluorescence images at individual wavelengths (top, wavelength number in yellow). Fluorescence spectra (bottom), mean ±SEM, n = 7.

a 2-exponential decay fit. Shorter fluorescence lifetime reached between 500–900 ps, the longer one between 1300–1900 ps (Fig. 4(c)). With increasing wavelength, some decrease in the fluorescence lifetime was noted.

Gathered results were then compared to fluorescence spectra and lifetimes of pigments isolated from Chlorella sp. Isolated pigments showed maximum spectra at 680 nm, which was blue-shifted of about 10 nm when compared to the spectrum of the Chlorella sp. (Fig. 4(a)). Analysis by a mono-exponential decay of the chlorophyll band in DMSO showed fluorescence lifetimes between 4800–5300 ps, while that of the carotenoid band was longer, between 4700–4800 ns (Fig. 4(b) and Fig. 4(c)). Data gathered for the chlorophyll band are in agreement with previously published results for chlorophyll a in ether, which was 4.9 ns [8]. Data showed significantly lower fluorescence lifetime in native Chlorella when compared to isolated pigments.

(a) Emission intensity.

(b) Normalized fluorescence decay at emission 700 (nm).

(c) Fluorescence lifetimes.

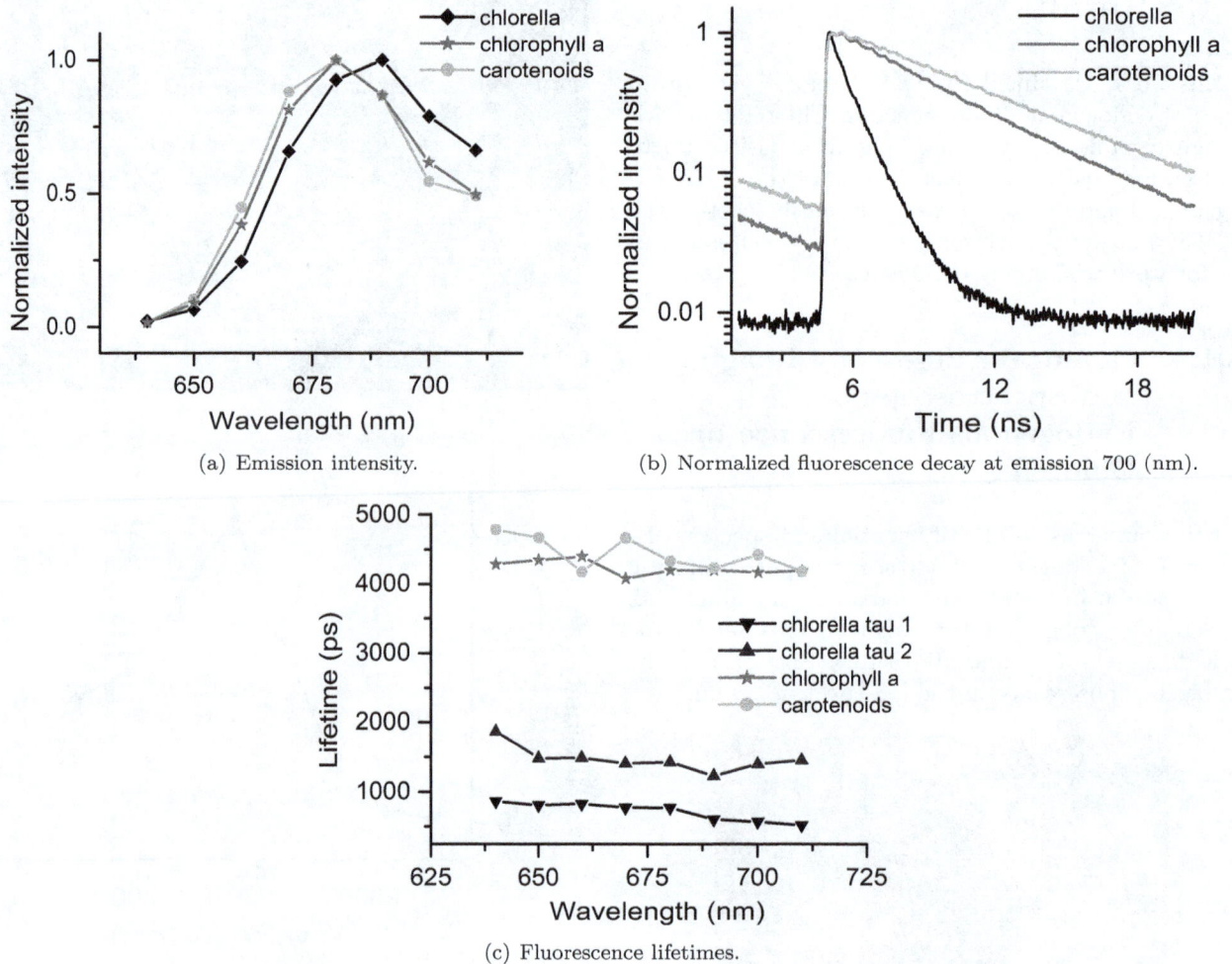

Fig. 4: Comparison of total photon counts of the fluorescence of Chlorella sp., chlorophyll a and carotenoids, excited at 635 (nm).

3.3. Fluorescence Lifetime Imaging of Chlorella sp.

Our last goal was to compare distribution of the fluorescence lifetimes in native algae, using FLIM. This unique approach allowed us to gather images of the lifetime distribution in individual algae with spatial resolution. FLIM images were recorded using TCSPC with 475 nm excitation and LP 500 nm emission. Fluorescence lifetimes were recorded from 150 to 2000 ps. Most algae showed short lifetime up to 200 ps, while some also exhibited lifetimes under 500 ps (Fig. 5). At the same time, we noted differences in the lifetime distribution within individual algae.

4. Discussion

The aim of this study is to characterize fluorescence properties of endogenous fluorescence in the Chlorella sp. algae, and compare them to that of their pigments. Gathered data showed endogenous fluores-

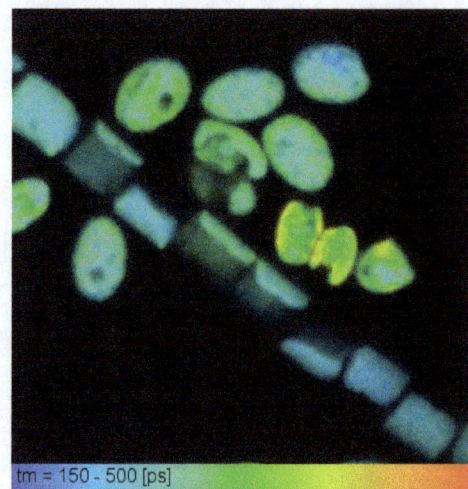

Fig. 5: FLIM image of Chlorella sp., ex. 475 (nm), LP 500 (nm), with mean lifetime 150–450 (ps) blue-red, Scale: 10 (μm).

cence of the algae peaking at 680–690 nm. This is in agreement with previously observed maximum value of 683 nm with half-width of about 20 nm [9], and/or the

660–680 nm peak range of fluorescence in green algae [10]. When compared to isolated pigments, fluorescence spectra were red-shifted of about 5–10 nm.

Fluorescence lifetimes of pigments, namely chlorophyll a, gives information about the primary photophysical events in photosynthesis [11]. Fluorescence of Chlorella sp. had double exponential decay with shorter fluorescence lifetime around 500–900 ps and longer one around 1300–1900 ps. Others demonstrated five exponential components in Chlorella, namely 53 ps, 89 ps, 174 ps, 535 ps and 1200 ps [10]. Our data are in agreement with the two longest lifetimes. In addition, FLIM recording uncovered the presence of a lifetime under 200 ps (blue at Fig. 5). However, applied methods did not allow us to record lifetimes at a picoseconds scale.

Importantly, obtained lifetimes in native Chlorella were much shorter than lifetimes of its isolated pigments that reached values longer than 4000 ps. This is expected, taking into consideration differences of the chlorophyll surroundings in live cells vs. in an artificial environment. We previously demonstrated [12], in agreement with others [1], longer fluorescence lifetime of the chlorophyll a compared to the chlorophyll b. Consequently, fluorescence lifetimes of all isolated pigments were clearly longer than endogenous fluorescence of algae. Further experiments are necessary to link these in vitro experiments to the data recorded in cells. Also, in the future, it would be interesting to resolve shortest lifetime components from individual algae using more advanced signal and data processing.

Overall, we can summarize that time-resolved endogenous fluorescence is a useful tool for monitoring the state of living systems and its changes due to modification of the cell environment [13]. In this work, we present, for the first time, the FLIM of the green algae, with distinct lifetimes, suggesting differences in their state. Further work is necessary to understand sensitivity of the recorded lifetimes to changing environment.

5.   Conclusion

Understanding fluorescence characteristics of endogenous fluorescence of algae and characterise its changes in different environments can help us to design appropriate monitoring systems, e.g. biosensors, capable of non-invasive identification of the algae presence and its state. This knowledge is valuable for evaluation of water pollution, allowing reliable monitoring of water quality and its cleaning, as well as better comprehension of the efficient solar energy capture mechanisms.

Acknowledgment

This publication was supported by the Slovak Research and Development Agency under the contract no. APVV-14-0716. Authors also acknowledge support from the Integrated Initiative of European Laser Infrastructures LASERLAB-EUROPE IV EU-H2020 grant no. 654148, the research support fund of the University of SS. Cyril and Methodius FPPV-52-2017 to T.T. and FPPV 18-2015 to M.V. We would like to thank S. Hostin from FPV UCM for precious advices.

References

[1] KARCZ, D., B. BORON, A. MATWIJCZUK, J. FURSO, J. STARON, A. RATUSZNA and L. FIEDOR. Lessons from chlorophylls: modifications of porphyrinoids towards optimized solar energy conversion. *Molecules*. 2014, vol. 19, iss. 10, pp. 15938–15954. ISSN 1420-3049. DOI: 10.3390/molecules191015938.

[2] MOTTEN, A. F. Diversity of photosynthetic pigments. In: *Tested studies for laboratory teaching*. Las Vegas: Association for Biology Laboratory Education (ABLE), 2004, pp. 159–177. ISBN 978-1890444075.

[3] SCRUTTON, N. G., M. L. GROOT and D. J. HEYES. Excited state dynamics and catalytic mechanism of the light-driven enzyme protochlorophyllide oxidoreductas. *Physical Chemistry Chemical Physics*. 2010, vol. 14, iss. 25, pp. 8818–8824. ISSN 1463-9076. DOI: 10.1039/C2CP23789J.

[4] MONICI, M. Cell and tissue autofluorescence research and diagnostic applications. *Biotechnology annual review*. 2005, vol. 11, iss. 1, pp. 227–256. ISSN 1387-2656. DOI: 10.1016/S1387-2656(05)11007-2.

[5] MARCU, L., P. M. W. FRENCH and D. S. ELSON. *Fluorescence Lifetime Spectroscopy and Imaging: Principles and Applications in Biomedical Diagnostics*. 1st ed. London: CRC Press, 2008. ISBN 978-1-4398-6168-4.

[6] VEDRINE, C., J.-C. LECLERC, C. DURRIEU and C. TRAN-MINH. Optical whole-cell biosensor using Chlorella vulgaris designed for monitoring herbicides. *Biophysical Journal*. 1969, vol. 9, iss. 4, pp. 586–591. ISSN 0956-5663. DOI: 10.1016/S0956-5663(02)00157-4.

[7] HOAGLAND, D. R. Optimum nutrient solutions for plants. *Science*. 1920, vol. 52, iss. 1325,

pp. 562–564. ISSN 0036-8075. DOI: 10.1126/science.52.1354.562.

[8] SINGHAL, G. S. and E. RABINOVITCH. Measurement of the Fluorescence Lifetime of Chlorophyll a In Vivo. *IEEE Transactions on Power Delivery.* 2015, vol. 30, iss. 3, pp. 1096–1103. ISSN 0006-3495. DOI: 10.1016/S0006-3495(69)86405-2.

[9] PEDROS, R., I. MOYA, Y. GOULAS and S. JACQUEMOUND. Chlorophyll fluorescence emission spectrum inside a leaf. *Photochemical and Photobiological Sciences.* 2008, vol. 7, iss. 4, pp. 498–502. ISSN 1474-905X. DOI: 10.1039/b719506k.

[10] RIZZO, F., G. ZUCCHELLI, R. JENNINGS and S. SANTABARBARA. Wavelength dependence of the fluorescence emission under conditions of open and closed Photosystem II reaction centres in the green alga Chlorella sorokiniana. *Biochimica et Biophysica Acta - Bioenergetics.* 2014, vol. 2014, iss. 6, pp. 726–733. ISSN 0005-2728. DOI: 10.1016/j.bbabio.2014.02.009.

[11] BRODY, S. S. Fluorescence lifetime, yield, energy transfer and spectrum in photosynthesis, 1950–1960. *Photosynthesis Research.* 2002, vol. 73, iss. 1–3, pp. 127–132. ISSN 0166-8595. DOI: 10.1023/A:1020405921105.

[12] DANISOVA, M., B. TOMIKOVA., T. TEPLICKY and A. M. CHORVATOVA. Comparison of fluorescence properties of Porphyrin IX and chlorophylls. In: *Advances in Electronic and Photonic Technologies (ADEPT).* Tatranska Lomnica: University of Zilina, 2016, pp. 43–46. ISBN 978-80-554-1226-9.

[13] TEPLICKY, T., J. HORILOVA, J. BRUNCKO, C. GLADINE, I. LAJDOVA, A. MATEASIK, D. CHORVAT and A. M. CHORVATOVA. Flavin fluorescence lifetime imaging of living peripheral blood mononuclear cells on micro and nanostructured surfaces. In: *Progress in Biomedical Optics and Imaging - Proceedings of SPIE, Multiphoton Microscopy in the Biomedical Sciences XV.* San Francisco: SPIE, 2015, pp. 1–10. ISBN 978-162841419-6. DOI: 10.1117/12.2076706.

About Authors

Tibor TEPLICKY was born in Poprad, Slovakia. He received his M.Sc. from Biomedical Physics in 2014. He is currently a doctoral student in Analytical Chemistry at the University of SS. Cyril and Methodius in Trnava, his research interests include preparation of scaffolds by 3D 2photon photopolymerisation and search for most appropriate sensors of metabolic state in living cells.

Miroslava DANISOVA was born in Bratislava, Slovakia. She received her M.Sc. degree in Biotechnology in 2017 from the SS. Cyril and Methodius University in Trnava. Her research included the study of endogenous fluorescence of algae and its pigments, namely chlorophylls.

Martin VALICA born in Myjava, Slovakia, received M.Sc. from Applied Chemistry and Biochemistry in 2014, and is currently a doctoral student in Analytical Chemistry at the University of SS. Cyril and Methodius in Trnava. His research interests include design, fabrication and testing of water quality biosensors.

Dusan CHORVAT JR. born in Bratislava, Slovakia in 1971, studied at the Faculty of Mathematics and Physics, Comenius University, Bratislava where he completed his master's degree in Biophysics and received his Ph.D. in Biophysics in 2004. He is a member of a team of founders of the International Laser Center (ILC) in Bratislava, Slovakia where he has worked as the head of the laboratory of Laser Microscopy and Spectroscopy since 1997. He was also the head of Dept. of Biophotonics between 2001 and 2009, and the vice-director since 2009.

Alzbeta MARCEK CHORVATOVA born in Bratislava, Slovakia, received M.Sc (1991) in Biophysics from the Comenius University in Bratislava, Slovakia, and Ph.D. in Physiology (1995), University Claude Bernard Lyon-1, France. She founded a lab at CHU Sainte-Justine Montreal, Canada. Currently, she is the head of the Dept. of Biophotonics at ILC and an associated professor at the University of SS. Cyril and Methodius in Trnava. She is a specialist in biophysics, physiology and biophotonics, specifically oriented towards evaluation of metabolic oxidative state by time-resolved autofluorescence.

16

Composition Related Electrical Active Defect States of InGaAs and GaAsN

Arpad KOSA[1], *Lubica STUCHLIKOVA*[1], *Ladislav HARMATHA*[1], *Jaroslav KOVAC*[1], *Beata SCIANA*[2], *Wojciech DAWIDOWSKI*[2], *Marek TLACZALA*[2]

[1]Institute of Electronics and Photonics, Faculty of Electrical Engineering and Information Technology, Slovak University of Technology, Ilkovicova 3, 812 19 Bratislava, Slovakia
[2]Division of Microelectronics and Nanotechnology, Faculty of Microsystem Electronics and Photonics, Wroclaw University of Science and Technology, Janiszewskiego 11/17, 50-372 Wroclaw, Poland

arpad.kosa@stuba.sk, lubica.stuchlikova@stuba.sk, ladislav.harmatha@stuba.sk, jaroslav.kovac@stuba.sk, beata.sciana@pwr.wroc.pl, wojciech.dawidowski@pwr.edu.pl, marek.tlaczala@pwr.wroc.pl

Abstract. *This paper discusses results of electrically active defect states - deep energy level analysis in InGaAs and GaAsN undoped semiconductor structures grown for solar cell applications. Main attention is focused on composition and growth condition dependent impurities and the investigation of their possible origins. For this purpose a widely utilized spectroscopy method, Deep Level Transient Fourier Spectroscopy, was utilized. The most significant responses of each sample labelled as InG2, InG3 and NG1, NG2 were discussed in detail and confirmed by simulations and literature data. The presence of a possible dual conduction type and dual state defect complex, dependent on the In/N composition, is reported. Beneficial characteristics of specific indium and nitrogen concentrations capable of eliminating or reducing certain point defects and dislocations are stated.*

Keywords

Deep energy levels, Deep Level Transient Fourier Spectroscopy, electrically active defects, GaAsN, indium, InGaAs, nitrogen, solar cells.

1. Introduction

Optimization of fabrication processes for achieving high quality semiconductor materials is one of the most important factors in the technical advancement of semiconductor structures. Concentrated effort is focused on the achievement of perfect crystal structures, although it is very hard to maintain and define specific growth conditions. Different growth parameters can affect the formation of defect states, such as temperature, pressure and high purity, and these can highly affect the function and electrical properties of final devices. From the energetic point of view, electrically active trap states, also called deep energy levels, are allowed states in the band gap of the semiconductor material, capable of capturing or emitting charge carriers, thereby directly influencing generation and recombination processes usually with non-beneficial effects.

In order to identify inadequacies of growth technologies, the manufacture of high quality semiconductor materials has to be supported by appropriate diagnostic approaches, in which Deep Level Transient Fourier Spectroscopy (DLTFS) has a key role [1] and [2]. Dilute-nitride InGaAsN based solar cells lattice matched to GaAs are continuously studied to realize higher efficiencies [3]. Using various indium and nitrogen concentrations in semiconductor layers, different band gaps can be achieved, suitable also for tandem solar cell applications [4].

Since a proper indium and nitrogen composition is the key element of these structures, the investigation of related deep energy levels is indispensable. In order to identify and state origins of indium or nitrogen generated deep energy levels and their behaviour at various compositions, it is essential to analyse InGaAs and GaAsN referent structures with various compositions.

The aim of this study is to assess and discuss findings from investigations of four $In_yGa_{1-y}As$ structures with varied indium concentrations from 3.4 to 12.8 %, and five $GaAs_{1-x}N_x$ samples with various nitrogen contents from 0.9 to 1.85 %, and according to these results to report most appropriate structures with lowest defect activities.

2. Experiment

The experimental part of this study was realised based on DLTFS measurements and evaluation procedures. This method is capable of measuring and processing a set of capacitance transient signals induced by charge carrier emission or capture for a set temperature range (85–550 K). Capacitance transient signals are then processed by Fourier transformation and calculated as DLTFS spectrums. The charge carrier emission or capture is ensured by electrical excitation, meaning different biased states of the examined structure applying reverse V_R and so called filling voltage V_P conditions. For a given time period T_W the transient signal changes exponentially and outlines a peak in the DLTFS spectrum. Each peak is a direct result of a possible defect state or defect complexes.

Advantages of this method like precision, high sensitivity and adaptability ensures accurate trap parameter outputs (activation energy ΔE_T, capture cross section σ_T and trap concentration N_T), calculated by Arrhenius curves - temperature dependence of the emission rate [1] and [2]. The accuracy of the method is mainly affected by complex situations when different defect complexes are interacting, which results in multi-level responses and complicated broad spectra. In such cases different evaluation approaches or mathematical deconvolution and simulation processes are used and compared by literature data.

2.1. Indium Related Defect States of InGaAs

Four InGaAs structures with variable indium compositions were investigated. All these samples were grown at the Wroclaw University of Science and Technology by Atmospheric Metal Organic Vapour Phase Epitaxy. As Tab. 1 lists the indium content was varied from 3.4 to 12.8 % by different flow rates of the indium dopant source trimethylindium (V_{TMIn}) [5]. Common flow rates of trimethylgallium (V_{TMGa}) and arsine (V_{ASH3}) were set to 7 and 50 ml·min^{-1}. Each InGaAs layer was deposited on a GaAs 450 nm buffer layer and n-GaAs:Si $n = \frac{1}{2 \cdot 10^{18}}$ cm^{-3} substrate. Structural properties were identical, only differences in layer widths and composition were observed at varied V_{TMIn} flow rates (Tab. 1). DLTFS investigation of such referent structures is quite interesting, since we are able to monitor and analyse the behaviour of electrically active defect states at different band gaps resulted from the In content. Deviation of activation energies is not unusual, meaning that not only slightly shifted DLTFS peaks but also fluctuating activation energies are assumed.

As Fig. 1 shows this prediction was confirmed by the DLTFS spectra comparison of each sample measured at same experimental conditions, where to clarify the investigation process group labels were introduced. If we closely examine e.g. the InG2 group (Fig. 1), we see that at this temperature range (300 K to 360 K) a positive peak was outlined in all cases but with shifted peak positions. We have assumed that these responses are originating from identical deep energy levels but significant at slightly different temperatures caused by the band gap difference [6]. Standard evaluation procedure resulted activation energies in range from 0.46 eV to 0.48 eV, hence similar values in each case. All the observed defect states were evaluated in a similar manner (InG1 - InG3).

Tab. 1: Distinct composition and structural parameters of the investigated GaAsN samples.

Structure	H$_2$ through V_{TMIn} (ml · min^{-1})	i In$_y$Ga$_{1-y}$As (nm)	y (%)
InGaAs i I	5	~ 110	3.4
InGaAs i II	20	~ 110	8.9
InGaAs i III	32	~ 120	10.5
InGaAs i IV	35	~ 110	12.8

By means of trap concentrations (peak amplitude) it is hard to conclude a definitive correspondence. Peak amplitudes are directly influenced by defect concentrations, locally distributed defects, and the investigated sample's contact area. Contact areas were more or less equal, although minor changes due to unevenly prepared contacts were expected, however, the number of significant trap states, eliminated or generated, can give a hint about the most pure sample and the proper indium content.

The DLTFS investigation indicates the following statements (see Fig. 1 and Tab. 2):

- higher indium content above 8.9 % reduces or entirely suppresses low temperature (below 250 K) point defects [7],

- two significant defect states or complexes were identified present in all samples (InG2 and InG3),

- InG2 was identified as a Cu ($\sim$0.48 eV) related hole trap [8] and [9],

- InG3 was evaluated as a probable dual state complex EL2 ($\sim$0.77 eV) and its meta stable double donor state EL2^{2+} ($\sim$0.57 eV) [8] and [9].

Tab. 2: Deep energy level parameters of the investigated GaAsN samples.

Trap	peak	$\Delta E_{Tn,p}$(eV)	$\sigma_{n,p}$ (cm^2)
ING1	positive	$\sim$0.25	$\sim$5.23·10^{-18}
	negative	$\sim$0.34	$\sim$2.29·10^{-16}
ING2	positive	$\sim$0.48	$\sim$3.3·10^{-17}
ING3	positive	$\sim$0.57	$\sim$2.0·10^{-14}
	negative	$\sim$0.77	$\sim$1.9·10^{-17}

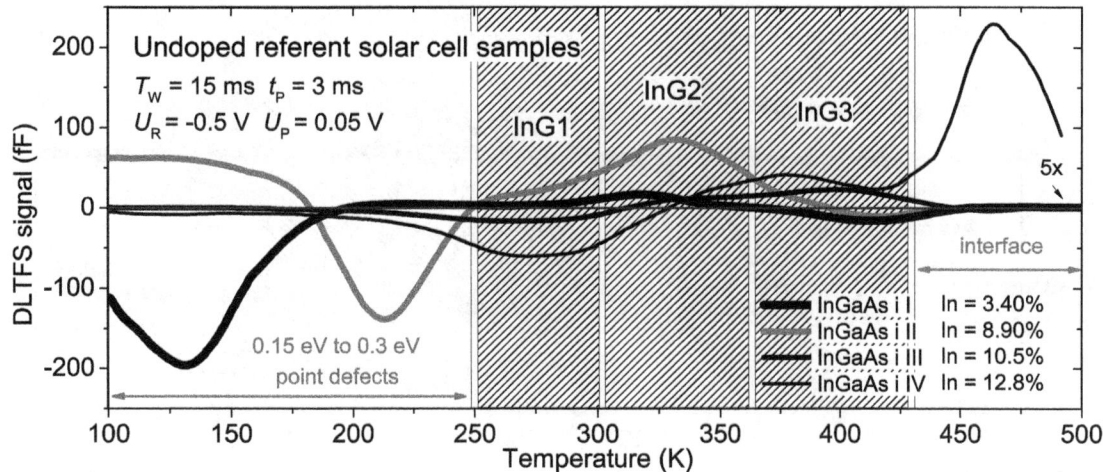

Fig. 1: DLTFS spectra of InGaAs undoped referent samples at different In concentrations, measured at identical experimental conditions.

These findings indicated that there is a high possibility that the observed formation of the identified EL2 and its meta-stable state is connected to the In concentration. A clear transition occurred around 9 to 10 % and was experimentally supported by the examined DLTFS results (Fig. 1 InG3 changing peak from negative to positive). The EL2 dominant native deep donor in GaAs is one of many intrinsic trap states investigated in III-V semiconducting compounds. It can be transformed to a meta-stable state e.g. by illumination of the crystal at low temperatures. It is not entirely understood what introduced this defect complex in the investigated samples, therefore further investigations are needed to fully evidence a possible In content relation.

At the analysed experimental conditions only samples InGaAs i I and II showed significant trap levels at lower (point defects) and InGaAs IV at higher temperatures (interface states).

Sample InGaAs i III with indium 10.5 % should be considered as the one with the lowest defect activity, where lowest peak amplitudes were observed, five times multiplied for spectral comparison.

The most unstable sample by these results was InGaAs IV with the highest In content, since it showed DLTFS peaks in the whole 200 K to 500 K range.

2.2. Nitrogen Related Defect States of GaAsN

The GaAsN DLTFS investigation included five referent samples with various nitrogen concentrations, addressing not only composition but also growth condition dependence of electrically active defect states. Each structure was prepared at different growth temperatures (585 °C, 605 °C, 595 °C, 565 °C and 575 °C)

to achieve a mixture of the nitrogen concentration. The following parameters were applied: growth precursor tertiarybutylhydrazine (V_{TBHy}) at flow rate of 1500 ml·min^{-1}, trimethylgallium $V_{TMGa} = 7$ ml·min^{-1} and arsine $V_{ASH3} = 50$ ml·min^{-1}. Each GaAsN layer was deposited on a GaAs 450 nm thick buffer layer and n-GaAs:Si $n = \frac{1}{2 \cdot 10^{18}}$ cm^{-3} substrate. Table 2 lists the main differences in composition and structural properties. This fabrication factor is also favourable for increasing or decreasing defect tendencies. Once again, shifting DLTFS curves had to be considered and group levels investigated (Fig. 2 NG1 and NG2) due to band gap differences.

As Fig. 2 shows, a broad peak with high trap concentration of sample GaAsN I (lowest N) was obviously dominating the spectrum comparison. Deep energy levels NG1 and NG2 were present in all samples, while this broad peak was only significant for this sample. It was assumed that higher nitrogen concentrations and/or temperature growth are capable of reducing/eliminating dislocation defects [10]. This reduction probably occurs at N = 1 % and is maintained at higher concentrations as well. Defect levels NG1 and NG2 were more relevant for these samples, although there was no exact correlation found between DLTFS amplitudes and the N concentration and growth temperature parameter pair. In overall view we can conclude that the DLTFS peaks were highest in two cases: for the sample GaAsN II with the highest growth temperature of 605 °C and concentration N = 1.15 %, and for GaAsN V with the highest nitrogen content N = 1.85 % and growth temperature of 575 °C (Fig. 2).

It could be speculated that both temperature and the nitrogen content tends to increase the defect concentration. Possible explanation on this behaviour is a complex defect, which could include a nitrogen induced defect state, increasing the signal at higher

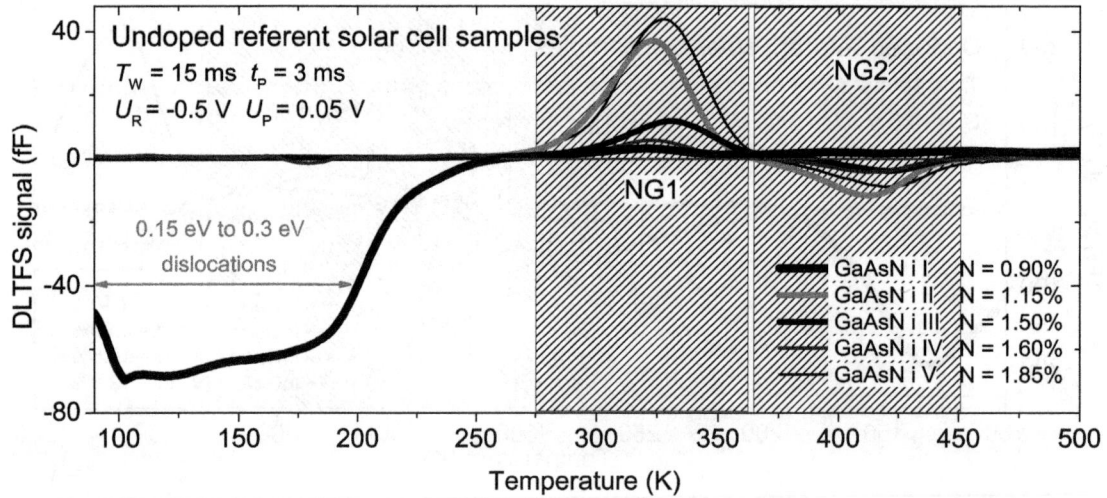

Fig. 2: DLTFS spectra of GaAsN undoped referent samples at different N concentrations, measured at identical experimental conditions.

Tab. 3: Distinct composition and structural parameters of the investigated GaAsN samples.

Structure	Growth temp. T_g (°C)	GaAs$_{1-x}$N$_x$ (nm)	x (%)
GaAsN I	585	~65	0.90
GaAsN II	605	~100	1.15
GaAsN III	595	~126	1.50
GaAsN IV	565	~100	1.60
GaAsN V	575	~130	1.85

nitrogen concentrations, co-existing with a GaAs defect becoming more dominant at higher temperatures (Tab. 4).

Tab. 4: Deep energy level parameters of the investigated GaAsN samples.

Trap	peak	$\Delta E_{Tn,p}$(eV)	$\sigma_{n,p}$ (cm^2)
ING1	positive	~0.48	~1.7·10^{-17}
	negative	~0.54	~3.9·10^{-16}
ING2	positive	~0.71	~8.0·10^{-16}
	negative	~0.66	~5.5·10^{-16}

This condition may be described for defect groups NG1 and NG2 as well, since NG1 was identified as a complex of a Cu related GaAs trap (NG1* $\cong$ 0.46 eV) together with a nitrogen induced defect (NG1** $\cong$ 0.54 eV) [11]. Cu was confirmed in InGaAs samples as well, strongly indicating that this assumption is correct. The NG2 interpretation is more complex since the spectrum transition occurred once again (see Fig. 2 NG2). In this case it was identified as a dual type complex (two different defect state responses as a complex spectrum) rather than a dual state as for InGaAs.

More precisely, as a possible nitrogen related negative signal (NG2 $\cong$ 0.66 eV) with a GaAs assumed positive one (NG2 $\cong$ 0.71 eV) [12]. Since the positive signal of NG2 at the investigated DLTFS measurement

conditions was only visible in GaAsN I (0.9 % of nitrogen, at growth temperature of 585 °C), it can be concluded that the growth temperature of 585 °C induces the most dominant state of the GaAs defect.

GaAsN IV with 1.60 % growth at 565 °C could be considered as the most balanced structure where NG1 and NG2 were the lowest and at the same time dislocations were suppressed. In means of N related defect concentration reduction, lower nitrogen content should be more relevant. An optimized growth temperature should be proposed avoiding the generation of growth temperature sensitive GaAs defects, but efficient enough to ensure a suitable nitrogen concentration eliminating dislocations. Since these disappeared already at 1.15 %, N content around this value should be considered with growth temperature lower than 585 °C. For these parameters the positive NG2 signal is probably dominant.

3.   Conclusion

The reported comprehensive InGaAs and GaAsN DLTFS investigation showed that each referent structure can be a source of several defects. Fundamental findings were observed related to composition and possible growth condition dependent defect states. New knowledge was acquired about emission and capture processes connected to indium and nitrogen, important to support the InGaAsN research and application in solar cells. According to the reported electrically active defect behavior, the InGaAs sample with 10.5 % can be stated as the most appropriate, since low temperature point defects were eliminated and trap concentrations were low. GaAsN referent structures showed several dislocations at low temperatures disappearing around

1 % of nitrogen, nevertheless, the growth of the structure with 1.6 % and 565 °C suppressing dislocations and reducing GaAs/N complex defect amplitudes can be stated as the most balanced one. Moreover, growth temperatures lower than 585 °C are arguable, at which significant trap states were reduced.

Acknowledgment

This work has been supported by the Scientific Grant Agency of the Ministry of Education of the Slovak Republic (VEGA 1/0651/16 and VEGA 1/0739/16). This work was co-financed by Wroclaw University of Science and Technology statutory grants and the Polish National Centre for Research and Development under the project No. PBS2/A3/15/2013 "PROFIT".

References

[1] LANG, D. V. Deep-level transient spectroscopy: A new method to characterize traps in semiconductors. *Journal of Applied Physics*. 1974, vol. 45, iss. 7, pp. 3023–3032. ISSN 1089-7550. DOI: 10.1063/1.1663719.

[2] WEISS, S. *Semiconductor Investigations with the DLTFS method* Hessen, 1991. Ph.D. thesis. University of Kassel.

[3] SABNIS, V., H. YUEN and M. WIEMER. High-efficiency multijunction solar cells employing dilute nitrides. In: *8th International conference on concentrating photovoltaic systems*. Toledo: AIP Publishing, 2012, pp. 14–19. ISBN 978-0-7354-1086-2. DOI: 10.1063/1.4753823.

[4] KOSA, A., L. STUCHLIKOVA, L. HARMATHA, M. MIKOLASEK, J. KOVAC, B. SCIANA, W. DAWIDOWSKI, D. RADZIEWITZ and M. TLACZALA. Defect distribution in InGaAsN/GaAs multilayer solar cells. *Solar Energy*. 2016, vol. 132, iss. 1, pp. 587–590. ISSN 0038-092X. DOI: 10.1016/j.solener.2016.03.057.

[5] DAWIDOWSKI, W., B. SCIANA, I. ZBOROVSKA-LINDERT, M. MIKOVASEK, M. LATKOWSKA, D. RADZIEWICZ, D. PUCICKI, K. BIELAK, M. BADURA, J. KOVAC and M. TLACZALA. AP-MOVPE technology and characterization of InGaAsN pin subcell for InGaAsN/GaAs tandem solar cell. *International Journal of Electronics and Telecommunications*. 2014, vol. 60, iss. 2, pp. 151–156. ISSN 2300-1933. DOI: 10.2478-eletel-2014-0018.

[6] TIXIER, S., S. E. WEBSTER, E. C. YOUNG, T. TIEDJE, S. FRANCOEUR, A. MASCARENHAS, P. WEI AND F. SCHITTEKATTE. Band gaps of the dilute quaternary alloys Ga N_x As$_{1-x-y}$ Bi$_y$ and Ga$_{1-y}$ In$_y$ N$_x$ As$_{1-x}$. *Applied Physics Letters*. 2005, vol. 86, iss. 11, pp. 112–113. ISSN 1077-3118. DOI: 10.1063/1.1886254.

[7] IOANNOU, D. E., Y. J. HUANG and A. A. ILIADIS. Deep states and misfit dislocations in indium-doped GaAs layers grown by molecular beam epitaxy. *Applied Physics Letters*. 1988, vol. 52, iss. 26, pp. 2258–2260. ISSN 1077-3118. DOI: 10.1063/1.99530.

[8] BREHME, S., P. KRISPIN and D. I. LUBYSHEV. Hole traps in indium-doped and indium-free GaAs grown by molecular beam epitaxy. *Semiconductor Science and Technology*. 1992, vol. 7, iss. 4, pp. 467–332. ISSN 0268-1242. DOI: 10.1088/0268-1242/7/4/005.

[9] NAZ, N. A., S. QURASHI, A. MAJID and M. Z. IQBAL. Doubly charged state of EL2 defect in MOCVD-grown GaAs. *Physica B: Condensed Matter*. 2007, vol. 401, iss. 1, pp. 250–253. ISSN 0921-4526. DOI: 10.1016/j.physb.2007.08.159.

[10] CHEN, K. M., Y. Q. JIA, Y. CHEN, A. P. LI, S. X. JIN and H. F. LIU. Deep levels in nitrogen implanted n-type GaAs. *Journal of Applied Physics*. 1995, vol. 78, iss. 6, pp. 4261–4263. ISSN 1089-7550. DOI: 10.1063/1.359889.

[11] POLYAKOV, A. Y., N. B. SMIRNOV, A. V. GOVORKOV, A. E. BOTCHKAREV, N. N. NELSON, M. M. E. FAHMI, J. A. GRIFFIN, A. KHAN, S. N. MOHAMMAD, D. K. JOHNSTONE, V. T. BUBLIK, K. D. CHSHERBATCHEV, M. I. VORONOVA and V. S. KASATOCHKIN. Optical properties and defects in GaAsN and InGaAsN films and quantum well structures. *Solid-State Electronics*. 2002, vol. 46, iss. 12, pp. 2147–2153. ISSN 0038-1101. DOI: 10.1016/S0038-1101(02)00178-8.

[12] YAMAGUCHI, M., B. BOUZAZI, H. SUZUKI, K. IKEDA, N. KOJIMA and Y. OHSHITA. (In)GaAsN materials and solar cells for super-high-efficiency multijunction solar cells. In: *Photovoltaic Specialists Conference*. Austin: IEEE, 2012, pp. 831–834. ISBN 978-1-4673-0066-7. DOI: 10.1109/PVSC.2012.6317732.

About Authors

Arpad KOSA was born in Dunajska Streda, Slovakia, in 1987. He is a researcher at the Institute of Electronics and Photonics, Slovak University of Technology. He received his Ph.D. degree in microelectronics from the Slovak University of Technology in 2016. His research interests include semiconductor defect analysis and investigation by experimental and mathematical methods, electronics and microelectronics, and mathematical evaluation algorithm programming.

Lubica STUCHLIKOVA was born in Humenne, Slovakia. She graduated from the Faculty of Electrical Engineering, Slovak University of Technology in Bratislava in 1990, received the Ph.D. degree in 1996 and was appointed associate professor in 2006. She works as a teacher at the Institute of Electronics and Photonics. From 2016 as full professor. Since 1990 she has been interested in semiconductor defects engineering and electrical characterization of semiconductor structures, devices and materials.

Ladislav HARMATHA (was born in Dobsina, Slovakia, in 1948. He graduated from the Faculty of Electrical Engineering, Slovak University of Technology Bratislava, in 1971, received the Ph.D. degree in 1984 and was appointed associate professor in 1996. Since 1988 he has worked as a senior scientist in the field of semiconductor defects engineering. His research is focused on defects in semiconductor structures and their characterization by electrical methods (capacitance transient spectroscopy).

Jaroslav KOVAC was born in Tornala, Slovakia, in 1947. He graduated from the Slovak University of Technology, Faculty of Electrical Engineering and Information Technology in 1970. Since 1971 he has been engaged in the research of optoelectronic devices technology at the Microelectronics Department of Faculty of Electrical Engineering, Slovak University of Technology Bratislava. He received a Ph.D. degree (1983) and professor degree (2001) from STU Bratislava. Since 1991 he has been the team leader of the Optoelectronic Group at the Institute of Electronics and Photonics.

Beata SCIANA was born in Wroclaw, Poland, in 1965. She graduated in Electronics from the Wroclaw University of Technology, Faculty of Electronics, in 1990. From 1990 to 1993 she worked at the Elwro Electronics Factory in Wroclaw. Since 1993 she has worked at Wroclaw University of Technology and pursued researches in epitaxial growth (MOVPE method) and material characterization of AIIIBV semiconductor compounds for application in advanced microelectronic and optoelectronic devices. She received her Ph.D. degree in Electronics from Wroclaw University of Technology in 2000.

Wojciech DAWIDOWSKI was born in Tomaszow Lubelski, Poland in 1987. He graduated from the Wroclaw University of Technology, Faculty of Microsystem Electronics and Photonics in 2011. He is a Ph.D. student and his research is focused on epitaxial growth of AIIIBV-N semiconductor compounds for solar cells, their fabrication and characterization.

Marek TLACZALA graduated in electronics from the Faculty of Electronics, Wroclaw University of Technology in 1972. In 1973-1976 he worked at the Electrotechnical University in Leningrad (now Sankt Petersburg). He received his Ph.D. degree in electronic engineering from the Electrotechnical University in Leningrad in 1976 and Dr.Sc. degree (habilitation) from the Faculty of Microsystem Electronics and Photonics, Wroclaw University of Technology in 2002. In 2009 he was appointed professor. At present he is a full professor and the head of the Semiconductor Devices Lab in the Faculty of Microsystem Electronics and Photonics, Wroclaw University of Technology.

17

Proximity Effect in Gate Fabrication Using Photolithography Technique

Joanna PRAZMOWSKA, Kornelia INDYKIEWICZ, Bogdan PASZKIEWICZ, Regina PASZKIEWICZ

Department of Microelectronics and Nanotechnology, Faculty of Microsystem Electronics and Photonics, Wroclaw University of Science and Technology, Wybrzeze Wyspianskiego 27, 503 70 Wroclaw, Poland

joanna.prazmowska@pwr.edu.pl, kornelia.indykiewicz@pwr.edu.pl, bogdan.paszkiewicz@pwr.edu.pl, regina.paszkiewicz@pwr.edu.pl

Abstract. *In the paper the technological factors influencing test structure gate length were described. The influence of test structure gate placement (Schottky metallization between ohmic contacts, on mesa and on GaN surface) was analyzed and discussed. Moreover, various distances between ohmic contacts paths were tested. Except for experimental investigations, simulations using finite elements method in COMSOL were performed for the same structure. The modelling results revealed crucial impact of a gap beyond the mask on the electric field distribution in photoresist layer. The smallest value of relative error of test finger lengths was observed for finger parts placed between ohmic paths on mesas. It was explained by thicker lift-off double layer between ohmic paths and the smallest Y-gap compared to test fingers placed on mesa and outside of it. Simulation did not bring an explanation of larger values of relative error for smaller distance between ohmic paths.*

Keywords

AlGaN/GaN transistors, h-line lithography, proximity effect.

1. Introduction

Continuous increase of scale of integration of electronic devices cause that the optical lithography faced its resolution limitation of used wavelength. According to Rayleigh's equation, enhancement of resolution could be assured by decrease of wavelength or higher numerical aperture of lens systems [1]. To obtain higher resolution, the wavelength was decreased from G-line (435 nm) to I-line (365 nm), further to 248 nm (excimer laser source with KrF) and to 193 nm (ArF) [2] and [3]. Also 8 various methods of image formation have been developed e.g. phase shifting method [1] and [2]. Moreover, there are efforts of development of superlenses [4], extreme ultraviolet and beyond extreme ultraviolet lithography [5], surface-plasmon polariton resonance [6].

Additionally, constants depending on resist material, process technologies and image formation techniques play an important role. The proximity effect defined as a variation in pattern width due to proximity of other nearby features is well known for electron-beam lithography [7]. The optical proximity effect was studied referring to features typical for transistors fabrication. In the paper the technological factors influencing test structure gate resolution were described. The influence of test structure gate placement was discussed. Observed phenomena were analyzed also based on computer simulations results.

2. Experimental Details

The dedicated test structures were made during AlGaN/GaN HEMT (High Electron Mobility Transistor) devices fabrication. The AlGaN/GaN heterostructures fabrication in metal-organic vapour phase epitaxy technique was described elsewhere [8]. Each transistor in the module on the wafer consisted of two test structures that differed in designed distance between ohmic contact paths:

- type 1 - designed distance of 3 μm plus designed length of test finger,

- type 2 - designed distance of 4 μm plus designed length of test finger).

Both types of dedicated test structures contained six fingers of various lengths in purpose of indirect analysis of gates lengths and chosen factors influencing its value. The designed fingers lengths were #1 – 0.6, #2 – 0.8, #3 – 1, #4 – 1.4, #5 – 2, #6 – 5 μm, respectively.

Additionally, the test structures embrace three different areas on which the test structures fingers were placed Fig. 1:

- area A - Schottky metallization on mesa between ohmic contacts,

- area B - Schottky metallization on mesa,

- area C - Schottky metallization on GaN surface (outside of the mesa).

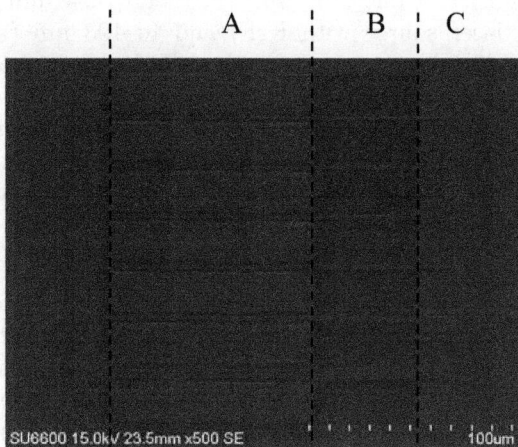

Fig. 1: SEM image of test structure.

The dedicated test structures were fabricated in AlGaN/GaN heterostructures by photolithography technique using Carl Suess MA56 mask aligner working in h-line mode. Mesas were etched through the SiO$_2$ mask (300 nm thick, deposited by plasma-enhanced chemical vapor deposition) in Reactive Ion Etching (RIE) system. For the RIE process a Cl$_2$:BCl$_3$ mixture of gasses was used. Time of etching equal to 70 s gave heights of mesas in the range from 70 to 87 nm. Metallization contacts were deposited in UHV system by thermal and e-beam evaporation. The metallization stack of Ti/Al/Mo/Au thermally formed in rapid thermal annealing system (at 820 °C for 60 s) was used as ohmic contact. Schottky contacts were of Ru/Au (30/150 nm) double layer.

Mesa structures patterns were made in standard lithography (using Microposit S1813 Photo Resist - Shipley) while ohmic and Schottky contacts were fabricated in lift-off technology using double layer - Shipley Microposit LOL 2000 and Megaposit SPR 700 – 1.0

(DOW). Pattern was transferred from chromium mask in vacuum contact (the vacuum seal inflates to form a chamber between mask and sample, which is then evacuated). The wavelength of exposure UV light was 405 nm and its intensity of 18 mW·cm^{-2}. The pre-bake time, time of exposure, LOL2000 and S1813 thicknesses, time of development as well as ultrasounds power during development were optimized for HEMTs gates fabrication. Additionally, step of edge bead remove to minimize the distance of the mask and sample during exposure was applied.

Fingers lengths of test structures were measured within a series of samples made in similar environmental conditions of lithography process. The yield of each sample exceeded 90 %. The lengths were measured repeatedly near the middle of each finger based on Scanning Electron Microscope (SEM) images.

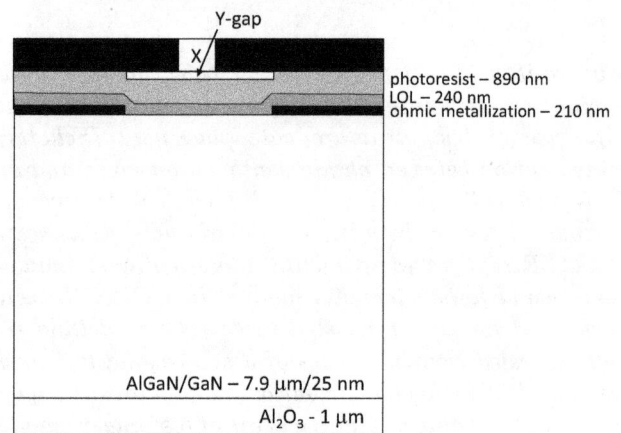

Fig. 2: The structure used for simulations in COMSOL.

Qualitative analysis of the electric field distribution in the Y-gap, photoresist and LOL (Fig. 2) during exposing was performed based on simulations using Finite Element Analysis (FEA) by COMSOL - the commercial software. The structure used for simulations was similar with that obtained in experiments. It contains (top-down):

- mask, with chromium areas and the X-gap (width as designed test finger length) - filled with vacuum during exposure,

- Y-gap, resulting from non-uniform spin coating of the photoresist due to ohmic contacts presence on the AlGaN/GaN surface - filled with vacuum during exposure,

- photoresist and LOL layer,

- ohmic metallization paths placed in the distances as in experiments (i.e. 3 and 4 μm plus designed gate length, depending on type of the structure),

- AlGaN/GaN structure.

The exposure parameters of vertically incident light in simulations were the same as for experimental part of the investigation. In the Tab. 1 the refractive index values of used materials for 405 nm wavelength are shown.

Tab. 1: Refractive index values of used materials.

SPR 700	LOL 2000	Al_2O_3	AlGaN	GaN	Glass	Metalli-zation
1.7	1.6	1.76	2.25	2.55	1.5	1.52

3. Results

In the first step the mean value of fingers lengths in three areas was estimated (Fig. 3). Additionally, the standard deviation was calculated. Its value was the smallest for fingers parts located between the ohmic contacts pads. Standard deviations of lengths of fingers designed for 1 μm, as gate length of HEMT structures, were similar.

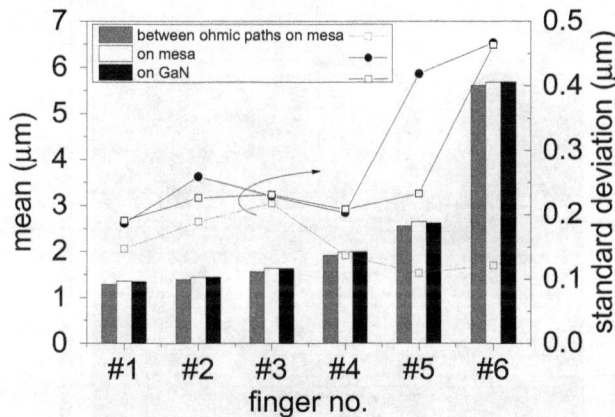

Fig. 3: Mean values and standard deviation of finger lengths of test structures.

The relative error of fingers lengths placed on three areas for both types of test structures is presented in Fig. 4(a) and Fig. 4(b).

The smallest value of relative error was obtained for finger parts placed between ohmic paths on top of mesas for structures of type 1 (depicted as area A in Fig. 1) as well as type 2 (depicted as area B in Fig. 1). Due to large height of ohmic contacts (Fig. 5) compared to lift–off double layer height the reflection on the metallization slope and its irregularities was expected to lengthen the fingers. The observed fingers length could be a consequence of thicker lift-off double layer between ohmic paths. The thicker lift-off double layer is an effect of the spin-off technique used for samples coating by resists. The time of exposure as well as time of development was equal for whole sample thus thicker layer of resists could give shorter fingers. For

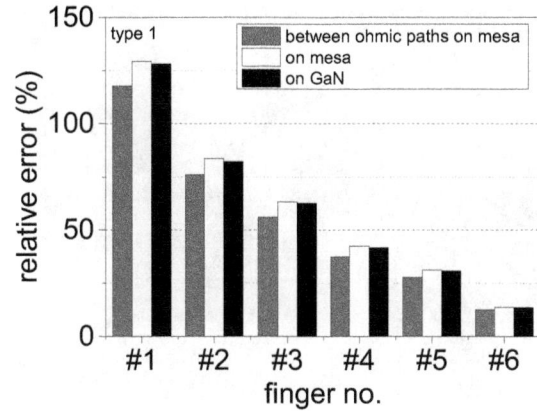

(a) Type 1.

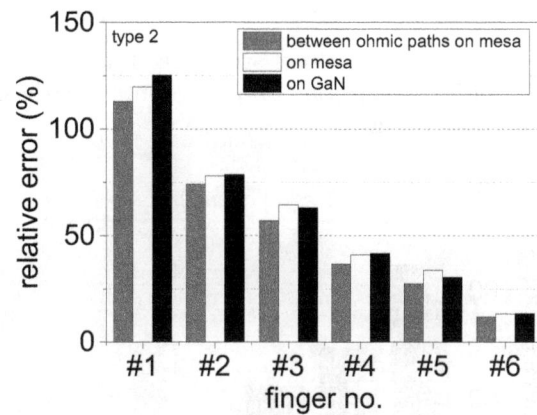

(b) Type 2.

Fig. 4: Relative error of fingers lengths placed on three areas for both types of test structures.

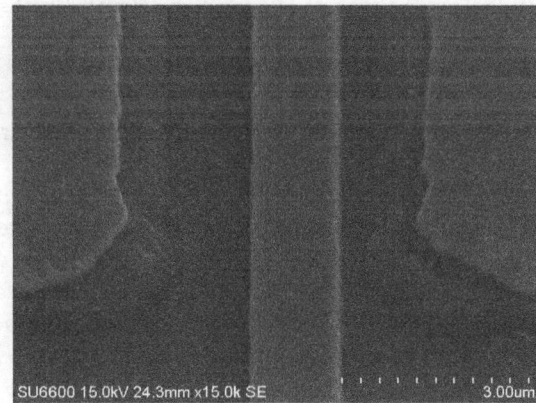

Fig. 5: SEM image of test structure finger between ohmic contacts paths.

fingers #1, #2 and #3 observed length was larger for parts placed on GaN surface compared to those placed on top of mesa. Lengthening of fingers within this area could be caused by reflecting of exposure UV-light on whiskers that occurred on GaN surface and further exposure of patterns Fig. 6.

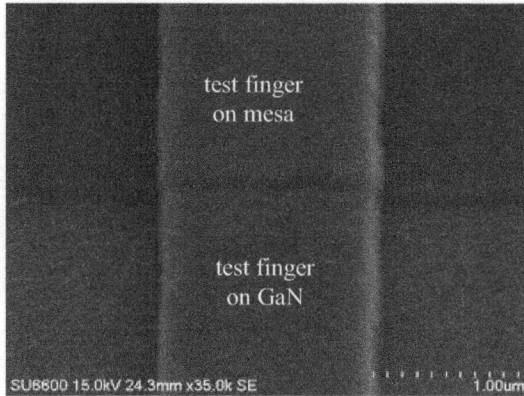

Fig. 6: SEM image of test structure finger on various areas.

The relative error of test structures fingers lengths (Fig. 7) indicated affection of ohmic contact paths distance on the lengths.

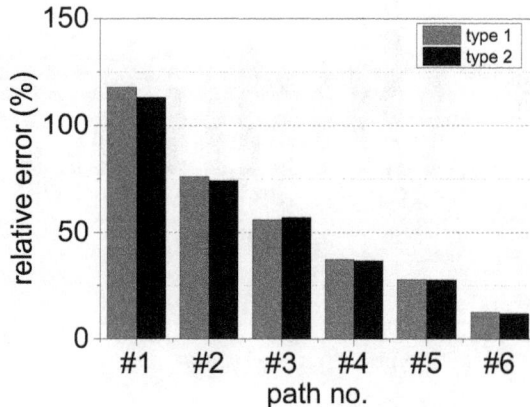

Fig. 7: The relative error of test structures fingers lengths for various ohmic contact paths distance.

Larger values of relative error were observed for smaller distance thus influence of UV-light scattering on ohmic contacts slopes could not be excluded. As a result of the simulations the electric field distribution in the structure was obtained. Only the issue of three designed finger lengths (i.e. 0.6 μm, 1 μm and 5 μm) were selected to further discussion. The case of the smallest designed finger length (0.6 μm) for both distances between ohmic contacts paths is presented in Fig. 8.

Additionally, electric field profiles in horizontal lines in four boundaries regions were estimated. The lines were located between:

- I - mask and photoresist layer or Y-gap,

- II - LOL and ohmic contact surface,

- III - ohmic contact and AlGaN/GaN,

- IV - under the Y-gap.

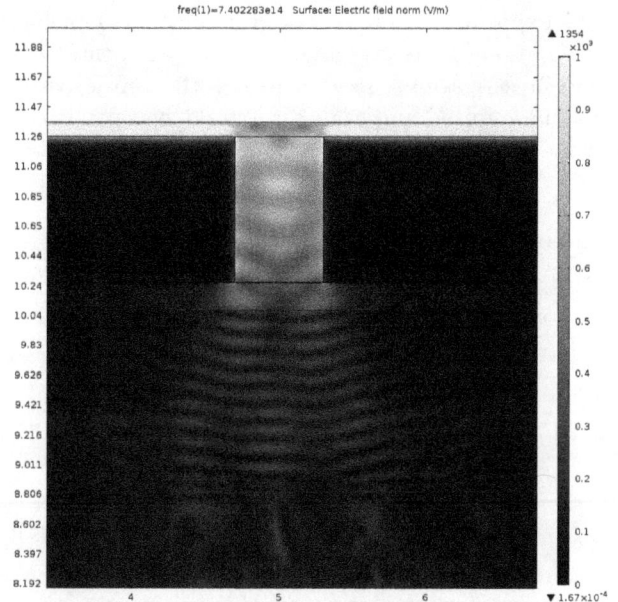

(a) 0.6 μm and 3.6 μm.

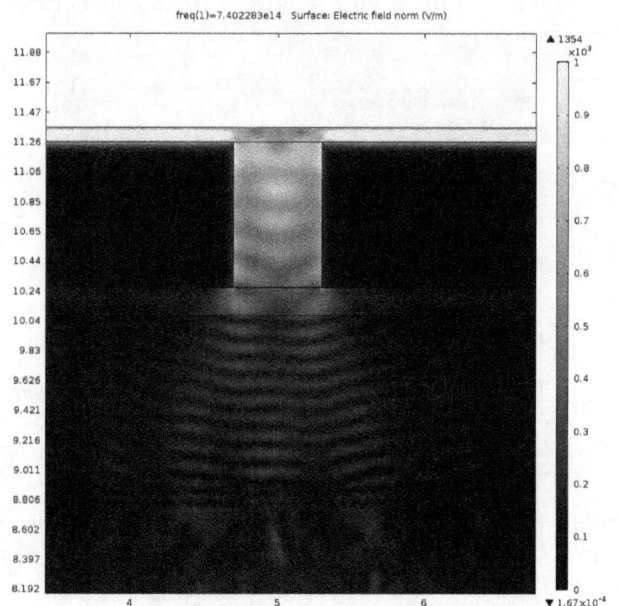

(b) 0.6 μm and 4.6 μm.

Fig. 8: Electric field distribution in structure for designed finger length and distance between ohmic contacts as depicted ($f = 7.402283 \cdot 10^{14}$, surface: Electric field norm (V/m)).

The electric field profiles for designed finger length and distance between ohmic contacts 0.6 μm and 3.6 μm in Fig. 9(a) and 0.6 μm and 4.6 μm in Fig. 9(b).

The analysis of profiles indicated expected diffraction on the edges of the chromium layer corners. The shadowing region for both cases was not evident as well as scattering on the ohmic contacts paths surfaces. A significant influence of Y-gap presence on the electric field distribution c ould be also observed for the investi-

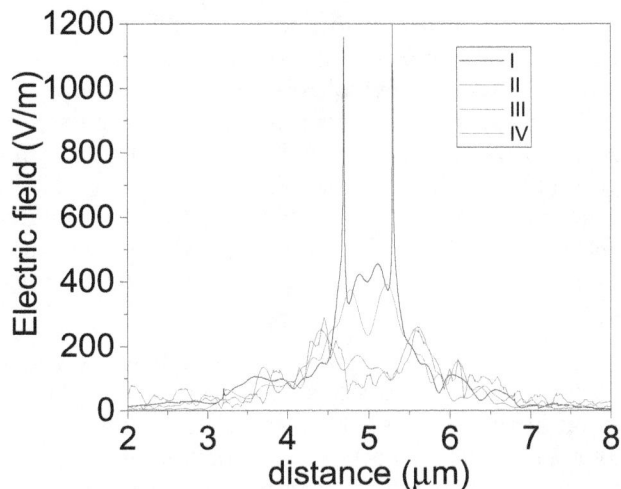

(a) 0.6 μm and 3.6 μm.

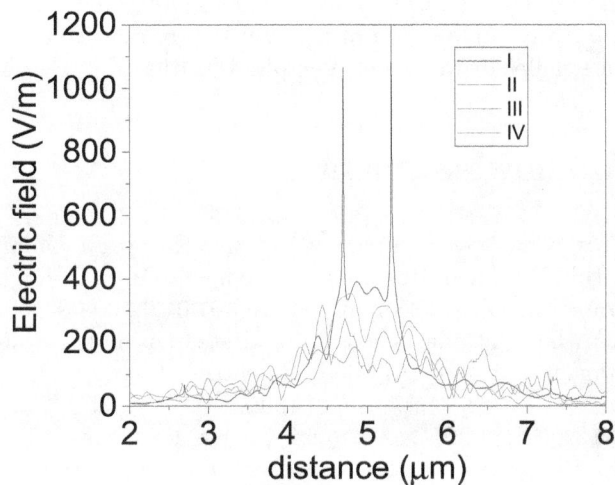

(b) 0.6 μm and 4.6 μm.

Fig. 9: Electric field profiles in structure for designed finger length and distance between ohmic contacts as depicted.

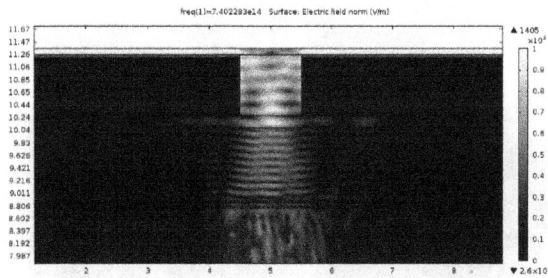

(a) 1 μm and 4 μm.

(b) 1 μm and 5 μm.

Fig. 10: Electric field distribution in structure for designed finger length and distance between ohmic contacts as depicted.

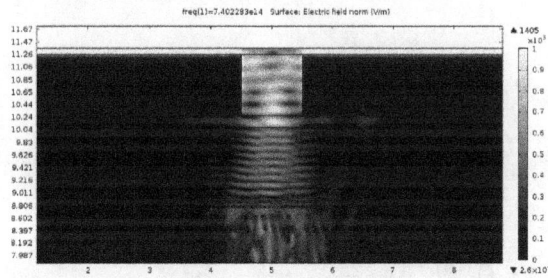

(a) Electric field distribution.

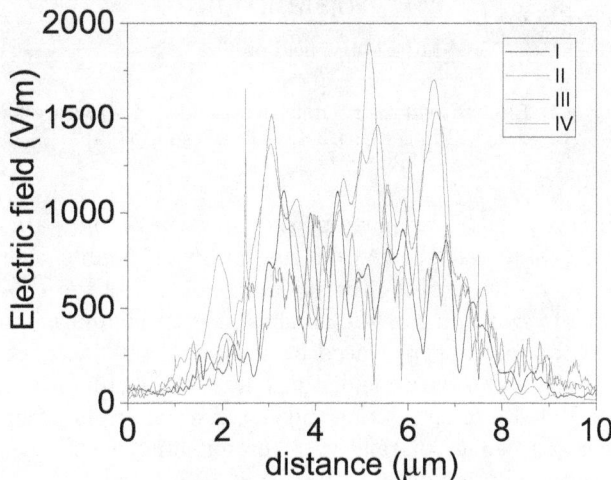

(b) Electric field profiles.

Fig. 11: Electric field distribution and profiles in structure for designed finger length and distance between ohmic contacts of 5 μm and 8 μm.

gated sample as well as for structure for designed finger length and distance between ohmic contacts 1 μm and 4 μm in Fig. 10(a) and 1 μm and 5 μm in Fig. 10(b).

Optical effects occurring during exposure were also remarkable for the samples of designed finger length and distance between ohmic contacts as 5 μm and 8 μm and 5 μm and 9 μm (not shown) presented in Fig. 11(a) and Fig. 11(b).

The desired electric field distribution and profiles in photoresist are presented in Fig. 12. From the simulation structure the Y-gap was excluded. Compared to previously shown electric field distributions that in Fig. 12 has regular shape what permitted to obtain designed finger length. The phenomena indicate great influence of Y-gap on resulting gate lengths.

The simulation results shown the significant influence of Y-gap on the electric field distribution in the photoresist. The smallest value of relative error was

freq(1)=7.402283e14   Surface: Electric field norm (V/m)

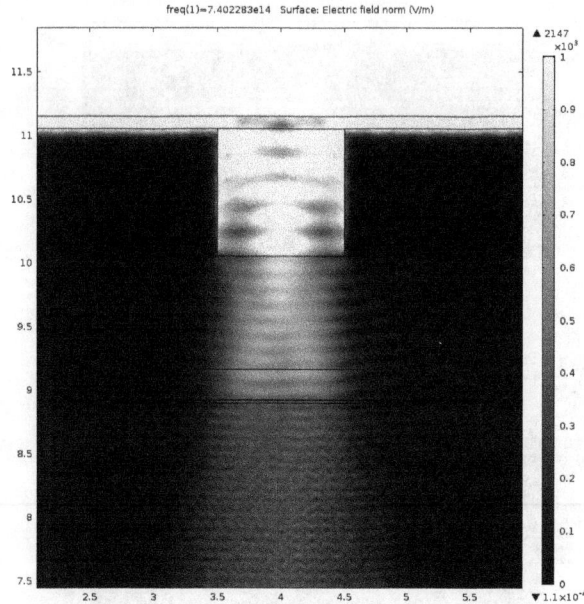

(a) Electric field distribution.

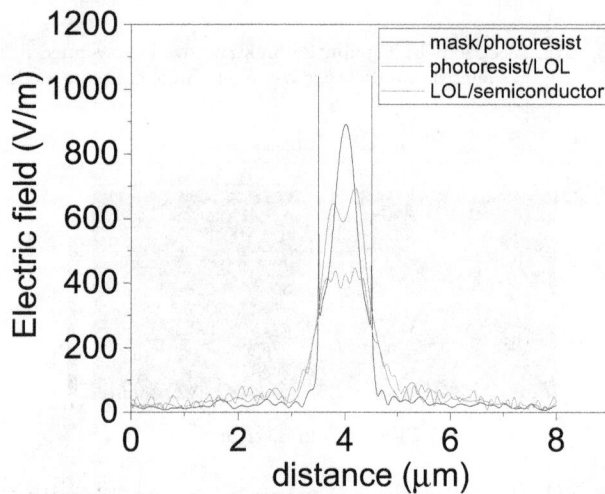

(b) Electric field profiles.

Fig. 12: Electric field distribution and profiles in structure for designed finger length of 1 μm without Y-gap.

observed for finger parts placed between ohmic paths on top of mesas. Apart from thicker lift-off double layer between ohmic paths the phenomenon could be a result of occurrence of the smallest Y-gap compared to that for test fingers placed on mesa and outside of it. The effect of separation distance was studied already in [8]. Results of simulation did not give an explanation of larger values of relative error for smaller distance between ohmic paths.

4.   Conclusions

In the paper, the technological factors influencing test structure gate length were described. The standard

deviation of fingers length was the smallest for fingers parts located between the ohmic contacts paths. Also the smallest value of relative error was obtained for finger parts placed between ohmic paths on top of mesas independently for both values of distance between ohmic contact paths. The relative error of test structures fingers lengths indicated affection of ohmic contact paths distance on the lengths. Larger values of relative error were observed for smaller distance between ohmic contacts.

The simulation results reveal great impact of Y-gap presence under the mask on the electric field distribution in the photoresist. The smallest value of relative error for finger parts placed between ohmic paths on top of mesas could be a result of occurrence of the smallest Y-gap compared to that for test fingers placed on and outside of mesa. Results of simulation did not bring any explanation of larger values of relative error for smaller distance between ohmic paths.

Acknowledgment

This work was co-financed by the European Union within European Regional Development Fund, through grant Innovative Economy (POIG.01.01.02-00-008/08-05), by National Centre for Research and Development through Applied Research Program grant no. 178782, program LIDER no. 027/533/L-5/13/NCBR/2014, National Centre for Science under the grants no. 2015/19/B/ST7/02494 and DEC-2012/07/D/ST7/02583, by Wroclaw University of Science and Technology statutory grants and Slovak-Polish International Cooperation Program.

References

[1] OKAZAKI, S. Resolution limits of optical lithography. *Journal of Vacuum Science & Technology B, Nanotechnology and Microelectronics: Materials, Processing, Measurement, and Phenomena.* 1991, vol. 9, iss. 1, pp. 2829–2833. ISSN 2166-2746. DOI: 10.1116/1.585650.

[2] HARRIOTT, L. R. Limits of lithography. *Proceedings of the IEEE.* 2001, vol. 89, iss. 3, pp. 366–374. ISSN 1558-2256. DOI: 10.1109/5.915379.

[3] ITO, T. and S. OKAZAKI. Limits of lithography. *Pushing the limits of lithography.* 2000, vol. 409, iss. 1, pp. 1027–1031. ISSN 0028-0836. DOI: 10.1038/35023233.

[4] FANG, N., H. LEE, C. SUN and X. ZHANG. Sub-Diffraction-Limited Optical Imaging with a Sil-

ver Superlens. *Science*. 2000, vol. 308, iss. 5721, pp. 534–537. ISSN 1095-9203. DOI: 10.1126/science.1108759.

[5] MOJARAD, N., J. GOBRECHT and Y. EKINCI. Beyond EUV lithography: a comparative study of efficient photoresists performance. *Scientific Reports*. 2015, vol. 5, no. 9235, pp. 1–5. ISSN 2045-2322. DOI: 10.1038/srep09235.

[6] LUO, X. and T. ISHIHARA. Subwavelength photolithography based on surface-plasmon polariton resonance. *Optics Express*. 2004, vol. 12, iss. 14, pp. 3055–3065. ISSN 1094-4087. DOI: 10.1364/OPEX.12.003055.

[7] LEUNISSEN, L. H. A., R. JONCKHEERE, U. HOFMANN, N. UNAL and C. KALUS. Experimental and simulation comparison of electron-beam proximity correction. *Journal of Vacuum Science and Technology B: Microelectronics and Nanometer Structures*. 2004, vol. 22, iss. 6, pp. 2943–2947. ISSN 1071-1023. DOI: 10.1116/1.1808742.

[8] WOSKO, M., B. PASZKIEWICZ, A. VINCZE, T. SZYMANSKI and R. PASZKIEWICZ. GaN/AlN superlattice high electron mobility transistor heterostructures on GaN/Si(111). *Physica Status Solidi (B)*. 2015, vol. 252, no. 5, pp. 1195–1200. ISSN 1071-1023. DOI: 10.1002/pssb.201451596.

[9] BAEK, S., G. KANG, M. KANG, C.-W. LEE and K. KIM. Resolution enhancement using plasmonic metamask for wafer-scale photolithography in the far field. *Scientific Reports*. 2016, vol. 6, no. 30476, pp. 1–8. ISSN 1071-1023. DOI: 10.1038/srep30476.

About Authors

Joanna PRAZMOWSKA received her M.Sc. degree in Electronics from Wroclaw University of Technology, Poland in 2005 and Ph.D. degree from Wroclaw University of Technology (WrUT) in 2011. Now she is assistant professor at WrUT. Her research interest embraces technology of semiconductor devices i.e. lithography process development of electronic, optoelectronic devices as well as gas sensors.

Kornelia INDYKIEWICZ born in Wyrzysk, Poland in 1986, is a Ph.D. student at Wroclaw University of Technology, Department of Microelectronics and Nanotechnology. She works in a multidisciplinary team doing research concerning AlGaN/GaN based transistors.

Bogdan PASZKIEWICZ received his M.Sc. degree in Electrical Engineering from St. Petersburg Electrotechnical University, St. Petersburg, Russia in 1979 and Ph.D. degree from the Wroclaw University of Technology in 1997. Now he is assistant professor at WrUT. His research is focused on the design and parameter evaluation of nitrides-based devices: HEMTs and sensors.

Regina PASZKIEWICZ received her M.Sc. degree in Electrical Engineering from St. Petersburg Electrotechnical University, St. Petersburg, Russia in 1982 and Ph.D. degree from the Wroclaw University of Technology in 1997. Now she is full professor at WrUT. Her research is focused on the technology of (Ga, Al, In) N semiconductors, microwave and optoelectronic devices technological processes development.

18

Femtosecond Laser Processing of Membranes for Sensor Devices on different Bulk Materials

Johann ZEHETNER[1], Gabriel VANKO[2], Jaroslav DZUBA[2], Tibor LALINSKY[2]

[1]Research Centre for Microtechnology, University of Applied Sciences,
Hochschulstrasse 1, 6850 Dornbirn, Austria
[2]Institute of Electrical Engineering, Slovak Academy of Sciences,
Dubravska cesta 9, 841 04 Bratislava, Slovakia

johann.zehetner@fhv.at, gabriel.vanko@savba.sk, jaroslav.dzuba@savba.sk, tibor.lalinsky@savba.sk

Abstract. *We demonstrate that diaphragms for sensor applications can be fabricated by laser ablation in a variety of substrates such as ceramics, glass, sapphire or SiC. However, ablation can cause pinholes in membranes made of SiC, Si and metals. Our experiments indicate that pinhole defects in the ablated membranes are affected by ripple structures related to the polarization of the laser. From our simulation results on light propagation in Laser-Induced Periodic Surface Structures (LIPSS) we find out that they are acting as a slot waveguide in SiC material. The results further show that field intensity is enhanced inside LIPSS and spreads out at surface distortions promoting the formation of pinholes. The membrane corner area is most vulnerable for pinhole formation. Pinholes funnel laser radiation into the bulk material causing structural damage and stress in the membrane. We show that a polarization flipping technique inhibits the formation of pin holes caused by LIPSS.*

Keywords

AlGaN/GaN HEMT, diaphragms, laser ablation, LIPSS, SiC MEMS, slot waveguide.

1. Introduction

Membranes are widely used as a structural unit for various sensor applications, MEMS, micro devices used in biology, medicine or life science [1] and [2]. Reactive Ion Etching (RIE) is an established and precise technology to produce them in materials (mainly Silicon) commonly used in micro technology. Only a few results were published in the field of bulk micromachining of wide band gap materials to fabricate 3D micromechanical structures. The widely used approaches employed laser tools working mostly with pulses in ns range, optionally in fs pulse duration [3], [4] and [5]. We showed that femtosecond laser ablation is a promising technology to fabricate micromechanical membrane structures in SiC, borosilicate glass, sapphire, Al_2O_3 and zirconium ceramic substrates [6] and [7]. Our main focus was on SiC, which can be machined by laser ablation faster than by RIE. On a 350 μm thick 4H-SiC substrate, we achieved by laser ablation an array of 275 μm deep and 1000 μm to 3000 μm in diameter blind holes without damaging the 2 μm GaN/AlGaN heterostructure layer grown on its opposite side. Recently we investigated a combination of ablation and RIE to produce thinner membranes without back side damage [6]. It is possible to use a substrate pre structured by ablation in a maskless RIE procedure when membrane fabrication and substrate thinning is intended.

The materials we tested (including diamond) revealed Laser-Induced Periodic Surface Structures (LIPSS) similar as investigated in [8]. On the other hand, we observed pinhole defects only in metals or high refractive index SiC and Si [7]. Comparing experimental and simulation results, we found that LIPSS are acting as slot waveguides in high refractive materials and can trigger the formation of pinholes. In the glass the formation of crater structures and spikes was explained previously by a different electronic damage mechanism due to nonlinear absorption of multiple incident ultrashort laser pulses [9]. The generation and orientation of LIPSS in different materials during fs-laser ablation using linear polarized light were extensively studied in [8]. Moreover, in materials we investigated, LIPSS were created perpendicular to the direction of the laser polarization. Such an angular orientated configuration is most suitable for a good performance of slot waveguides [10]. Consequently, steady rotation or frequent angular flipping of the polarization is an effective measure firstly to avoid a pronounced formation of LIPSS and secondly to reduce the performance of existing LIPSS in their waveguide function. The slot waveguide approach does not explain the formation of LIPSS and small pinholes some hundreds of nanometer in diameter. However, our qualitative simulation results in [10] pointed towards an existing feedback mechanism promoting the growth from nanometer size to typically 2 to 5 micrometer by interconnecting with adjacent bores.

In this paper, we report on consequences for the diaphragm surface texture caused by the structural stress induced from the radiation funnelling effects of the pinholes in the corner zone of the structures. We give a brief possible reason why the formation of High Spatial Frequency LIPSS (HSFL) is often observed at the edges of pinholes and scratches. In a comparison of two different substrate materials, we demonstrate the relevance of back reflections from a thin membrane for the ablation quality limits.

2. Experiments

For laser ablation, we used a Spectra Physics SPIRIT delivering 350 fs pulses at 200 kHz with an average power output of 4000 mW at 1040 nm and 1600 mW at 520 nm. This laser is part of the laser work station microSTRUCTvario from 3D-MICROMAC and the used 100 mm focal length scanner optics provides a focal spot diameter of about 13 μm, and 25 μm at 520 nm and 1040 nm wavelength, respectively. In all experiments we used 1000 mm·s^{-1} scan speed and 5 μm hatch distance between lines in the scan mesh.

To evaluate the influence of pinholes and structural stress on the surface texture and membrane quality the scan pattern was rotated by 15° increments between consecutive scans at a constant polarization direction parallel to the x-scan axis. Alternatively, the specimen was rotated. Rotating the sample automatically changes the polarization and the spatial orientation of the scan-pattern relative to the sample continuously. For practical use, the specimen rotation is only suitable for circular structures but not feasible at a general design pattern. Therefore, we conducted experiments with 90° flipping of the polarization after several consecutive scans by frequently inserting and removing a half-wave plate into the beam path of the laser setup (Fig. 1). This method preserves a constant spatial scan-pattern on the substrate, only the polarization direction is altered in 90° increments. We compared the three methods with our standard ablation procedure which is xy-scanning at constant special scan-pattern orientation and constant polarization direction. This

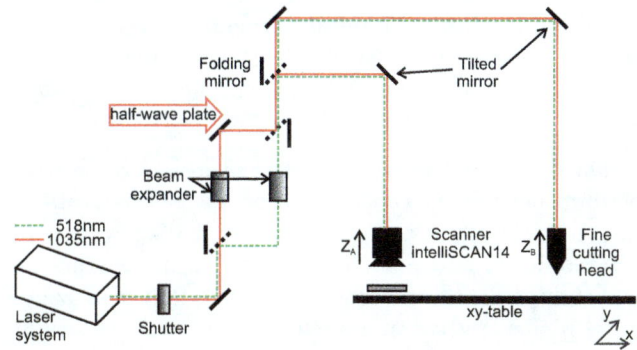

Fig. 1: Schematic of laser setup, to supress pinhole formation in the membrane a half-wave plate was used to flip the polarization 90° after several consecutive scans.

offered a possibility to distinguish whether the scan pattern rotation or the polarization change is the dominant parameter for a proper surface quality of the obtained diaphragm.

3. Results and Discussion

3.1. HSFL Formation at Pinhole Edges

In SiC we often observed HSFL structures at the edges of pinholes, deep scratches and slopes. In such locations LIPSS show the tendency to split into structures with a periodicity of only a fraction of the used laser wavelength. Figure 2 depicts the edge of a pinhole in SiC. The horizontal area of the membrane shows LIPSS

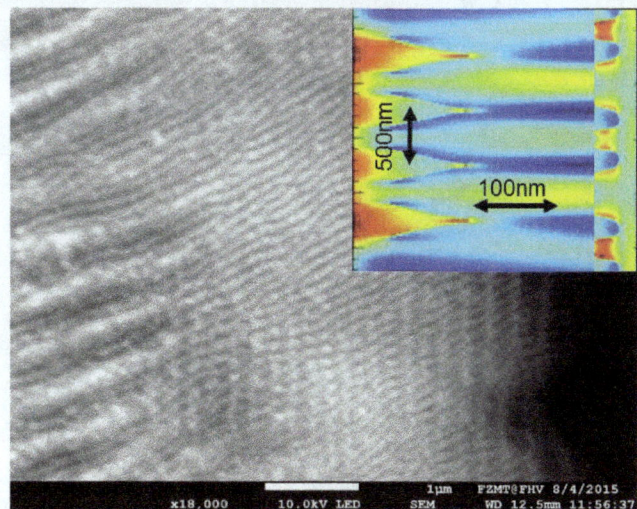

Fig. 2: At the edges of bores, pinholes or scratches we observed often a transition from LIPSS into HSFL. The insert shows how the light field below the ridges of the LIPSS can couple to the groves of HSFL structures. The splitting is initiated at the edge and continues down the wall of the pinhole.

separated in about 500 nm increments. The transition from LIPSS to HSFL is visible at the intersection of the horizontal bottom of the membrane and the slope section of the pinhole.

The calculated intensity distribution in LIPSS with 500 nm periodicity can be seen on the left side of the insert in Fig. 2. Calculating the light distribution of a periodic structure at 250 nm spatial increments on the right side of the insert revealed that the zones with higher radiation match exactly with the light distribution of higher intensity from the 500 nm LIPSS inside the bulk material. We assumed for all simulations a horizontal plane and a vertically oriented laser beam for the ablation. The calculated distribution of the laser intensity deeper inside the bulk material cannot interfere with the radiation inside the HSFL as long as both are in the same plane. However, apparently there is a chance to get sufficient intensity overlap at the edge of a pinhole or laser structured bore (Fig. 3 and Fig. 3 upper right insert, respectively). In this locally confined area the inner bulk radiation field from the LIPSS surfaces at the side wall where it can positively interfere with the light distribution of the HSFL structure. The small radius of a contour or pinhole edge and the high refractive index of SiC make such an edge to a lens of very short focal length concentrating laser light intensity sufficient high for the LIPSS to HSFL transition. We visualized this concentration effect in the lower right insert of Fig. 3. The lens effect of the pinhole edge causes a permanent change of the refractive index in an acrylic sample causing intensity related colour changes when observed between polarizers. In addition, the sidewall of a bore or pinhole is not perpendicular due to the Gaussian intensity distribution of a laser beam. Light refraction at the sloped sidewall directs further laser radiation into the bulk material as can be seen in the upper right insert of Fig. 3. The intensity of the light entering the bulk via the side wall was at least intense enough to create darker colour centres in a glass substrate, which we used to visualize leaking radiation from the ablation zone.

We observed that the pinhole formation starts predominately at scratches and corners. One reason can be seen in already mentioned focusing effects of sharp edges, a second in the spatial spreading of laser intensity at distortions and interruptions of slot waveguides. For a smoothly aligned LIPSS array the laser radiation, which hits the ablation surface perpendicularly, is well confined inside the slots (simulation results in upper insert Fig. 4). In the area between them where ridges from the already formed LIPSS are located, the laser radiation is penetrating deep into the substrate material. If such a waveguide arrangement is interrupted by a scratch like in Fig. 4 the confinement of the laser radiation is broken as the calculated intensity distribution depicts in the lower insert of Fig. 4. Neighbouring nanometer sized bores are now in conditions to join and grow towards pinholes with several micrometers in diameter (Fig. 3 and Fig. 4). Such pinholes funnel the laser light directly into the membrane and create extremely high intensities at the edges and tips causing excessive thermal load and structural stress (visualized at the tip of the pinhole by polarizing microscopy in Fig. 3 lower left insert).

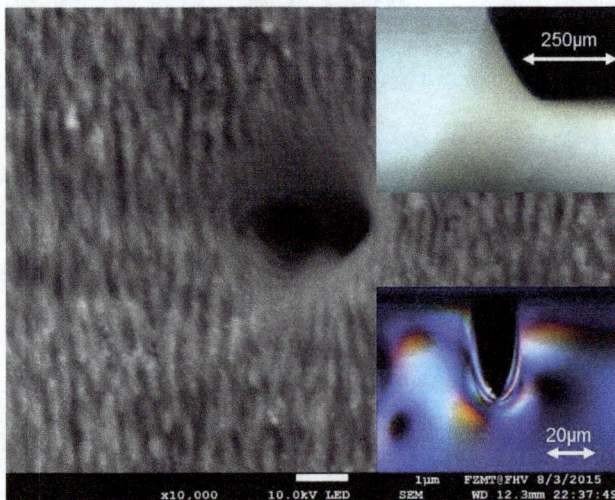

Fig. 3: LIPSS in the horizontal plane of a SiC membrane alter beginning at the edge of the generated pinhole from 500 nm spatial increments to 250 nm over the entire sloped sidewall.

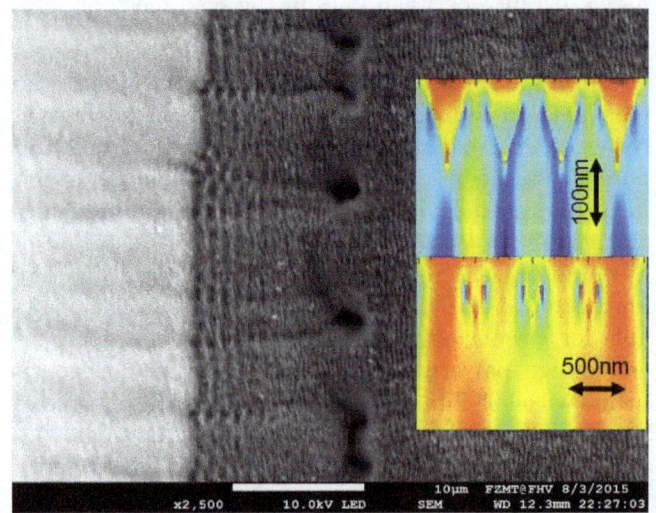

Fig. 4: LIPSS and pinholes are generated in the size of several hundreds of nanometer. Predominately at positions of scratches radiation is no longer well confined inside the waveguide (insert) and adjacent bores can combine to pinholes with several micrometers in diameter.

3.2. Pinhole and Stress Related Surface Structures at the Membrane

Induced stress at the diaphragm is accumulating at the corner area of the bore. The side wall quality is dependent mainly on the scanning method, laser parameter, light polarization and laser delay time. Structural distortions at the side wall due to delay issues in laser on/off timing can result in membrane stress influencing the surface texture of the membrane.

In Fig. 5, a standard xy-scan procedure was used (scan pattern and orientation of the laser polarization spatially fixed for all consecutive scans). The first scan moved in the x-direction from left to right and back from right to left shifting the parallel scan line 5 μm upward in the y-direction until the total surface area was covered by the scans. Next, the same procedure was performed in the y-direction from up and down. At the start and return positions the scanner has to accelerate and slow down. If the delay time is not properly adjusted with respect to the scan speed the first pulse at the start position and the last pulse at the return position will not hit the correct x-coordinates (in the case of an x direction scan). In Fig. 5, this caused a positioning error at the side wall of 30 μm and scratch-like distortions at the side wall prolonging down to the corner between wall and membrane. According to our simulations and experimental observations, LIPSS interrupted by scratches provide good conditions for pinhole growth (Fig. 4). The funnelling of laser power via the pinholes into the substrate causes thermal and mechanical stress locally influencing the ablation ratio. This can generate a stress induced surface structure on the whole membrane surface typical for the specific scan strategy. Figure 5 shows such a wave-like pattern on a SiC membrane produced by an xy-scanning strategy at constant polarization in the x-direction.

Rotating the scan pattern between the consecutive xy-scans by e.g. 15° as depicted in the insert of Fig. 6, changes the side wall damage pattern causing a different membrane stress profile and surface texture. The distortions are now distributed more evenly, fewer scratches and pinholes are generated at the corner of the membrane. Mechanical stress is consequently reduced resulting in a much smoother surface texture with respect to the standard xy-scanning method. However, also in the pattern rotation procedure the polarization direction was constant in the x-direction and we observed pinhole formation all over the membrane when the laser power was increased beyond 470 mW for SiC [10]. Spots of higher stress levels are generated and the membrane quality degenerates predominately at the corner zone where an increased ablation ratio can be observed. In one experiment we rotated the sample with about 30 revolutions per minute while performing a standard xy scanning procedure with constant polarization orientated in the machine's x-direction. By that means as well scanning direction and polarization are permanently changed relative to the ablation area. The laser power was set to such a high level that pinholes were created all over the membrane area when the ablation was made without sample rotation. The results with rotating sample can be seen in Fig. 7. The geometrical surface quality at the side wall, at the critical corner zone and the membrane itself was very smooth. All the pinholes were successfully supressed. Unfortunately, this method is not useful when an array of

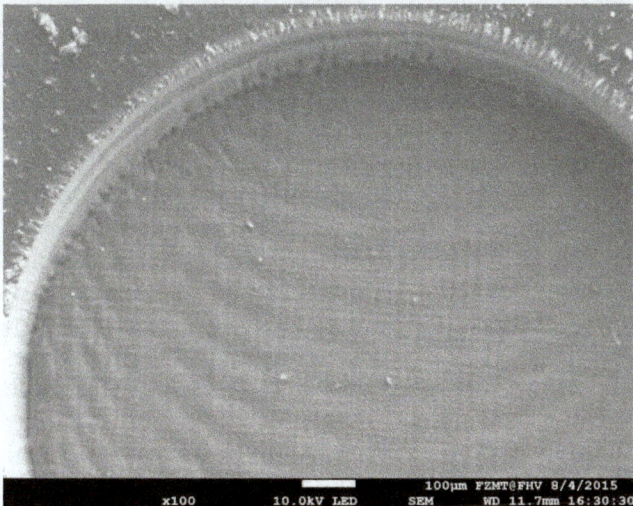

Fig. 5: Wrong delay time setting of the xy-scanner causes poor side wall quality. Scratches at the corner area promote pinhole formation, generates stress and wave patterns at the membrane.

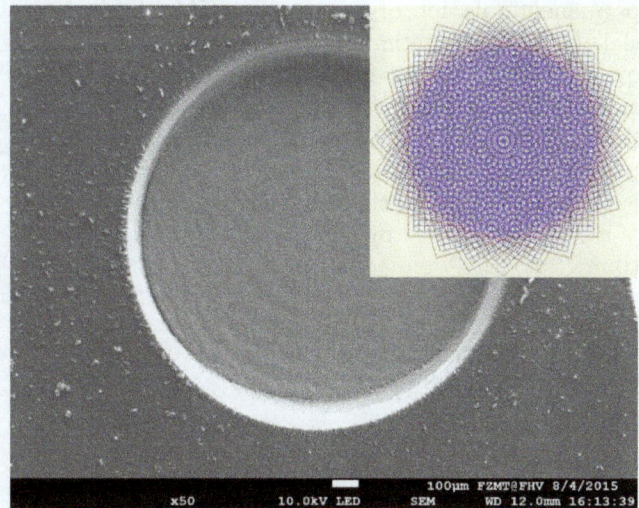

Fig. 6: Structural distortions at the sidewall and in the corner area of the SiC membrane are reduced by 15° rotation of the scan direction between consecutive scans (insert). This causes also a smoother surface pattern at the membrane with respect to a standard xy-scan as shown in Fig. 5.

membranes or contour pattern other than circular shall be produced. We obtained very similar high quality membranes without pinholes and a sufficiently reduced stress related surface texture when we combined the pattern rotation approach with a 90° polarization flip between consecutive scans.

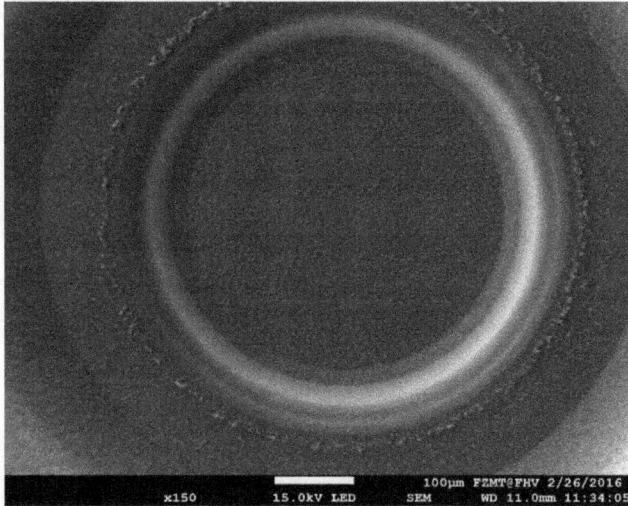

Fig. 7: A rotating sample (30 rpm) generates less scratches, smoother side walls and membranes without pinholes and surface wave structures. Pinhole formation is most effectively suppressed by simultaneous rotation of the polarization direction and the smoothening effect by the scan pattern rotation.

At the membrane the formation of pinholes was already suppressed by the 90° flipping of the polarization direction, while keeping the xy-scan pattern orientation unchanged. The most effective single measure to avoid pinhole formation is the flipping of the laser polarization direction between consecutive scans by 90° or the permanent rotation of the sample. This disables the slot waveguide function of the LIPSS and as a consequence interrupts the growth process of pinholes. It is not possible to achieve the same effect by changing the scan direction only as such an action does not suppress the waveguide performance of the LIPSS. But it is possible to smooth the irregularities at the side wall of a structure caused by delay issues of the laser on/off timing.

3.3. Back Reflection Related Damage

In Fig. 7, one can see four shallow lines on the bottom of the bore. This surface modification at the ablation side of the membrane was caused by an existing structure on the back side of the sample. On the SiC sample with an approximately 250 nm thick step structure on the back side of the diaphragm (the laser ablation is made at the front side), we realized that reflection and focusing effects contribute significantly to the quality

of ablated membranes. The insert in Fig. 8 depicts this structure on the backside of the sample. The corners and edges of a step are comparable to a lens of very short focal length. Light from back reflections is collected and accumulates in a small area inside the material close to the surface of the thin membrane causing the surface damage along the step edge at the front side (ablation side). This damage is shown enlarged in Fig. 8 and at the complete membrane area in Fig. 7. A significant circular distortion on the sample backside is also visible in the insert of Fig. 8. Its causes are the corresponding focusing and light funneling effect of small pinholes in the corner area of the membrane depicted in Fig. 5 and Fig. 6. We visualized these focusing and laser light funneling effect of pinholes in the lower right insert of Fig. 3 using a transparent polymer sample. During the membrane fabrication, the focused funneling of laser radiation causes a permanent stress and temperature related change of the refractive index in the zone with higher light intensity and we made this visible in a polarizing microscope. This type of damage is a focusing effect and not influenced by the light polarization, but using a fixed polarization ablation procedure, the distortions from focused back reflections become likely an activator for pinhole formation. Figure 8 shows the damage at the front side of the membrane where the regular ablation is performed. During ablation the sample was rotating, changing scanning track and polarization continuously. However, the pattern of the structure on the back side of the 68 μm thick membrane is clearly reproduced at the ablation plane. The direct laser radiation and the back reflected focused portion add up to a locally higher light intensity. Besides damage on the back side, this causes an increased ablation ratio at the front side along the reflection pattern depicted in Fig. 8. Pinhole formation starts at a certain threshold what we demonstrated in [10]. Consequently, pinholes at the membrane will first occur along this back reflection induced distortion pattern. Without flipping or rotation of the laser polarization this problem appears already at thicker membranes or lower laser power. The most vulnerable area for front and backside damage remains the corner zone of the membrane. The highest impact to improve the membrane quality and reduce the damage there was provided by the permanent change of the polarization direction by 90° flipping or constant rotating to suppress the waveguide function of the LIPSS. A better delay time adjustment additionally reduces sharp grooves and obstacles at the side wall of the membrane cavity. Also, other measures like the scan pattern rotation at 15° increments (Fig. 6) or permanent sample rotation are smoothing the side wall and corner of a membrane. All measures in combination gave the best result depicted in Fig. 7 but the main contribution was clearly by the elimination of the waveguide function of the LIPSS. In SiC we are

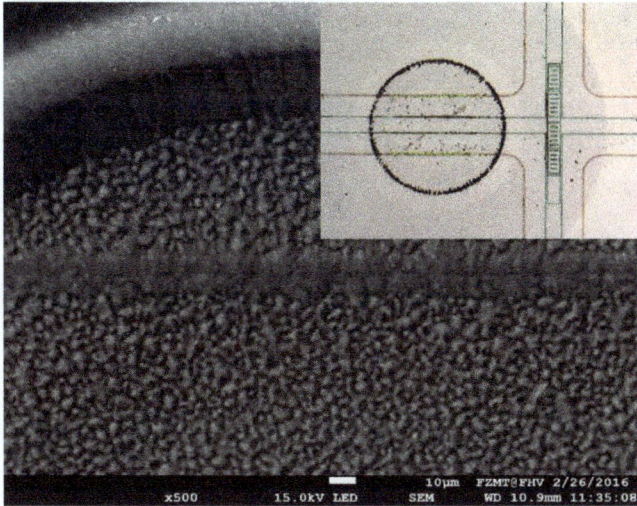

Fig. 8: A 250 nm step structure is the cause of damage at the back side and the front side of the membrane. Focused laser back reflections are too intense to be neglected.

currently able to produce diaphragms in the range of 70 μm to 80 μm using 90° polarization flipping after consecutive scans. At constant polarization direction, the limits are between 100 μm and 150 μm. Substrate purity and structural quality additionally influence the threshold for pinhole formation and consequently the minimum thickness achievable for a diaphragm. As soon as micrometer sized pinholes are formed the focused laser radiation funneling process into the substrate starts and within a few consecutive scans, the membrane is destroyed. In lower refractive index material (glass, sapphire, diamond) we observed LIPSS too, but there was no pinhole formation like in SiC or Si. We could produce much thinner diaphragms than in SiC, the thinnest one of 14 μm thickness was obtained in glass without polarization flipping. This is a further indication that the waveguide function is the main cause for damage, in low refractive index material the performance of slot waveguides is reduced.

In the attempt to produce a thin membrane in a 100 μm borosilicate glass sheet, we could maintain a constant ablation ratio of precisely 1.5 μm per scan without readjusting the focus position. The focus was set at the beginning of the structuring process onto the front surface (entrance side of the laser beam) of the sample. However, as soon as we obtained a diaphragm thickness of 12 μm the laser caused damage on the glass surface at the backside of the diaphragm (Fig. 9). In the very next scan the membranes cracked and disintegrated. One scan before glass particles started chipping off the surface the approximately 14 μm thick membrane was totally intact (insert of Fig. 9). Sharp edges of the chipped spots in combination with back reflections generate focusing effects of the same type as described for SiC and contribute to a higher ablation speed. In addition, the absorption of the laser

light is increased at the now rough backside and in fact, front and back side of the diaphragm is located at the same distance with respect to the focal plane (88 μm and 100 μm respectively) leading to a double side ablation. Mechanical stress generated by the laser ablation, focused radiation back reflection and double side ablation destroyed the membrane within one scan. The slot waveguide issue is not relevant in glass and polarization flipping has a lower impact on the membrane quality (only radiation losses via the side wall of the bore depicted in the upper insert of Fig. 3 are influenced by the orientation of the laser polarization).

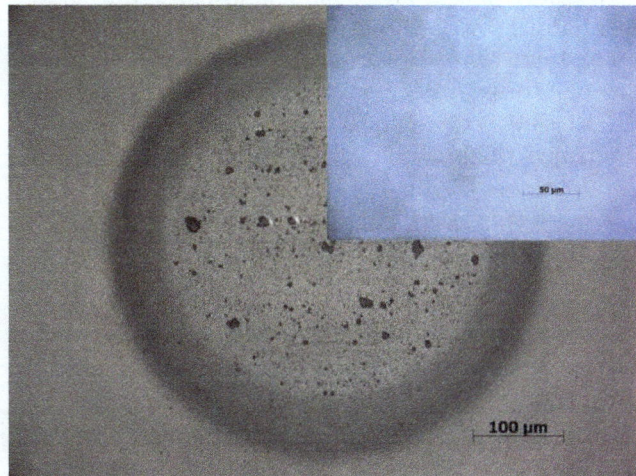

Fig. 9: Laser induced backside damage on a 12 μm glass membrane. The next consecutive scan destroyed the membrane. At a 14 μm thick membrane the glass surface is still intact (insert).

4. Conclusion

Surface homogeneity at the overall ablation area including side walls of the bore and backside surface of the membrane is essential to produce thin membranes especially in high refractive index materials (for example SiC). In such materials, LIPSS in combination with surface irregularities (grooves or scratches) promote the formation of pinholes. Pinholes funnel laser radiation into the bulk material causing structural damage and stress in the membrane. The edges of grooves, scratches and pinholes provide focusing structures with very short focal length. Hot spots of intense focused laser radiation can destroy a thin membrane. Especially the membrane corner area is vulnerable for pinhole formation. We developed an effective counter measure against pinhole formation which is the permanent change of the laser polarization direction after consecutive scans. In lower refractive index materials such as glass, sapphire or diamond we observed no threatening pinhole generation by laser ablation.

In these materials focussing effects of sharp structural edges or scratches are likely the main cause for structural damage in the membrane and measures to generate a smooth ablation area contribute most to the membrane quality. In high refractive materials LIPSS function very well as slot waveguides and consequently the ablation result is dominated by the laser polarization direction, focussing effects contribute secondary for the final optimization of the surface quality. This observation is in accordance with our simulation results for SiC. The calculated laser radiation distribution confirmed the experimental findings qualitatively that LIPSS in high refractive index material act as slot waveguides and promote pinhole formation at spatial disruptions. At locations of scratches and grooves interrupting the slot waveguides, the growth of pinholes is promoted by dislocated laser radiation prior confined in the slots. Changing the direction of polarization interrupts the growth cycle of the pinholes effectively and increases the quality of the fabricated membrane. We observed one effect, the transition of LIPSS into HSFL which requires both, a strong focusing effect at sharp edges and the effective slot waveguide function in high refractive index material. We demonstrated that femtosecond laser ablation is a powerful tool to generate a variety of structures for possible applications e.g. in MEMS sensor devices, microfluidic and biotechnology. Controlling the laser polarization is an effective method to influence the interaction of the laser radiation with the material used. For the laser radiation we demonstrated three different cases of main interacting mechanism responding very specifically to the substrate materials or geometric conditions. For good quality ablation results one has to consider micro focussing effects in glass, the slot waveguide function in SiC and the combination of focussing and waveguide function for the LIPSS to HSFL transition in SiC pinholes.

Acknowledgment

This work was supported in part by the Slovak - Austrian bilateral project SK-AT-0019-10, ICM OeAD SK05/2011 and Slovak Research and Development Agency under contract APVV-0450-10.

References

[1] VANKO, G., M. DRZIK, M. VALLO, T. LALINSKY, V. KUTIS, S. STANCIK, I. RYGER and A. BENCUROVA. AlGaN/GaN C-HEMT structures for dynamic stress detection. *Sensors and Actuators A: Physical.* 2011, vol. 172, iss. 1, pp. 98–102. ISSN 0924-4247. DOI: 10.1016/j.sna.2011.02.049.

[2] PEARTON, S. J., B. S. KANG, S. KIM, F. REN, B. P. GILA, C. R. ABERNATHY, L. LIN and S. N. G. CHU. GaN-based diodes and transistors for chemical, gas, biological and pressure sensing. *Journal of Physics: Condensed Matter.* 2004, vol. 16, no. 29, pp. 961–994. ISSN 0953-8984. DOI: 10.1088/0953-8984/16/29/R02.

[3] DESBIENS, J. G. and P. MASSON. ArF excimer laser micromachining of Pyrex, SiC and PZT for rapid prototyping of MEMS components. *Sensors and Actuators A: Physical.* 2007, vol. 136, iss. 2, pp. 554–563. ISSN 0924-4247. DOI: 10.1016/j.sna.2007.01.002.

[4] SAVRIAMA, G., V. JARRY, L. BARREAU, C. BOULMER-LEBORGNE and N. SEMMAR. Laser micro cutting of wide band gap materials. In: *9th International Symposium on High Power Laser Ablation.* Santa Fe: AIP, 2012, pp. 169–178. ISBN 978-073541068-8. DOI: 10.1063/1.4739871.

[5] ZOPPEL, S., M. FARSARI, R. MERZ, J. ZEHETNER, G. STANGL, G. A. REIDER, C. FOTAKIS. Laser micro machining of 3C-SiC single crystals. *Microelectronic Engineering.* 2006, vol. 83, iss. 4–9, pp. 1400–1402. ISSN 0167-9317. DOI: 10.1016/j.mee.2006.01.064.

[6] VANKO, G., P. HUDEK, J. ZEHETNER, J. DZUBA, P. CHOLEVA, V. KUTIS, M. VALLO, I. RYGER and T. LALINSKY. Bulk micromachining of SiC substrates for MEMS sensor applications. *Microelectronic Engineering.* 2013, vol. 110, iss. 1, pp. 260–264. ISSN 0167-9317. DOI: 10.1016/j.mee.2013.01.046.

[7] ZEHETNER, J., G. VANKO, P. CHOLEVA, J. DZUBA, I. RYGER and T. LALINSKY. Using of laser ablation technique in the processing technology of GaN/SiC based MEMS for extreme conditions. In: *10th International Conference on Advanced Semiconductor Devices & Microsystems.* Smolenice: IEEE, 2014, pp. 259–262. ISBN 978-1-4799-5474-2. DOI: 10.1109/ASDAM.2014.6998693.

[8] BONSE, J., J. KRUEGER, S. HOEHM and A. ROSENFELD. Femtosecond laser induced periodic surface structures. *Journal of Laser Applications.* 2012, vol. 24, iss. 4, pp. 042006-1–7. ISSN 1938-1387. DOI: 10.2351/1.4712658.

[9] SUN, M., U. EPPELT, S. RUSS, C. HARMANN, C. SIEBERT, J. ZHU and W. SCHULZ. Numerical analysis of laser ablation and damage in

glass with multiple picosecond laser pulses. *Optics Express*. 2013, vol. 21, iss. 7, pp. 7858–7867. ISSN 1094-4087. DOI: 10.1364/OE.21.007858.

[10] ZEHETNER, J., S. KRAUS, M. LUCKI, G. VANKO, J. DZUBA and T. LALINSKY. Manufacturing of membranes by laser ablation in SiC, sapphire, glass and ceramic for GaN/ferroelectric thin film MEMS and pressure sensors. *Microsystem Technologies*. 2016, vol. 4, iss. 22, pp. 1883–1892. ISSN 0946-7076. DOI: 10.1007/s00542-016-2887-2.

About Authors

Johann ZEHETNER was born in Amstetten, Austria in 1958. He received his Ph.D. from Technical University Vienna in 1992 for his research on diode-pumped and mode locked laser systems. He has acquired experience in research and industry at the Osaka National Research Institute, New Cosmos Electric Co. Ltd. Osaka, Emitec Japan in Tokyo and Trotec Laser Japan Ltd. At present he is senior researcher and lecturer at the Research Centre for Microtechnology, University of Applied Sciences Vorarlberg. His research interests include femtosecond laser ablation and materials processing for medical, fluidic and sensor technology.

Gabriel VANKO was born in Velky Krtis, Slovakia in 1981. He received the degree in engineering from Faculty of Electrical Engineering and Information Technology, Slovak Technical University, Bratislava in 2006. From 2006 to 2010 he was a Ph.D. student in Department of Microelectronic Structures at Institute of Electrical Engineering of Slovak Academy of Sciences, Bratislava. Since that he is the member of the same department and his work is focused on the technology and characterizationof the AlGaN/GaN based HEMT devices and their applications in the MEMS sensors for harsh environment.

Jaroslav DZUBA was born in Rimavska Sobota, Slovakia in 1987. He received his Ph.D. degree at the Faculty of Electrical Engineering and Information Technology, Slovak Technical University, Bratislava in 2014. Currently, he is a research fellow in the Department of Microelectronics and Sensors at Institute of Electrical Engineering, Slovak Academy of Sciences, Bratislava, Slovakia and his research interests are focused on Finite Element Method (FEM) simulations of MEMS sensors, thermo-mechanic characterization of MEMS, verification of simulated models in direct iteration with experimental results.

Tibor LALINSKY was born in Bratislava, Slovakia in 1951. He received M.Sc. degree from Slovak University of Technology, Bratislava in 1974, Ph.D. degree on GaAs FETs technology from the Institute of Electrical Engineering, Slovak Academy of Sciences, Bratislava in 1981 and Doctor of Science degree on MEMS device technology from the same Institute in 2007. Since 1985, he is the head of Department of Microelectronic Structures at the above Institute. Since 1995, he focus on design and developmentof III-V and III-N compound semiconductor M(N)EMS devices. He has contributed more than 150 papers in referred journals and conference proceedings.

19

Biometric Image Recognition Based on Optical Correlator

David SOLUS, Lubos OVSENIK, Jan TURAN,
Tomas IVANIGA, Jakub ORAVEC, Michal MARTON

Department of Electronics and Multimedia Communications, Faculty of Electrical Engineering and
Informatics, Technical University of Kosice, Park Komenskeho 13, 042 01 Kosice, Slovak Republic

david.solus@tuke.sk, lubos.ovsenik@tuke.sk, jan.turan@tuke.sk, tomas.ivaniga@tuke.sk,
jakub.oravec@tuke.sk, michal.marton@tuke.sk

Abstract. *The aim of this paper is to design a biometric images recognition system able to recognize biometric images-eye and DNA marker. The input scenes are processed by user-friendly software created in C# programming language and then are compared with reference images stored in database. In this system, Cambridge optical correlator is used as an image comparator based on similarity of images in the recognition phase.*

Keywords

Biometrical images, Cambridge optical correlator, JTC, MF, optical Fourier transform, recognition.

1. Introduction

Nowadays, identification of individual person based on biometrics ispopular research field. For example, it helps to tackle various criminal activity, to obtain related information about origin of the person and it is possible to use it almost anywhere. In order to identify someone, it is necessary to store the information about biometric features into a specific database. Such databases are among the most extensive databases of the world. They contain a huge amount of information that must be processed as quickly as possible. There are many ways and algorithms for faster and also more reliable searching. One way is the optical data processing.

At present, biometrics is an area that is developing rapidly. The development also involves greater demands on information processing speed and fast growing database size. Optical correlator has the potential to quickly search in these huge databases. The biometrics database is an example of such massive analysis databases of DNA data. Every individual has different DNA structure and thus different genes. Each of these genes may affect the health of an individual. This means that the presence or absence of a gene strongly influences the onset of the disease. It is necessary to search for a relationship between a gene and expression of phenotypes. The discovery of such associations is a complex process that passes through billions of database elements in thousands of iterations and it's expensive for electronic computers. A potential solution is optical correlator, which is appropriate in the biomedicine not only for finding a DNA data, but also in identification of fingerprints and facial features. The advantage of biometric facial features, unlike other methods of identifying people is their universality, uniqueness and by their comparison the individual can be clearly identified.

The aim of this paper is to explore the application of optical correlator for biometric image recognition. Specifically, it deals with the design of eye recognition system based on the iris of the eye and also design of DNA markers recognition system. Designed user-friendly software for image pre-processing and Cambridge optical Correlator (CC) are parts of these systems, which can recognize the image based on their similarity.

In Sec. 2. , the theoretical principles of optical data processing are described. Section 3. contains a basic description of the optical correlator and distribution of its components based on the method of processing information. The proposed systems for biometrical images recognition - eyes and DNA markers, are involved in Sec. 4. Experiments and results are presented in Sec. 5. and conclusion is summarized in Sec. 6.

2. Fourier Optics

The Fourier Transform (FT) is used for the analysis and processing of digital images and enables image description in other variables via integral transformation. Therefore, it can be used for various purposes, such as detection of edges or image compression. This method can also provide information about the location of edges in images, colour scheme and image through frequency and amplitude spectrum receiving assistance FT [1] and [2].

2.1. Introduction to Fourier Optics

Fourier optics provides a description of the propagation of light waves based on harmonic analysis and linear systems. Harmonic analysis is based on the expansion of an arbitrary function of time as a sum of harmonic functions of time with different frequencies and different amplitudes:

$$f(t) = \int_{-\infty}^{+\infty} \tilde{f}(v)\exp(-j2\pi vt)\mathrm{d}v, \qquad (1)$$

where the function $\tilde{f}(v)\exp(-j2\pi vt)$ is a harmonic function with frequency v and complex amplitude $\tilde{f}(v)$.

This approach is useful for the description of linear systems. If the response of the system to each harmonic function is known, the response to an arbitrary input function can be determined. The response of a linear system can be therefore determined by using harmonic analysis at the input and superposition at the output. Consider now an arbitrary function $f(x,y)$ of the two variables x and y. We suppose that x and y represent the spatial coordinates in a plane, $f(x,y)$ can also be written as a sum of harmonic functions of x and y:

$$\begin{aligned}f(x,y) = \\ = \int\int \tilde{f}(v_x, v_y)\exp\left[-j2\pi(v_x x + v_y x)\right]\mathrm{d}v_x\mathrm{d}v_y,\end{aligned} \qquad (2)$$

where $\tilde{f}(v_x, v_y)$ is the complex amplitude of the harmonic functions, v_x and v_y are the spatial frequencies in the x and y directions, respectively [1] and [3].

2.2. Optical Fourier Transform Implementation

One of the Fourier optics properties is that the lens produces amplitude distribution of light at its back focal plane. This distribution is proportional to the FT

amplitude in a plane with the displayed object (object plane).

Special case, advantageous for the study of the optical FT and the spatial filtering, is a two-lens system (Fig. 1). According to Fourier optics, such system constitutes a cascade of two subsystems implementing the FT. The first subsystem creates a direct FT in the focal plane of the first lens (also called the Fourier plane). A second subsystem (between the Fourier plane and the focal plane of the system) performs inverse FT and we can immediately deduce the inter changeability of running light beams. The result is a picture that is an exact replica of the object.

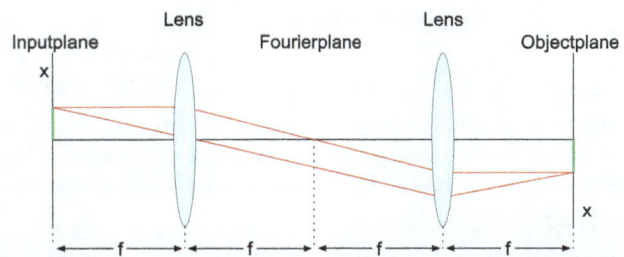

Fig. 1: 4-f two lens system.

This two-lens imaging system is called 4-f system or coherent optical processor. Its use as a spatial filter is clearly shown in the case, where the amplitude of the light wave $f(x,y)$ is a function of two variables, see Fig. 2. As follows from the above analogy - every "point" in the Fourier plane corresponds to one spatial frequency. Put in a Fourier plane suitable mask (i.e., impervious shield having a defined spaced slot = filter), which will block some spatial frequencies and leak others. We get in the image plane an image with an amplitude $F(x,y)$ which is a filtered version of the subject of the amplitude $f(x,y)$. This optical spatial frequency filtering can selectively suppress or completely eliminate some geometric features of the image, change the contrast etc. [1], [3] and [4].

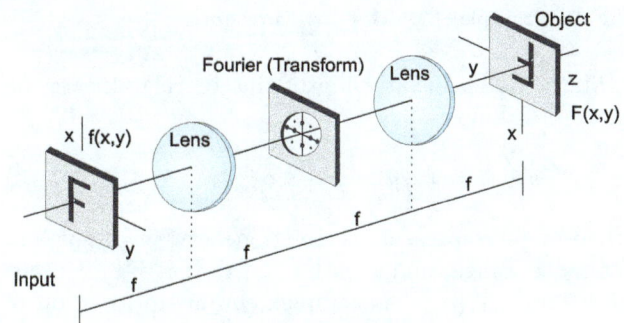

Fig. 2: 4-f system configuration for implementing the optical Fourier transform.

3. Optical Correlator

Optical correlator is a device executing comprehensive comparison with high speed. Optical correlator takes advantage of Fourier and Hilbert transforms. There are several types of correlators that achieve comparable results despite the fact that their information processing is different in each of these types. In general, the Optical correlators are divided into two basic types according to method of processing information - Optical correlators using Matched Filter (MF), and Optical correlators using Joint Transformation (JTC) [5].

3.1. Matched Filter

The basis of MF is Vanderlugt's filter which was invented by A. Vanderlugt in 1964. The MF is currently regarded as the most frequently used correlator. Its architecture is based on a 4-f system described above (Fig. 2).

The correlation of this type of correlator consists of two independent correlations. The first is the correlation of the input image $s(x, y)$, and the second is correlation of the reference image $r(x, y)$. Input scene is shown on Spatial Light Modulator (SLM). Then this scene undergoes FT. Reference image is used to create Matched filter which is applied as reference filter. After that, input scene is multiplied by reference filter. Finally, product of multiplication undergoes FT which produces plane with correlation peaks. The process of correlation is shown in Fig. 3.

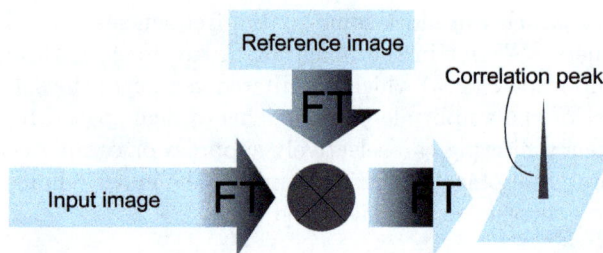

Fig. 3: Principle of Match Filter correlator.

Mathematical expression of this correlation can be written as follows:

$$s \oplus r = c(x,y) = F\left\{S(\alpha,\beta)R^*(\alpha,\beta)\right\}, \quad (3)$$

where $s \oplus r$ represents correlation between input and reference image, and F is FT. $S(\alpha,\beta)$ is FT of input image and $R^*(\alpha,\beta)$ is complex conjugate function of $R(\alpha,\beta)$.

The advantages of this method are high space - bandwidth product and extremely fast process time, but disadvantage is the need to create reference filter,

Matched filter from reference image, so also reference image is needed for start of the correlation process [5] and [6].

3.2. Joint Transform Correlator

JTC was invented in 1966 by Weaver and Goodman. The principle of the correlator based on the fact that reference filter is not required before starting correlation is shown in the Fig. 4. It can be seen that JTC has an input and reference image aligned and displayed together on the SLM. Subsequently, these images are transformed by Fourier transformation and a non-linear camera then captures the intensity distribution of transform to produce Joint Power Spectrum (JPS).

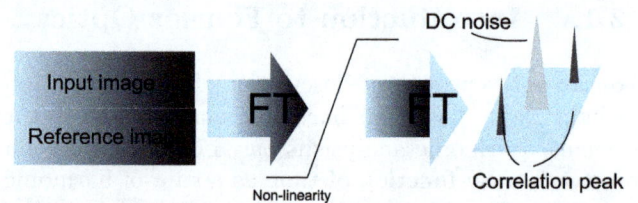

Fig. 4: Principle of Joint Transform Correlator.

JPS is binary or threshold processed and this processed image enters to transform process as input image of second FT. Output of this transform is the correlation plane that includes correlation peaks per match. Found matches are shown as highly localized intensities. Peak intensities provide a measure of similarity of compared images. Positions of peaks denote relative align of the images in the input scene.

Mathematical expression of the input scene can be written as follows:

$$f(x,y) = s(x,y) + r(x,y), \quad (4)$$

where $s(x,y)$ and $r(x,y)$ represents input and reference images shown in input plane. The input scene $f(x,y)$ is then transformed by FT and produces $F(\alpha,\beta)$.

$$F(\alpha,\beta) = F\left\{s(x,y)\right\} + F\left\{r(x,y)\right\} = \\ = S(\alpha,\beta) + R(\alpha,\beta). \quad (5)$$

Using next equations, the square of absolute value of transformed input scene can be produced.

$$J(\alpha,\beta) = |F(\alpha,\beta)|^2 = |S(\alpha,\beta) + R(\alpha,\beta)|^2 = \\ = |S(\alpha,\beta)|^2 + |R(\alpha,\beta)|^2 + 2|S(\alpha,\beta) \cdot R(\alpha,\beta)|. \quad (6)$$

The square of absolute value of complex function is equal to multiplication of function with its complex conjugated function and then we can write:

$$J(\alpha,\beta) = S(\alpha,\beta)S^*(\alpha,\beta) + R(\alpha,\beta)R^*(\alpha,\beta) + \\ + S(\alpha,\beta)R(\alpha,\beta) + R(\alpha,\beta)S^*(\alpha,\beta). \quad (7)$$

It is important to note that the distance of correlation peaks located in the correlation plane is half of the distance of the input and the reference image. This type of optical correlator is less widely used than MF [5], [6] and [7].

3.3. Cambridge Correlator

Optical correlator of Cambridge type (Fig. 5) was designed and constructed at the University of Cambridge. Bearers of the main idea of the technology are Nick New and Tim Wilkinson. It belongs to a group of JTC also referred to as 1/f Phase-Only Joint Transform Correlator. The expression 1/f indicates only a simple Optical Fourier Transform in both stages of the JTC process. This means using the same optics twice in the process of correlation. The term phase only refers to the images being displayed on the SLM in phase, which greatly improves system performance [5], [8] and [9].

1) Optical System

The basis of CC is a unique Fourier Transform Engine©. Compact and powerful data processing system, which is built on the principle of optical diffraction FT. Thanks to the unique distribution of optoelectronic components into a shape resembling the letter "W", the system can use their full potential. This minimizes the distance by which the electrical signals are transmitted, and also optimizes the time needed for processing. In Fig. 5, the main parts of CC are shown [5], [8], [10] and [11].

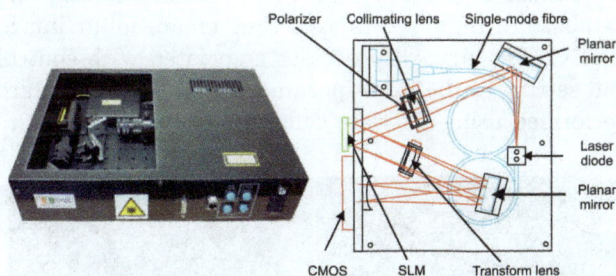

Fig. 5: Main parts of Cambridge optical correlator.

CC consists of several basic optical elements. The first part of elements contains a laser diode and single-mode fibre. Its dimensions are advantages in terms of integration into various devices that are used as CC. The collimating lenses are used to regulate the scattering of the light beam. Polarizer is intended to polarization of the light beam and then it is reflected by planar mirrors. The SLM is located in the system because of storing spatial modulation of the light beam. Currently, there are two kinds of SLM, the Electrically Addressable SLM (EASLM) and Optically Addressable SLM (OASLM). An analyser deals with the analysis of

the light beam before it enters the beam transformation lens. The transform lens is obtained by FT of the input scene, which is then processed by a CMOS sensor. Mentioned sensor includes an integrated circuit, consisting of several pixel sensor, while photodetector and active amplifiers are components of each pixel in the sensor. The camera type Leutron PicSight with Gigabit Ethernet interface is used in this device. Frame rate of camera is up to 200 Hz and it has VGA resolution(640 × 480 pixels) [5], [8], [9], [10] and [11].

2) Fourier Optics Experimenter

CC described above is coupled with simulation software, "Fourier Optics Experimenter" (FOE) for learning and easy understanding of Fourier optics, especially Optical Correlation. FOE can perform Optical Correlation based on CC between two or more images (videos), see Fig. 6.

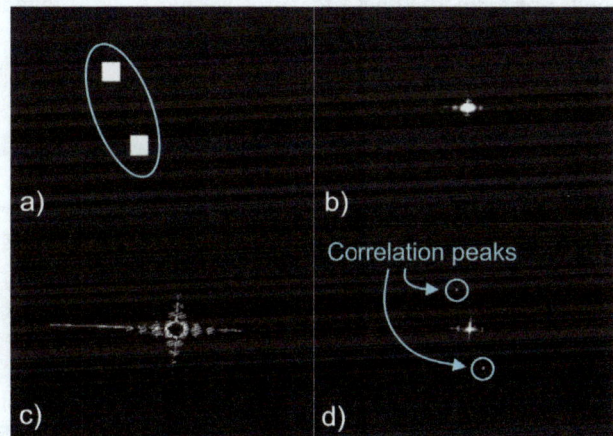

Fig. 6: Process of correlation between same images.

In Fig. 6 we can see process of Optical Correlation between two samples. The input scene (Fig. 6(a)) is created by images mentioned above. We can see JPS in Fig. 6(b) and this JPS binary or threshold processed in Fig. 6(c). The optical output is shown in Fig. 6(d). As it was mentioned above, the optical output contains highly localized intensities and their value might be in within range <0;255> where value "255" refers to total match and value "0" refers to mismatch. The advantage of JTC is ability to do more than one optical correlation at the same time. So, we can compare more images in one correlation process and save process time of recognizing any patterns [8], [9], [10] and [11].

4. Design of Biometric Image Recognition System

Design of systems for biometric image recognition is created because of faster and more efficient processing of the vast amount of information from a database. The term biometric image recognition, in that case expresses the comparison of two images of the eyes or comparing two DNA markers. At first, the image is pre-processed by user-friendly software created in the programming language C#. The obtained image is then compared with the reference image by CC.

4.1. Principles of the Eye Recognition System

There are several methods that can be used for the eye recognition. These methods include the identification according to iris and its structure, respectively. Iris is a pigmented internal structure of the eye, where you can notice a number of clearly visible external signs. These features include rings, scratches, spills, stains, and pigment. Each iris is specific thanks to the construction of tissues (muscle fibres, connective tissue, pigment), which are formed. These features should be stabilized in the first year of a person's life.

This method is unique in several respects. One of them is the fact that finding two identical irisesby random selection is as less likely asfinding two identical fingerprints. Another fact is that two identical twins can be distinguished from each other by the structure of their irises. Each individual has a unique iris and they are distinguishable from each other. And so from that point of view there is no other outside biometric characteristic of person which would be able to distinguish more people than the iris. While fingerprints can change (growth, burns, cuts, etc.) irises of the eye in humans throughout their life does not change (except for some diseases) [12] and [13].

Block diagram of the eye recognition system is shown in Fig. 7.

In the process of recognition, an image of the eye is scanned at the first stage. This image is then sent to the computer, where it is processed. Image processing consists of multiple processes that are loading, filtration, cropping and creating experimental databases of eyes.

The procedure is shown in Fig. 8. First, the image is:

- loaded Fig. 8(a),
- subsequently inverted Fig. 8(b),
- image is converted by a filter to grayscale Fig. 8(c),

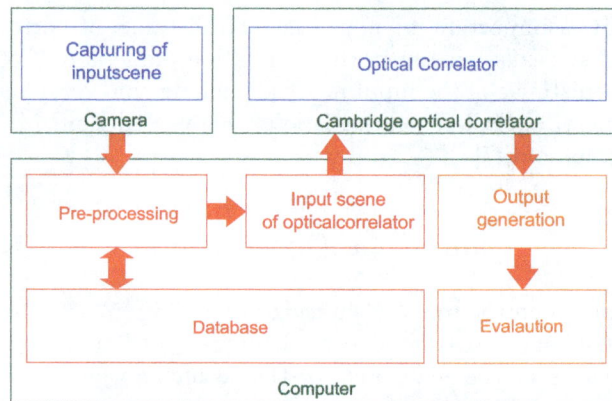

Fig. 7: Block diagram of the eye recognition system.

- pupil of the eye is found using a binary image Fig. 8(d),
- the result of the eye image pre-processing is cropped surrounding of the pupil Fig. 8(e).

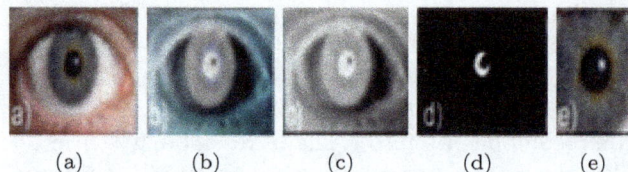

(a)   (b)   (c)   (d)   (e)

Fig. 8: Pre-process of image of eye.

The obtained image is stored in the experimental database and together with the image from the database of reference images they create input image for CC (Fig. 9). All processes associated with control and setting of the input parameters for CC images are performed using software called FOE.

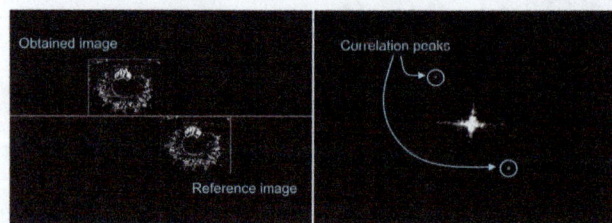

Fig. 9: Input image of CC.

The input image is adjusted by Roberts edge detector. These two images are then transformed and the FT forms their JPS. Then the cross-correlation is executed, which results in correlation peaks with certain values of intensity depending on the degree of similarity of the two images displayed in the correlation plane. Values of intensities obtained by using CC are recorded in a text file that is sent to a computer for evaluation.

4.2. Principles of the DNA Marker Recognition System

DNA profile, also called DNA fingerprinting, is a technique for determining the nucleotide sequence of certain areas of DNA that are unique. Each person has a unique fingerprint DNA. Unlike conventional fingerprint that occurs only on the fingertips and can be altered surgically, DNA fingerprinting is a same for every cell, tissue and organ of a person. It is the ideal way to distinguish the individual from others. DNA fingerprint can be obtained by DNA-extracting from cores of white blood cells or hair follicle cells at the root of the hair falling out or pulling out hair [14].

Applications of DNA fingerprints:

- individuality,

- establishing paternity, maternity,

- detection of hereditary diseases,

- the human race,

- a forensic point of view,

- sociology.

Block diagram of the DNA marker recognition system is similar to previous system (see Fig. 7), but preprocessing is different. The first step is loading an image of DNA markers from the database. Subsequently, a colour filtration is performed as it is shown in Fig. 10.

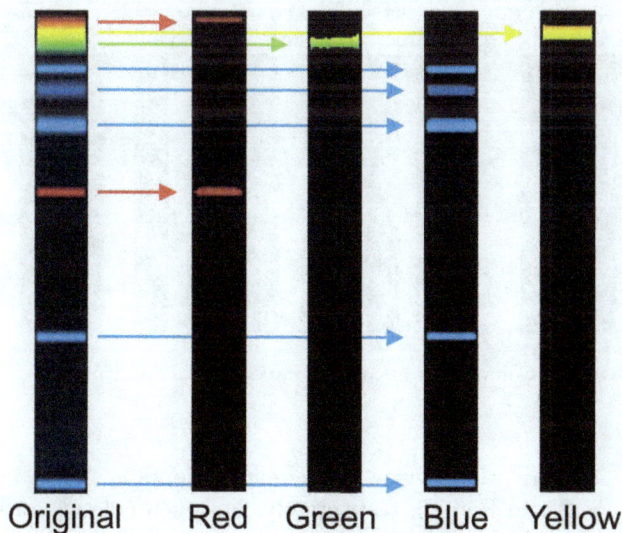

Fig. 10: Colour filtration of DNA marker.

Colour filtration separates colours that are located in the DNA markers (red, blue, green, and yellow) from each other. It is possible to set various different colour combinations depending on the type of DNA markers. From these components of DNA markers, a new

database is created. The next step is creation of an input image for the CC from the acquired and reference images. Further steps are in principle the same as it was in the previous subsection in the eye recognition. JPS is created from the input image by FT. The result contains correlation peaks with certain values of brightness depending on the similarity of the two images on the input scene. These values are then saved to a text file which is transferred to a computer for evaluation of these correlation peaks.

5. Experiments and Results

Verification of system for biometric images recognition consists of two parts-verification of eye recognition system and DNA marker recognition system. Both systems contain user-friendly software created by program language C# for input images pre-processing. Pre-processed images are verified by CC. The verification also includes software that enables usage of the correlator-FOE, which provides access to a variety of manual settings of CC.

5.1. Experiments of Eye Recognition System

For verification of system for the eye recognition, an experimental image database of eight different eyes is created. Each image of the eye is pre-processed by created software. The results of this program are eight images of iris of eye (see Fig. 11).

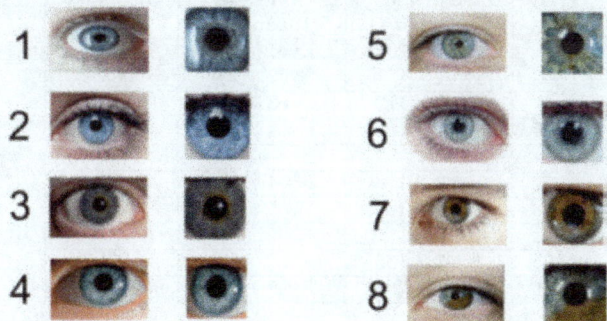

Fig. 11: Experimental image database of eyes.

Pairs of irises are then hundred times compared by FOE and CC. Images of two irises are inserted to input plane of the FOE for determining the similarity, which is shown in Fig. 12.

The first input plane a) contains two identical images; the second input plane b) contains different images. Match of the input images can be detected from the resulting brightness values of the correlation peaks. These values are then entered in a text file.

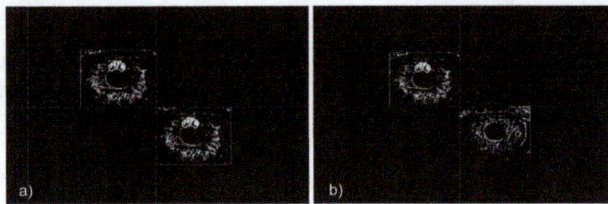

Fig. 12: Input plane of a) same images, b) different images.

A complete overview of the experiments is shown in Tab. 1 and Fig. 13. The average percentages of intensity of compared samples are calculated by the following equation:

$$I_{ID}(\%) = \frac{100 \cdot \bar{I}_{ID}}{255}, \qquad (8)$$

where I_{ID} is average value of intensity of two compared samples. This value is calculated by these equations:

$$\bar{I}_{N \cdot sample} = \frac{\sum\limits_{n=1}^{100} I_{N \cdot n \cdot sample}}{100}, \qquad (9)$$

$$\bar{I}_{ID} = \frac{\bar{I}_{1 \cdot sample} + \bar{I}_{2 \cdot sample}}{2}, \qquad (10)$$

where $I_{N \cdot n \cdot sample}$ isintensity value of sample and $\bar{I}_{N \cdot sample}$ is average intensity of this sample.The threshold value is set to 80%. When the average percentage intensity value of compared samples reaches at least this threshold, the samples are identical.

Tab. 1: Overview of the experiments.

ID	1. Sample / Average Intensity	2. Sample / Average Intensity	Average Intensity (%)
1	Eye1 / 236	Eye1 / 236	93
2	Eye1 / 151	Eye2 / 171	63
3	Eye1 / 172	Eye3 / 160	65
4	Eye1 / 140	Eye4 / 161	59
...	...	...	...
17	Eye3 / 176	Eye4 / 138	62
18	Eye3 / 165	Eye5 / 146	61
...	...	...	...
35	Eye7 / 135	Eye8 / 156	57
36	Eye8 / 235	Eye8 / 235	92

5.2. Experiments of DNA Marker Recognition System

Verification of systems for DNA markers recognition consists of experimental verification of the DNA markers (Fig. 14).

The basis of verification is filtration of DNA markers colour components and subsequently comparing the colour components of one marker with second marker (hundred times), see Fig. 15.

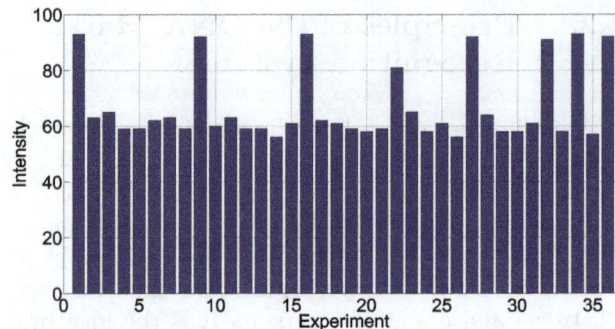

Fig. 13: Overview of the experiments.

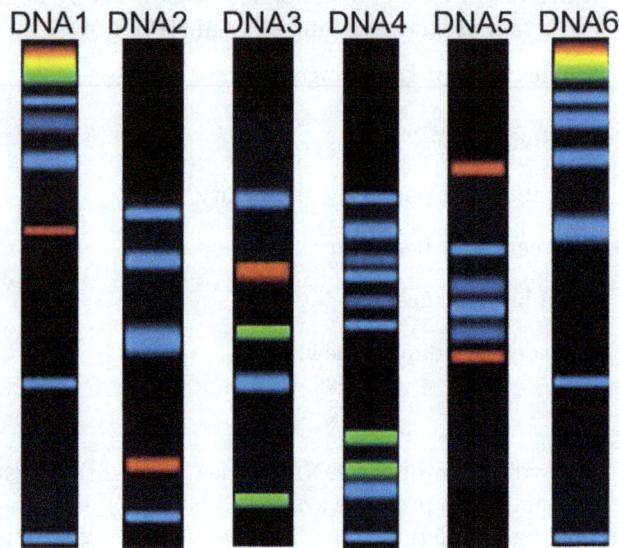

Fig. 14: Overview of the experiments.

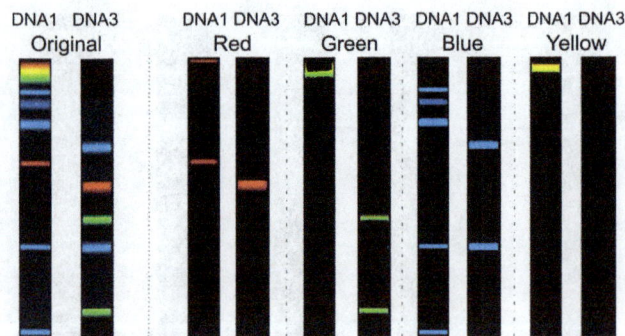

Fig. 15: Comparison of colour components of DNA markers.

A complete overview of the experiments is shown in Tab. 2 and Fig. 16, respectively. For each colour component of DNA marker, the average value of intensity is calculated by Eq. (10). A positive evaluation is based on the fact that intensity value of each colour components is above than set value of 220. In this case, the markers are considered as identical. If at least one colour components intensity value is below a set value, then compared DNA markers are considered to be different.

Tab. 2: Overview of the experiments.

ID	1. Sample	2. Sample	Intensity (R)	Intensity (G)	Intensity (B)	Intensity (Y)
1	Marker1	Marker1	235	234	235	233
2	Marker1	Marker2	110	0	134	0
3	Marker1	Marker3	118	67	71	0
4	Marker1	Marker4	0	43	74	0
...	...	...	...	...	...	...
13	Marker3	Marker4	0	52	39	255
14	Marker3	Marker5	98	0	57	255
...	...	...	...	...	...	...
20	Marker5	Marker6	46	0	73	0
21	Marker6	Marker6	231	234	233	232

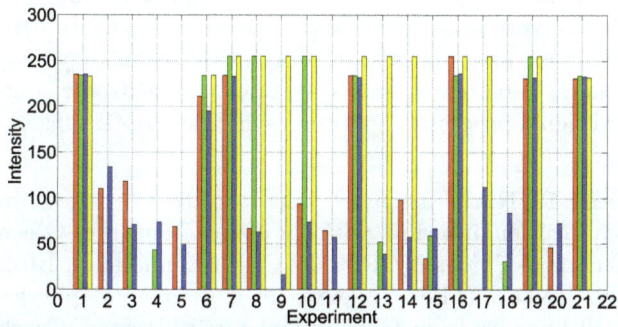

Fig. 16: Overview of the experiments.

6.   Conclusion

In this paper, system for recognition of biometric images has been described. This system contains of two systems - eye recognition system and DNA marker recognition system. Both systems use user-friendly software created in programming language C# to extract and pre-process input images. These images are compared with reference stored in experimental database by CC. Software FOE is used for comparing, which is inseparable component of CC. The result of comparison was intensity of correlation peaks. The average values of intensities and percentage match of compared images were obtained.

Eight images of iris of the eye were created and then compared. 3600 experiments have been done and the functionality of eye recognition system has been verified. Intensity value of the same images was always greater than 200 (80 %) and for different images, this value was not greater than 166 (65 %).

The system for DNA marker recognition is based on comparison of DNA marker colour components. 2100 experiments have been done and the functionality of DNA marker recognition system has been verified. The same images had intensity value of each colour component higher than 220. In the case, when both DNA markers did not contain a specific colour component, these components were considered as identical.

The system for biometric images recognition - eye and DNA marker, was verified experimentally. Optical image processing done by CC has some potential in biometrics. Also, CC could be implemented in many other areas where the speed and accuracy is important. The system can be extended to the fingerprint and palm recognition system in the future.

Acknowledgment

This publication arose thanks to the support of the Operational Programme Research and development for the project "Centre of Information and Communication Technologies for Knowledge Systems (ITMS code 26220120020), co-financed by the European Regional Development Fund".

References

[1] DUBOIS, A. *Fourier Optics*. Paris: ParisTech. September, 2014.

[2] GOODMAN, J. W. *Introduction to Fourier Optics*. 1st ed. Colorado: Roberts and Company, 2005. ISBN 0-9747077-2-4.

[3] ANGELSKY, O. V. *Optical Correlation Techniques and Applications*. 1st ed. Washington: SPIE Publications, 2007. ISBN 978-0819465344.

[4] LEHAR, S. An Intuitive Explanation of Fourier Theory. *Boston University* [online]. 2017. Available at: http://cns-alumni.bu.edu/~slehar/fourier/fourier.html.

[5] http://cns-alumni.bu.edu/ Optical processing products. *Cambridge Correlators* [online]. 2014. Available at: http://www.cambridgecorrelators.com.

[6] AMBS, P. Optical Computing: A 60-Year Adventure. *Advances in Optical Technologies*. 2010, vol. 2010, Article ID 372652, pp. 1–15. ISSN 1687-6393. DOI: 10.1155/2010/372652.

[7] LAYTON, A. and R. MARSH. Object distance detection using a joint transform correlator. In: *IEEE International Conference on Electro Information Technology*. Grand Forks: IEEE, 2016, pp. 707–709. ISBN 978-1-4673-9985-2. DOI: 10.1109/EIT.2016.7535326.

[8] Cambridge Correlators. *Fourier Optics Experimenter*. User Guide.

[9] HARASTHY, T., L. OVSENIK and J. TURAN. Current summary of the practical using of optical correlators. *Acta Electrotechnica et Informatica*. 2012, vol. 12, no. 4, pp. 30–38. ISSN 1335-8243. DOI: 10.2478/v10198-012-0042-2.

[10] TURAN, J., L. OVSENIK and T. HARASTHY. Traffic Sign Recognition System based onCambridge Correlator Image Comparator. *Carpathian Journal of Electronic and Computer Engineering*. 2012, vol. 5, no. 1, pp. 127–132. ISSN 1844-9689.

[11] SOLUS, D. Optical Correlator in Image and Video Processing Systems. In: *15th Scientific Conference of Young Researchers*. Kosice: TU, 2015, pp. 96–99. ISBN 978-80-553-2130-1.

[12] SOLUS, D., L. OVSENIK and J. TURAN. Signal processing - object detection methods with usage of optical correlator. In: *Radioelektronika 2016*. Danvers: IEEE, 2016, pp. 315–318. ISBN 978-1-5090-1673-0.

[13] SOKAL, R. R. and F. J. ROHLF. *The Principles and Practice of Statistics in Biological Research*. 3rd ed. New York: W.H. Freeman and Company, 2012. ISBN 978-0-7167-8604-4.

[14] KIAH, H. M., G. J. PULEO and O. MILENKOVIC. Codes for DNA Sequence Profiles. *IEEE Transactions on Information Theory*. 2016, vol. 62, iss. 6, pp. 3125–3146. ISSN 0018-9448. DOI: 10.1109/TIT.2016.2555321.

About Authors

David SOLUS received Ing. (M.Sc.) degree in 2014 at Department of Electronics and Multimedia Telecommunications, Faculty of Electrical Engineering and Informatics of Technical University of Kosice. Since September 2014 he has been at University of Technology, Kosice as Ph.D. student. His research interests include Optical correlator in image and video processing systems.

Lubos OVSENIK received Ing. (M.Sc.) degree in radioelectronics from the University of Technology, Kosice, in 1990. He received Ph.D. Degree in electronics from University of Technology, Kosice, Slovakia, in 2002. Since February 1997, he has been at the University of Technology, Kosice as Associate Professor for electronics and information technology. His general research interests include optoelectronic, digital signal processing, photonics and fiber optic sensors.

Jan TURAN received Ing. (M.Sc.) degree in physical engineering with honours from the Czech Technical University, Prague, Czech Republic, in 1974, and RNDr. (M.Sc.) degree in experimental physics with honours from Charles University, Prague, Czech Republic, in 1980. He received a C.Sc. (Ph.D.) and Dr.Sc. degrees in radioelectronics from University of Technology, Kosice, Slovakia, in 1983, and 1992, respectively. Since March 1979, he has been at the University of Technology, Kosice as Professor for electronics and information technology. His research interests include digital signal processing and fibre optics, communication and sensing.

Tomas IVANIGA is currently Ph.D. student at University of Technology, Kosice. His research interests include Mitigation of degradation mechanism in all optical WDM systems.

Jakub ORAVEC is currently Ph.D. student at University of Technology, Kosice. His research interests include steganography and digital image processing.

Michal MARTON is currently Ph.D. student at University of Technology, Kosice. His research interests include optical fiber gyroscopic systems and optical communication systems.

20

Non-Invasive Fiber-Optic Biomedical Sensor for Basic Vital Sign Monitoring

Jan NEDOMA[1], Marcel FAJKUS[1], Radek MARTINEK[2], Vladimir VASINEK[1]

[1]Department of Telecommunications, Faculty of Electrical Engineering and Computer Science, VSB–Technical University of Ostrava, 17. listopadu 15/2172, 708 33 Ostrava, Czech Republic
[2]Department of Cybernetics and Biomedical Engineering, Faculty of Electrical Engineering and Computer Science, VSB–Technical University of Ostrava, 17. listopadu 15/2172, 708 33 Ostrava, Czech Republic

jan.nedoma@vsb.cz, marcel.fajkus@vsb.cz, radek.martinek@vsb.cz, vladimir.vasinek@vsb.cz

Abstract. *This article focuses on the functionality verification of a novel non-invasive fibre-optic sensor monitoring basic vital signs such as Respiratory Rate (RR), Heart Rate (HR) and Body Temperature (BT). The integration of three sensors in one unit is a unique solution patented by our research team. The integrated sensor is based on two Fiber Bragg Gratings (FBGs) encapsulated inside an inert polymer (non-reactive to human skin) called PolyDiMethylSiloxane (PDMS). The PDMS is beginning to find widespread applications in the biomedical field due to its desirable properties, especially its immunity to ElectroMagnetic Interference (EMI). The integrated sensor's functionality was verified by carrying out a series of laboratory experiments in 10 volunteer subjects after giving them a written informed consent. The Bland-Altman statistical analysis produced satisfactory accuracy for the respiratory and heart rate measurements and their respective reference signals in all test subjects. A total relative error of 0.31 % was determined for body temperature measurements. The main contribution of this article is a proof-of-concept of a novel noninvasive fiber-optic sensor which could be used for basic vital sign monitoring. This sensor offers a potential to enhance and improve the comfort level of patients in hospitals and clinics and can even be considered for use in Magnetic Resonance Imaging (MRI) environments.*

Keywords

Basic vital sign monitoring, biomedical instrumentation, body temperature, heart rate, fiber Bragg gratings, fiber-optic sensor, noninvasive, polydimethylsiloxane, respiration rate.

1. Introduction

The emerging trends in biomedical instrumentation development clearly show that the immediate future of vital sign monitoring favors the utilization of sophisticated diagnostic tools and devices which integrate more diagnostic parameters into one universal device. More precisely stated: the integration of a variety of basic sensors into one measurement unit with the purpose to increase safety as well as comfort levels of patients is of great research interest. This article reports our contribution to the field and shares our recent research findings on biomedical applications of fiber-optic sensors. For a review of this emerging field please see articles [1] and [2]. Our research team has developed a patented fiber-optic sensor that allows monitoring of mechanical vibrations of the human body evoked by life activities such as breathing and cardiac rhythms as well as body temperature [3].

A number of research articles which used one or more FBGs, have presented results of measurements of respiration rate, heart rate or both simultaneously [4], [5], [6], [7], [8] and [9]. For example, Chethana et al. have presented very interesting results based on the design and construction of an FBG-based sensor attached to the patient's chest that enables respiratory and heart rate monitoring [4]. In this design, it is essential to pay special attention to the tension of the optical fiber so that adequate sensitivity is achieved. The detailed design of this FBG-based sensor for monitoring respiratory and heart rates in human subjects is presented in [5]. In this work, the sensor consists of an FBG embedded inside a single-mode optical fiber that operates with the wavelength of approximately 1550 nm with a maximum relative measurement error of 12 %. The experimental results reported in article [6] describe an FBG-based sensor prototype designed for monitoring the respiratory rate. In this work, the FBG sensor is encapsulated inside a PDMS enclosure. The sensor assembly is mounted on an elastic contact strap that encircles the patient's chest. The tension in the chest caused by breathing leads to a spectral shift of the reflected light from the FBG. In [7], Dziuda et al. present results obtained from monitoring the respiration and heart rates of a patient in a Magnetic Resonance Imaging (MRI) environment using a fiber-optic FBG-based sensor. This sensor was proposed by its developers to specifically acquire BallistoCardioGraphic (BCG) signals from a patient positioned inside a dynamic magnetic field. The authors in [8] report a fiber-optic-based smart textile sensor for respiratory rate monitoring capable of operating in MRI environments. In this work, two FBGs placed on the thorax enable the conversion of chest wall movements during respiration to measurable signals. Interestingly, article [9] focuses on an MRI-friendly fiber-optic sensor for monitoring the heart and respiration rates simultaneously. In this design, the sensor employs a Plexiglas springboard to which an FBG is attached to convert the patient's body movements to mechanical strain while lying on the springboard.

Current research by our team substantiates that our novel sensor based on Mach-Zehnder interferometers along with adaptive signal processing methods can find applications in a variety of fields including noninvasive monitoring of basic vital signs in obstetrics and gynecology (uterine contractions, fetal heart rate monitoring and others) [10], [11], [12] and [13].

Recent literature in the field of noninvasive maternal and fetal vital sign monitoring provides ample evidence that the integration of several diagnostic measures in one all-purpose instrument or sensor, although an appealing concept, is facing many challenges; consequently, there is a vast need for improvement and research in this area. Our sensor, which allows the measurement of body temperature in addition to monitoring the heart and/or respirations rates as described in the articles above, offers an innovative solution in noninvasive, basic vital sign monitoring and is thus a step forward.

2. Methods

FBGs are currently the most frequently used single-point fiber-optic sensors due to their desirable properties such as: small size with high tensile strength, immunity to electromagnetic interference, and minimal aging effect with regard to the components from which they are assembled [14], [15] and [16]. Basically, FBGs function by means of the periodical change of the refractive index in their optical core, selectively filtering certain wavelengths that are reflected back, while allowing the remaining part of the spectrum to pass through. All the reflected light signals combine coherently to form one large reflection at a particular wavelength when the grating period is approximately 1/2 of the input light's wavelength. This is referred to as the Bragg Condition, and the wavelength at which this reflection occurs is called the Bragg Wavelength. An example of the FBG structure and its working principle is shown in Fig. 1. As FBGs are sensitive to strain and temperature changes, they are suitable for many biomedical measurements. Single-point FBG sensors can be connected together in cascade, thereby producing a multi-point sensor within one optical fiber. The easiest method for enhancing the resolution of individual sensors is to use wavelength-division multiplexing. We can integrate tens of sensors within the wavelength-division multiplex, whose capacity is given by the type of a measured value, the size of measuring ranges and the size of the protection zone, see [17] and [18].

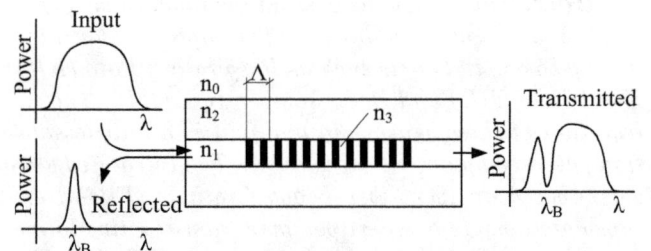

Fig. 1: An example of FBGs structure and working principle.

The size of Fiber Bragg wavelength is given by the following relationship:

$$\lambda_B = 2n_{eff}\Lambda, \tag{1}$$

where n_{eff} is the effective refractive index of the used optical fiber with Bragg grating, and Λ is the period of changes in the refractive index of the core of the used optical fiber. Deformation and temperature dependence are given by the central Fiber Bragg Wavelength and parameter values (where λ_B is the Bragg wavelength, $\Delta\lambda_B$ is the shift of the Bragg wavelength, $\Delta\varepsilon$ is the change of deformation and ΔT represents a change in temperature). To determine individual sensitivities, normalized deformation and temperature coefficients are used [19]. The normalized deformation coefficient is given by the following relationship:

$$\frac{1}{\lambda_B}\frac{\Delta\lambda_B}{\Delta\varepsilon} = 0.78 \cdot 10^{-6} \ \mu\text{strain}^{-1}, \tag{2}$$

and the normalized temperature coefficient is given by the following relationship:

$$\frac{1}{\lambda_B}\frac{\Delta\lambda_B}{\Delta T} = 6.678 \cdot 10^{-6} \ °\text{C}^{-1}. \tag{3}$$

3. Results

Our sensor with the following dimensions: 70 mm (length) × 40 mm (width) × 4 mm (thickness), and the weight of 50 grams, is used for measuring the respiratory rate, heart rate and body temperature in the human body. It is based on two FBGs which are encapsulated inside a polydimethylsiloxane polymer [20] and [21], see (Fig. 2). Our results reported elsewhere [22] indicate that this type of encapsulation does not affect the structure of the FBG.

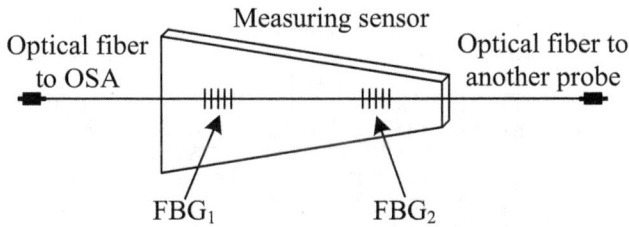

Fig. 2: The design of our noninvasive basic vital sign sensor.

Measurements were carried out in ten volunteer subjects of both sexes (5 men: M1–M5, and 5 women: F1–F5), after obtaining their written informed consents, in a research laboratory with the temperature of 24 °C. The subjects were between 21 and 47 years of age, their heights were between 156 and 197 cm, and their weights were between 47 and 108 kg. No significant differences were found in the quality of the recorded data based on the subjects' age, height, and weight. The sensor probe was placed on the chest (around the pulmonic area) and fixed by a contact elastic strap, see Fig. 3. The subjects were tested in the supine position in a relaxed state.

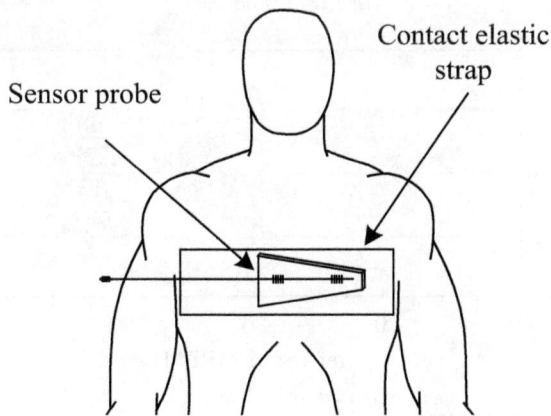

Fig. 3: Implementation sensor with contact elastic strap on human body.

An Optical Interrogator (OI) system developed by our team was used to further process the sensed vital signs acquired by the sensor [3]. The OI system is composed of a wideband spectral light source from a Light Emitting Diode (LED) with the central wavelength of 1550 nm and the output power of 1 mW. Furthermore, it is composed of an Optical Spectrum Analyzer (OSA) unit using the sampling frequency of 250 Hz, an optical circulator, a Digital Signal Processing (DSP) Unit, and an Electronic Control Unit (ECU) for each individual optical element. The vital sign information (comprised of heart rate, respiratory rate, and body temperature) was displayed in a graphical user interface in an application created in LabVieW (2015, National Instruments, Austin, Texas, USA) by our research team.

The heart rate (expressed in beats per minute BPM) and respiratory rate (expressed in respiration per minute RPM) were obtained by a spectral evaluation of the measured signals. Based on signal peak detection and the calculation of time intervals between these peaks, the heart and respiratory rates were determined. The reference ECG and respiratory signals were acquired by using a real-time monitoring system with standard bioelectrodes, a respiratory sensing module fixed to a subject's chest along with a real-time ECG and respiratory signal monitoring system based on a virtual instrumentation system (NI ELVIS, II Series, National Instruments, Austin, TX, USA). The body temperature (expressed in degrees Celsius °C) was obtained by the mathematical relationships shown in Eq. (4). A digital thermometer (Greisinger, Prague, Czech Republic) was used to acquire the reference temperature signal and recordings.

To determine the body temperature, we used two FBGs with different temperature and deformation sensitivities. Different sensitivities were within the proposed sensor range given by a specific form and shape of encapsulation. It is established that if the sensor is affected by deformation or temperature, the size of both of these impacts could be determined by using the following relationship [23]:

$$
\begin{pmatrix} \Delta T \\ \Delta \varepsilon \end{pmatrix} = \frac{1}{K_{1T}K_{2\varepsilon} + K_{2T}K_{1\varepsilon}} \cdots
$$
$$
\cdots \begin{pmatrix} K_{2\varepsilon} & -K_{1\varepsilon} \\ -K_{2T} & K_{1T} \end{pmatrix} \begin{pmatrix} \Delta \lambda_{B1} \\ \Delta \lambda_{B2} \end{pmatrix},
\tag{4}
$$

where $\Delta \varepsilon$ is deformation, ΔT is the temperature change, $K_{n\varepsilon}$ is the deformation coefficient, and K_{nT} is the temperature coefficient belonging to the first or second FBG. $\Delta \lambda_{B1}$ and $\Delta \lambda_{B2}$ represent the shift of the Bragg Wavelength for the first FBG_1 and the second FBG_2, respectively.

Figure 4 shows a 30-second long record of changes in Fiber Bragg Wavelength during the measurement of breathing activity in test subject M1 as an example.

For comparison, Fig. 5 shows a 10-second long record of changes in the Fiber Bragg Wavelength during the measurement of heartbeat activity in a female test subject (F4).

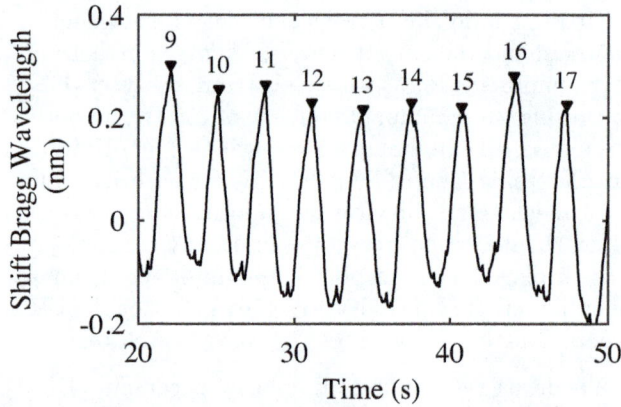

Fig. 4: A 30-second record of changes in the Fiber Bragg Wavelength during the measurement of breathing activity in a male test subject (M1).

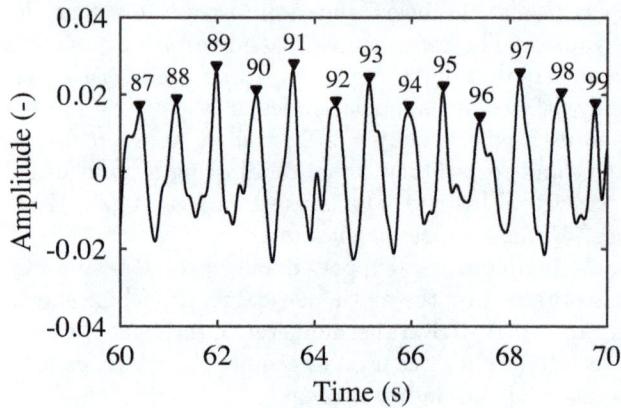

Fig. 5: A 10-second record of changes in the Fiber Bragg Wavelength during the measurement of heartbeat activity in a female subject (F4).

Figure 6 shows a 60-second-long record of temperature measurement in test subject M1.

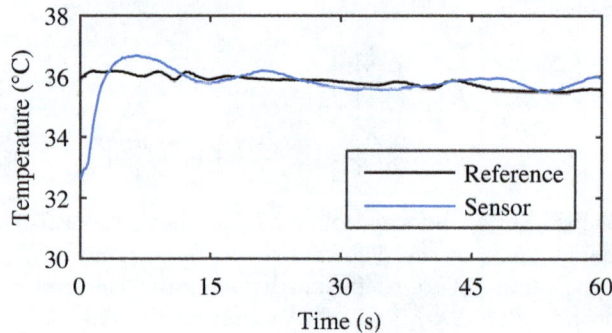

Fig. 6: A 60-second-long record of measurement and determination of temperature in test subject M1.

To compare the differences between the reference signals and those acquired from our novel sensor, the Bland-Altman plot was utilized [24]. The differences between the sensor and the reference traces, $x_1 - x_2$, are plotted against the average, $(x_1 + x_2)/2$. The reproducibility is considered to be good if 95 % of the

results lie within the ± 1.96 SD (Standard Deviation) range.

The key experimental results for the heart and respiratory rate measurements are summarized in Tab. 1. In the case of heart rate measurements for the entire data set, 95.34 % (95.45 % for men and 95.24 % for women) of the values lie within the ± 1.96 SD range for the HR determination and no significant differences were found between observed individuals (Fig. 7).

Tab. 1: Summary of respiratory and heart rates measurements.

		RR		HR	
Sub.	Rec. time (s)	NoS sensor	Samples in ± 1.96 SD (%)	NoS sensor	Samples in ± 1.96 SD (%)
M1	720	204	94.61	816	95.47
M2	530	133	94.38	636	95.60
M3	680	197	95.53	884	94.57
M4	540	141	93.80	675	95.56
M5	440	122	95.18	535	96.07
F1	340	71	94.57	459	96.08
F2	450	135	95.56	630	95.71
F3	420	106	96.33	553	94.39
F4	490	139	94.54	760	96.05
F5	530	133	95.59	616	93.99

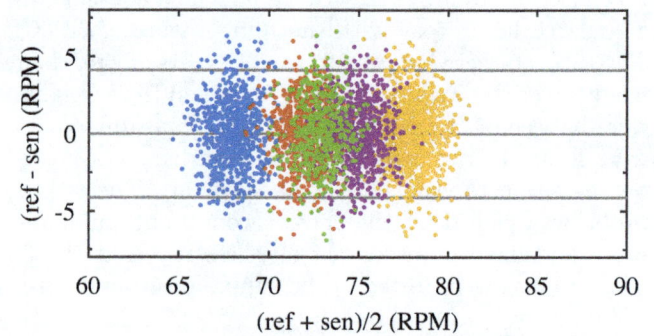

(a) Five tested men.

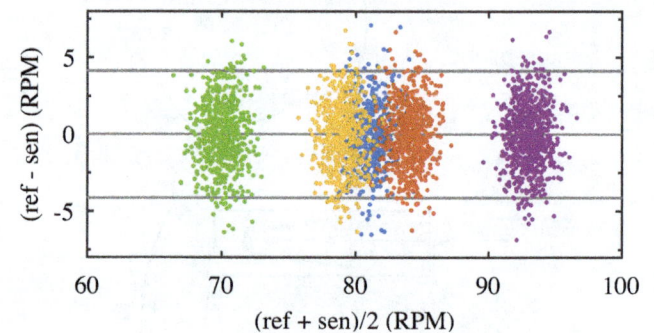

(b) Five tested women.

Fig. 7: Statistical analysis of heart rate using the Bland-Altman plot.

In the case of respiratory rate measurements for the entire data set, 95.01 % (94.71 % for men and 95.31 % for women) of the values lie within the ± 1.96 SD range (Fig. 8).

(a) Five tested men.

(b) Five tested women.

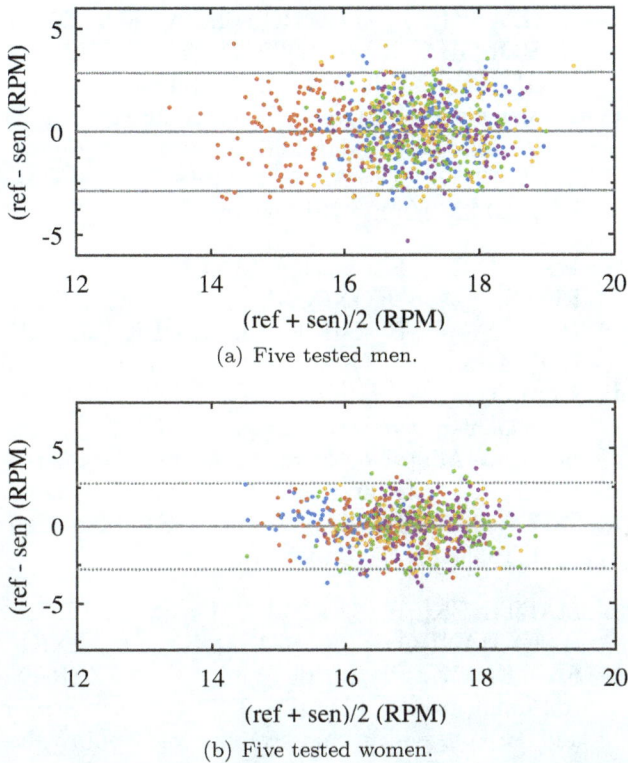

Fig. 8: Statistical analysis of the respiratory rate using the Bland-Altman plot.

The key experimental results of body temperature measurements are summarized in Tab. 2. This table shows temperature values obtained after a measurement time interval of 60 seconds. The maximum relative error of temperature measurement was 0.55 %.

Tab. 2: Summary of body temperature measurements.

Subject	Rec. time (s)	60	Relative error (%)
M1	ref. ($^\circ$ C)	35.7	0.28
	sensor ($^\circ$ C)	35.8	
M2	ref. ($^\circ$ C)	35.6	0.28
	sensor ($^\circ$ C)	35.7	
M3	ref. ($^\circ$ C)	36.3	0.55
	sensor ($^\circ$ C)	36.5	
M4	ref. ($^\circ$ C)	36.8	0.27
	sensor ($^\circ$ C)	36.7	
M5	ref. ($^\circ$ C)	36.9	0.27
	sensor ($^\circ$ C)	37.0	
F1	ref. ($^\circ$ C)	36.7	0.27
	sensor ($^\circ$ C)	36.8	
F2	ref. ($^\circ$ C)	36.6	0.27
	sensor ($^\circ$ C)	36.5	
F3	ref. ($^\circ$ C)	36.3	0.28
	sensor ($^\circ$ C)	36.4	
F4	ref. ($^\circ$ C)	36.6	0.27
	sensor ($^\circ$ C)	36.7	
F5	ref. ($^\circ$ C)	37.1	0.27
	sensor ($^\circ$ C)	37.2	

4. Conclusion

Here we described the functionality verification of a novel non-invasive fiber-optic sensor for the monitoring of human basic vital signs: Respiratory Rate (RR), Heart Rate (HR), Body Temperature (BT). Experiments were carried out in a research laboratory condition on 10 volunteer test subjects after obtaining their written informed consents. At the completion of the experiments all test subjects were asked whether they had sensed or encountered any feelings of discomfort, especially at the moment of fixing the measurement sensor into a contact strap. None of the test subjects expressed any sense of discomfort. The total time for carrying out all experiments was 85 minutes 39 seconds. The Bland-Altman Statistical Analysis for the respiratory rate (95.01 %) and heart rate (95.34 %) measurements showed satisfactory accuracy for all data acquired from the test subjects. The maximum relative error for temperature measurement was 0.55 %. The outcomes of these experiments have unambiguously proved the functionality of our novel sensor. We are hoping that our contribution reported here paves the way for researchers in this fast developing and emerging field and facilitates their efforts in expanding the applications of fiber-optic sensors and devices in sophisticated medical diagnostic instrumentation in the near future.

Acknowledgment

This article was supported by the Technology Agency of the Czech Republic TA04021263 Project and by the Ministry of Education of the Czech Republic (Projects Nos. SP2017/128 and SP2017/79). This research was partially supported by the Ministry of Education, Youth and Sports of the Czech Republic through Grant Project no. CZ. 1.07/2.3.00/20.0217 within the framework of the Operation Programme Education for Competitiveness financed by the European Structural Funds and from the state budget of the Czech Republic. The Ministry of the Interior of the Czech Republic (Project No. VI20152020008) also provided support for this article.

References

[1] RORIZ, P., L. CARVALHO, O. FRAZAO, J. L. SANTOS and J. A. SIMOES. From conventional sensors to fibre optic sensors for strain and force measurements in biomechanics applications: A review. *Journal of Biomechanics*. 2014, vol. 47, iss. 6, pp. 1251–1261. ISSN 0021-9290. DOI: 10.1016/j.jbiomech.2014.01.054.

[2] DZIUDA, L. Fiber-optic sensors for monitoring patient physiological parameters: a review of applicable technologies and relevance to use during magnetic resonance imaging procedures. *Journal of Biomedical Optics*. 2015, vol. 20, iss. 1, pp. 1–23. ISSN 1083-3668. DOI: 10.1117/1.JBO.20.1.010901.

[3] FAJKUS, M., J. NEDOMA, R. MARTINEK, V. VASINEK, H. NAZERAN and P. SISKA. A Non-invasive Multichannel Hybrid Fiber-optic Sensor System for Vital Sign Monitoring. *Sensors*. 2016, vol. 17, iss. 1, pp. 1–17. ISSN 1424-8220. DOI: 10.3390/s17010111.

[4] CHETHANA, K., A. S. GURU PRASAD, S. N. OMKAR and S. ASOKAN. Fiber bragg grating sensor based device for simultaneous measurement of respiratory and cardiac activities. *Journal of Biophotonics*. 2016, vol. 10, iss. 2, pp. 278–285. ISSN 1864-063X. DOI: 10.1002/jbio.201500268.

[5] DZIUDA, L., F. W. SKIBNIEWSKI, M. KREJ and J. LEWANDOWSKI. Monitoring respiration and cardiac activity using fiber Bragg grating-based sensor. *IEEE Transactions on Biomedical Engineering*. 2012, vol. 59, iss. 7, pp. 1934–1942. ISSN 0018-9294. DOI: 10.1109/TBME.2012.2194145.

[6] FAJKUS, M., J. NEDOMA, P. SISKA and V. VASINEK. FBG sensor of breathing encapsulated into polydimethylsiloxane. In: *Proceedings of SPIE: Optical Materials and Biomaterials in Security and Defence Systems Technology XIII*. Edinburg: SPIE, 2016, pp. 1–6. ISBN 978-151060392-9. DOI: 10.1117/12.2241663.

[7] DZIUDA, L., M. KREJ and F. W. SKIBNIEWSKI. Fiber bragg grating strain sensor incorporated to monitor patient vital signs during MRI. *IEEE Sensors Journal*. 2013, vol. 13, iss. 12, pp. 4986–4991. ISSN 1530-437X. DOI: 10.1109/JSEN.2013.2279160.

[8] CIOCCHETTI, M., C. MASSARONI, P. SACCOMANDI, M. A. CAPONERO, A. POLIMADEI, D. FORMICA and E. SCHENA. Smart textile based on fiber bragg grating sensors for respiratory monitoring: Design and preliminary trials. *Biosensors*. 2015, vol. 5, iss. 3, pp. 602–615. ISSN 2079-6374. DOI: 10.3390/bios5030602.

[9] DZIUDA, L., F. W. SKIBNIEWSKI, M. KREJ and P. M. BARAN. Fiber Bragg grating-based sensor for monitoring respiration and heart activity during magnetic resonance imaging examinations. *Journal of Biomedical Optics*. 2013, vol. 18, iss. 5, pp. 1–14. ISSN 1083-3668. DOI: 10.1117/1.JBO.18.5.057006.

[10] JEZEWSKI, J., J. WROBEL, A. MATONIA, K. HOROBA, R. MARTINEK, T. KUPKA and M. JEZEWSKI. Is abdominal fetal electrocardiography an alternative to Doppler ultrasound for FHR variability evaluation? *Frontiers in Physiology*. 2017, vol. 8, iss. 5, pp. 1–14. ISSN 1664-042X. DOI: 10.3389/fphys.2017.00305.

[11] MARTINEK, R., R. KAHANKOVA, H. NAZERAN, J. KONECNY, J. JEZEWSKI, P. JANKU, P. BILIK, J. ZIDEK, J. NEDOMA and M. FAJKUS. Non-invasive Fetal Monitoring: A Maternal Surface ECG Electrode Placement-based Novel Approach for Optimization of Adaptive Filter Control Parameters Using the LMS and RLS Algorithms. *Sensors*. 2017, vol. 17, iss. 5, pp. 1–31. ISSN 1424-8220. DOI: 10.3390/s17051154.

[12] MARTINEK, R., J. NEDOMA, M. FAJKUS, R. KAHANKOVA, R. KONECNY, P. JANKU, S. KEPAK, P. BILIK and H. NAZERAN. A Phonocardiographic-Based Fiber-Optic Sensor and Adaptive Filtering System for Noninvasive Continuous Fetal Heart Rate Monitoring. *Sensors*. 2017, vol. 17, iss. 4, pp. 1–26. ISSN 1424-8220. DOI: 10.3390/s17040890.

[13] NEDOMA, J., M. FAJKUS, P. SISKA, R. MARTINEK and V. VASINEK. Non-invasive fiber optical probe encapsulated into polydimethylsiloxane for measuring respiratory and heart rate of the human body. *Advances in Electrical and Electronic Engineering*. 2017, vol. 15, no. 1, pp. 93–100. ISSN 1336-1376. DOI: 10.15598/aeee.v15i1.1923.

[14] BEDNAREK, L., O. MARCINKA, F. PERECAR, M. PAPES, L. HAJEK, J. NEDOMA and V. VASINEK. The ageing process of optical couplers by gamma irradiation. In: *Proceedings of SPIE: Photonic Fiber and Crystal Devices: Advances in Materials and Innovations in Device Applications IX*. San Diego: SPIE, 2015, pp. 1–8. ISBN 978-162841752-4. DOI: 10.1117/12.2187044.

[15] BEDNAREK, L., L. HAJEK, A. VANDERKA, J. NEDOMA, M. FAJKUS, O. ZBORIL and V. VASINEK. The influence of thermal aging on the optical coupler. In: *Photonic Fiber and Crystal Devices: Advances in Materials and Innovations in Device Applications X*. San Diego: SPIE, 2016, pp. 1–7. ISBN 978-151060307-3. DOI: 10.1117/12.2236294.

[16] BEDNAREK, L., R. POBORIL, A. VANDERKA, L. HAJEK, J. NEDOMA and V. VASINEK. Influence of load by high power on the optical coupler. In: *20th Slovak-Czech-Polish Optical Conference on Wave and Quantum Aspects of Contemporary*

Optics. Jasna: SPIE, 2016, pp. 1–8. ISBN 978-151060733-0. DOI: 10.1117/12.2256824.

[17] FAJKUS, M., I. NAVRUZ, S. KEPAK, A. DAVIDSON, P. SISKA, J. CUBIK and V. VASINEK. Capacity of wavelength and time division multiplexing for quasi-distributed measurement using fiber bragg gratings. *Advances in Electrical and Electronic Engineering.* 2015, vol. 13, no. 5, pp. 575–582. ISSN 1336-1376. DOI: 10.15598/aeee.v13i5.1508.

[18] FAJKUS, M., J. NEDOMA, S. KEPAK, L. RAPANT, R. MARTINEK, L. BEDNAREK, M. NOVAK and V. VASINEK. Mathematical model of optimized design of multi-point sensoric measurement with Bragg gratings using wavelength divison multiplex. In: *Proceedings of SPIE: Optical Modelling and Design IV.* Brussels: SPIE, 2016, pp. 1–9. ISBN 978-151060134-5. DOI: 10.1117/12.2239551.

[19] KERSEY, A. D., M. A. DAVIS, H. J. PATRICK, M. LEBLANC, K. P. KOO, C. G. ASKINS, M. A. PUTHAM and E. J. FRIEBELE. Fiber grating sensors. *Journal of Lightwave Technology.* 1997, vol. 15, no. 8, pp. 1442–1463. ISSN 0733-8724. DOI: 10.1109/50.618377.

[20] FENDINGER, N. J. *Polydimethylsiloxane (PDMS): Environmental Fate and Effects.* Weinheim: Wiley Verlag, 2005. ISBN 978-3527620777.

[21] NEDOMA, J., M. FAJKUS and V. VASINEK. Influence of PDMS encapsulation on the sensitivity and frequency range of fiber-optic interferometer. In: *Proceedings of SPIE: Optical Materials and Biomaterials in Security and Defence Systems Technology XIII.* Edinburg: SPIE, 2016, pp. 1–10. ISBN 978-151060392-9. DOI: 10.1117/12.2243170.

[22] NEDOMA, J., M. FAJKUS, L. BEDNAREK, J. FRNDA, J. ZAVADIL and V. VASINEK. Encapsulation of FBG sensor into the PDMS and its effect on spectral and temperature characteristics. *Advances in Electrical and Electronic Engineering.* 2016, vol. 14, no. 4, pp. 460–466. ISSN 1336-1376. DOI: 10.15598/aeee.v14i4.1786.

[23] MOKHTAR, M. R., T. B. SUN and K. T. V. GRATTAN. Bragg grating packages with nonuniform dimensions for strain and temperature sensing. *IEEE Sensors Journal.* 2012, vol. 12, iss. 1, pp. 139–144. ISSN 1530-437X. DOI: 10.1109/JSEN.2011.2134845.

[24] BLAND, J. M. and D. G. ALTMAN. Measuring agreement in method comparison studies. *Statistical methods in laboratory medicine.* 1999, vol. 8, iss. 2, pp. 135–160. ISSN 0962-2802.

About Authors

Jan NEDOMA was born in 1988 in Prostejov. In 2012 he received a Bachelor's degree from VSB–Technical University of Ostrava, Faculty of Electrical Engineering and Computer Science, Department of Telecommunications. Two years later, he received his Master's degree in the field of Telecommunications in the same workplace. He is currently an employee and a Ph.D. student of Department of Telecommunications at VSB–Technical University of Ostrava. He works in the field of optical communications and fiber optic sensor systems.

Marcel FAJKUS was born in 1987 in Ostrava. In 2009 he received a Bachelor's degree from VSB–Technical University of Ostrava, Faculty of Electrical Engineering and Computer Science, Department of Telecommunications. Two years later, he received a Master's degree in the field of Telecommunications in the same workplace. He is currently an employee and a Ph.D. student of Department of Telecommunications at VSB–Technical University of Ostrava. He works in the field of optical communications and fiber optic sensor systems.

Radek MARTINEK was born in 1984 in Czech Republic. In 2009 he received Master's degree in Information and Communication Technology from VSB–Technical University of Ostrava. Since 2012 he has worked here as a research fellow. In 2014 he successfully defended his dissertation thesis titled "The use of complex adaptive methods of signal processing for refining the diagnostic quality of the abdominal fetal electrocardiogram". He has worked as an assistant professor at VSB–Technical University of Ostrava since 2014.

Vladimir VASINEK was born in Ostrava. In 1980 he graduated in Physics, specialization in Optoelectronics, from the Science Faculty of Palacky University. He was awarded the title of RNDr. at the Science Faculty of Palacky University in the field of Applied Electronics. The scientific degree of Ph.D. was conferred upon him in the branch of Quantum Electronics and Optics in 1989. He became an associate professor in 1994 in the branch of Applied Physics. He has been a professor of Electronics and Communication Science since 2007. He pursues this branch at the Department of Telecommunications at VSB–Technical University of Ostrava. His research work is dedicated to optical communications, optical fibers, optoelectronics, optical measurements, optical networks projecting, fiber optic sensors, MW access networks. He is a member of many societies: OSA, SPIE, EOS, Czech Photonics Society.

21

Validation of a Novel Fiber-Optic Sensor System for Monitoring Cardiorespiratory Activities During MRI Examinations

Jan NEDOMA[1], Marcel FAJKUS[1], Martin NOVAK[1],
Nela STRBIKOVA[2], Vladimir VASINEK[1], Homer NAZERAN[3],
Jan VANUS[2], Frantisek PERECAR[1], Radek MARTINEK[2]

[1]Department of Telecommunications, Faculty of Electrical Engineering and Computer Science,
VSB–Technical University of Ostrava, 17. listopadu 15, 708 33 Ostrava, Czech Republic
[2]Department of Cybernetics and Biomedical Engineering, Faculty of Electrical Engineering and Computer
Science, VSB–Technical University of Ostrava, 17. listopadu 15, 708 33 Ostrava, Czech Republic
[3]Department of Electrical and Computer Engineering, College of Engineering,
University of Texas El Paso, 500 W University Ave, El Paso, TX 79968, United States of America

jan.nedoma@vsb.cz, marcel.fajkus@vsb.cz, martin.novak.st@vsb.cz, nela.strbikova.st@vsb.cz
vladimir.vasinek@vsb.cz, hnazeran@utep.edu, jan.vanus@vsb.cz, frantisek.perecar@vsb.cz,
radek.martinek@vsb.cz

Abstract. *In this article we report on the validation of a novel fiber-optic sensor system suitable for simultaneous cardiac and respiration activity monitoring during Magnetic Resonance Imaging (MRI) examinations. This MRI-compatible Heart Rate (HR) and Respiration Rate (RR) measurement system is based on the Fiber-optic Bragg Grating (FBG) sensors. Using our system, we performed real measurements on 4 test subjects (2 males and 2 females) after obtaining their written informed consents. The sensor was encapsulated inside a Polydimethylsiloxane polymer (PDMS), as this material does not react with the human skin and is unresponsive to Electromagnetic Interference (EMI). The advantage of our design is that the sensor could be embedded inside a pad which is placed underneath a patient's body while lying in the supine position. The main feature of our system design is to maximize patient's safety and comfort while assisting the clinical staff in predicting and detecting impending patient's hyperventilation and panic attacks. To further validate the efficacy of our system, we used the Bland-Altman statistical analysis test on data acquired from all test subjects to determine the accuracy of cardiac and respiratory rate measurements. Our satisfactory results provide promising means to leverage the advancement of research in the field of noninvasive vital sign monitoring in MRI environments. In addition, our method and system enable the clinical staff to predict and detect patient's hyperventilation and panic attacks while undergoing an MRI examination.*

Keywords

Electromagnetic interference, fiber Bragg grating, fiber-optic sensor, heart rate, magnetic resonance imaging environment, noninvasive, polydimethylsiloxane, respiration rate, vital sign monitoring.

1. Introduction

Hyperventilation and panic attacks in patients undergoing an MRI examination are major concerns for clinicians and MRI machine operators due to their very frequent occurrence [1], [2] and [3]. Hyperventilation is a state of abnormally fast and deep breathing, preceded by a feeling of shortness of breath or "air hunger" (dyspnea). Symptoms of hyperventilation in an anxious patient include faintness or impaired consciousness, sometimes along with a felling of chest tightness, a sensation of smothering, fast heart palpitation, and dizziness. As a result, the patient may exhibit muscle cramps and panic attacks. Therefore, monitoring a patient's respiration and heart rates during an MRI examination can prove very useful and assist the clinical staff in predicting impending hyperventilation and panic attacks.

A possible method to monitor a patient's vital signs during an MRI examination is to use Fiber-Optic Sensors (FOS). These sensors are finding increased applications in many fast developing biomedical areas including performing measurements in MRI environments [4].

Recent advancements in this field can be summarized as follows [5], [6] and [7]. Chethana et al. [5] report interesting results on the design and construction of an FBG-based sensor suitable for heart rate and respiratory rate monitoring. As this sensor is applied to a patient's chest, it is very important to pay special attention to the tension developed in the optical fiber to ensure that adequate sensitivity is reached during measurements. It should be mentioned that this sensor has not been tested in an MRI environment. Dziuda et al. [6] report their results obtained from an FBG-based optical strain sensor used for monitoring respiration and cardiac activities during an MRI examination. These authors offer a solution based on Ballistocardiography (BCG), which uses a different measurement principle compared to the sensor described in [5]. In article [7], Dziuda et al. describe the validation of a fiber-optic sensor for monitoring respiratory and cardiac activities under laboratory conditions and report a maximum relative measurement error of 12 %.

From the brief review above, it is evident that noninvasive vital sign monitoring in MRI environments still faces some challenges and there is more room for research and improvement in this field. Recognizing this demand, our research team has developed a novel sensor system that allows monitoring the mechanical vibrations in the human body which are evoked by living activities such as breathing and cardiac rhythms [8], [9] and [10]. Our main aim here is to report the evaluation results of our small size, low-cost fiber-optic sensor solution with minimal weight in the form of a pad, which could be placed underneath a patient's body while lying in a supine position, thereby enabling simultaneous monitoring of cardiac and respiratory activities with an accuracy exceeding 95 %. The relative error level of approximately 5 % in our sensor system is clinically acceptable as our system is designed for cardiorespiratory activity monitoring rather than for performing accurate diagnosis of cardiopulmonary conditions. In addition, our sensor offers the advantage of enabling the clinical staff to detect and predict impending hyperventilation and panic attacks during MRI examinations.

2. Methods

Our novel measurement probe (weight: 150 g, dimensions: first layer - $75 \times 75 \times 4$ mm, second layer - $10 \times 10 \times 1$ mm) is based on a FBG encapsulated inside a PDMS polymer. A FBG is formed by a periodic change of refractive index in the core of optical fiber (n_1, n_3) and n_2 represents the refractive index of the fiber cladding, see Fig. 1.

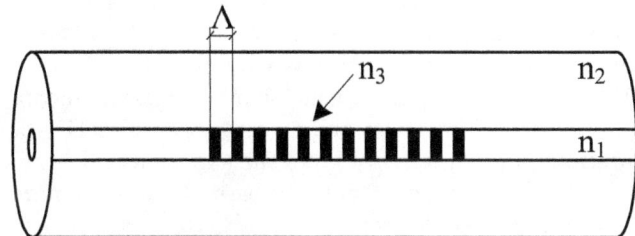

Fig. 1: Structure of fiber Bragg grating.

Dependent on the grating period, the light of a specific wavelength called the Bragg wavelength λ_B is reflected, and the other wavelengths are transmitted. FBG is one of the most widely used types of the fiber-optic sensors [4], [11], [12] and [13]. The Bragg wavelength is given by:

$$\lambda_B = 2n_{eff}\Lambda, \qquad (1)$$

where n_{eff} is the effective refractive index of the used optical fiber with Bragg grating and Λ is the period of changes in the refractive index of the core of the used optical fiber.

The primary use of FBG is based on the deformational and temperature sensitivities. According to dependencies on the mechanical stress and temperature, size of Bragg wavelength change $\Delta\lambda_B$ can be defined by:

$$\frac{\Delta\lambda_B}{\lambda_B} = k\varepsilon + (\alpha_\Lambda + \alpha_n)\Delta T, \qquad (2)$$

where k is the deformational coefficient, α_n the optical temperature coefficient, α_Λ the coefficient of thermal expansion, ΔT the temperature change, and ε the applied deformation. Deformational and temperature dependence are determined both by the parameter values and the central Bragg wavelength. Normalized deformational and temperature coefficients are determined based upon their individual sensitivities [14], [15] and [16].

The normalized FBG strain response at constant temperature is:

$$\frac{1}{\lambda_B}\frac{\Delta\lambda_B}{\Delta\varepsilon} = 0.78 \cdot 10^{-6} \; \mu\varepsilon^{-1}, \qquad (3)$$

where $\Delta\varepsilon$ is the applied deformation change and the normalized temperature sensitivity at constant strain is:

$$\frac{1}{\lambda_B}\frac{\Delta\lambda_B}{\Delta T} = 6.678 \cdot 10^{-6} \; {}^\circ\mathrm{C}^{-1}. \qquad (4)$$

The Polydimethylsiloxane polymer does not react with the human skin and is resistant to EMI. The cumulative results presented in our published works [17], [18], [19], [20] and [21] indicate that this type of encapsulation does not affect the structure of the FBG or interferometer. Figure 2 shows the experimental setup and the positioning of our sensor system to acquire data from a test subject.

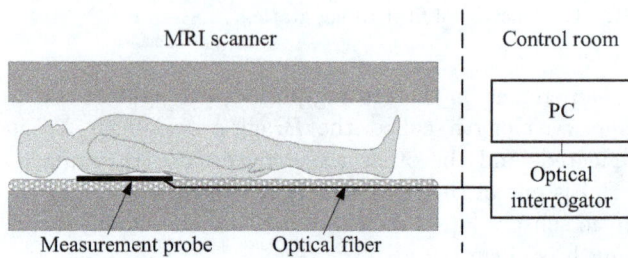

Fig. 2: Positioning of the sensor system and the experimental setup.

Data obtained from our FBG-based sensor were transferred to a control room by using an optical fiber based on the G.652.D Standard. The Optical Interrogator is composed of a wideband spectral light source from a Light-Emitting Diode (LED) with a central wavelength of 1550 nm, a spectral width of 40 nm and an output power of 1.5 mW. Furthermore, our data collection system is composed of an Optical Circulator, an Optical Spectrum Analyzer (OSA) using a sampling frequency of 100 Hz, in addition to a Digital Signal Processing (DSP) as well as an Electronic Control Unit (ECU) for each individual optical element. The cardiac and respiratory signals sensed by our sensor were further processed by our data acquisition system and finally displayed on a PC screen in a graphical user interface as part of an application created in LabVIEW (2015, National Instruments, Austin, Texas, USA).

To obtain the Heart Rate (HR) and Respiratory Rate (RR) in our tested subjects, we first performed a spectral evaluation of the measured signals and then implemented peak detection to calculate the time intervals between these peaks. The RR values were expressed in respiration per minute (rpm) and the HR values were expressed in beat per minute (bpm). These were calculated as the inverse of the detected time intervals in these signals and were then multiplied by 60.

3. Results

Data collection was carried out in a clinical setup (Private Clinic Prostejov) on 4 test subjects (2 males: age: 26 and 28 year, height: 174 and 179 cm, weight: 78 and 84 kg; and 2 females: age: 21 and 26 year, height: 162 and 167 cm, and weight: 52 and 58 kg) after obtaining their written informed consents. A Signa HDxt 1.5T MRI Scanner [22] was used in our experiments (Please see Fig. 4). To obtain reference heart and respiratory rate information we made use of the Scanner's built-in features (Please see the lower part of Fig. 3). The total data acquisition time for all of the four test subjects was 87 minutes and 43 seconds.

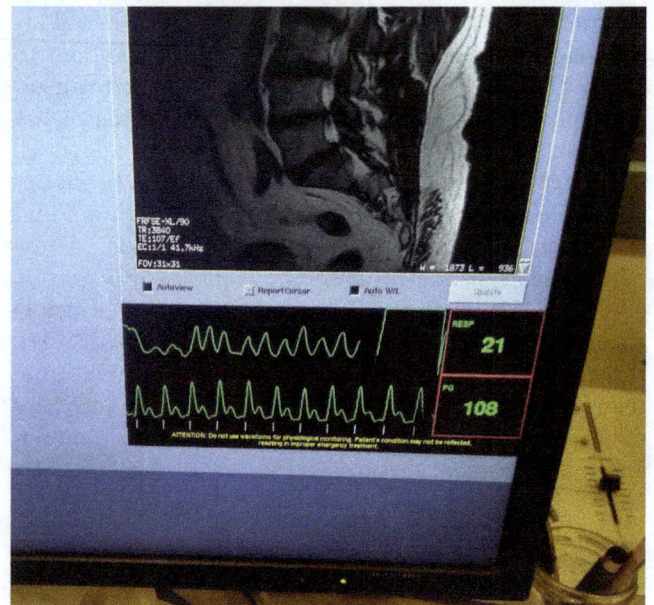

Fig. 3: Estimated reference heart and respiratory signals and their rates displayed by the Signa HDxt 1.5T MRI Scanner (Please see lower part of this figure).

For better visualization of the experimental setup and positioning of the cardiorespiratory sensor within the MRI Scanner's bed, Fig. 4 shows a photo with the sensor pad encircled with red color.

To compare the differences between the HRs and RRs estimated from the cardiac and respiratory signals acquired from our sensor pad with their corresponding reference values (HRs and RRs) determined from the Signa HDxt 1.5T MRI Scanner, the Bland-Altman Plots were used [23]. In these plots, the differences between the sensor and the reference data (reference - sensor), are plotted against their average, (reference + sensor)/2 values. The reproducibility is considered to be good if 95 % of the results lie within a ± 1.96 SD (Standard Deviation) range. Please see Fig. 6.

Figure 5 shows an example of a 60-second recording of the respiratory activity in a male subject (MAN2) sensed by our FBG-based sensor.

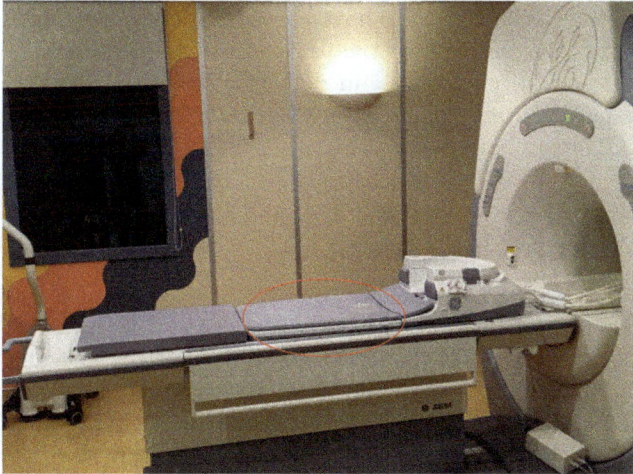

Fig. 4: A photo of the sensor pad positioned within the MRI Scanner's bed (encircled with red color).

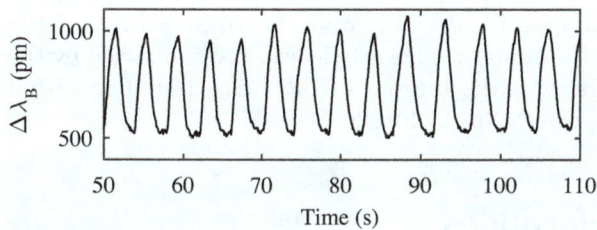

Fig. 5: The recording from our FBG sensor representing the time course of the respiratory activity in a male test subject (MAN2).

The key results of the respiratory rate measurements are summarized in Tab. 1. Recording time represents the subject's total data acquisition time and NoS sensor represents the number of measured samples from the FBG sensor. The maximum relative error was 4.41 %. For the entire data set, 96.10 % (96.29 % for males and 95.91 % for females) of the values lied within the ±1.96 SD range for the respiratory rate accuracy determination. Figure 6 shows the Bland-Altman Plot of respiratory rate measurements for two male (left) and for two female (right) subjects.

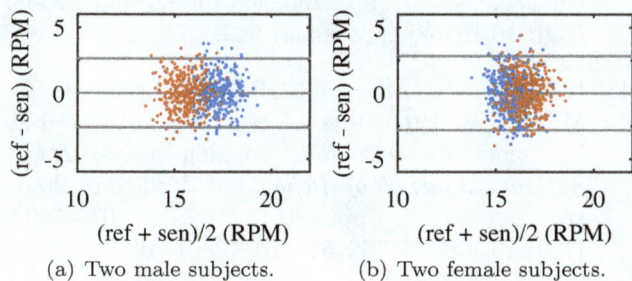

(a) Two male subjects.   (b) Two female subjects.

Fig. 6: Statistical analysis using the Bland-Altman Plots for respiratory rate measurements.

Figure 7(a) shows an example of a 60-second recording of cardiac activity in a female subject (F1)

Tab. 1: Statistical data for respiratory rate measurements in 4 test subjects.

Subject	Rec. time (s)	Respiratory Rate (RR)	
		NoS sensor	Samples in ±1.96 SD (%)
MAN1	1032	295	96.27
MAN2	1157	298	96.31
FEMALE1	1876	497	95.77
FEMALE2	1198	329	96.05

sensed by our FBG-based sensor. Figure 7(b) shows a 4-second expanded version of the recording shown in part Fig. 7(a).

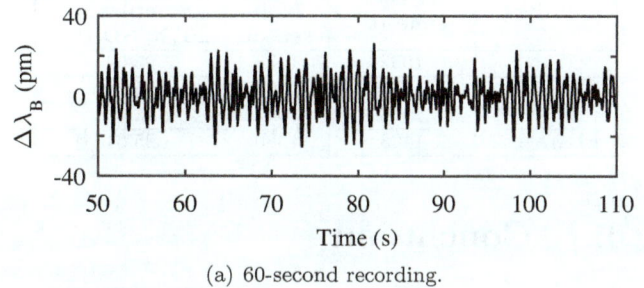

(a) 60-second recording.

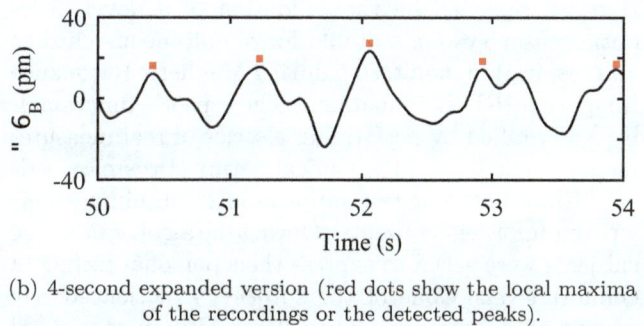

(b) 4-second expanded version (red dots show the local maxima of the recordings or the detected peaks).

Fig. 7: The recording of cardiac beat activity by using our FBG sensor in a female test subject (F1).

The key results of the heart rate measurements are summarized in Tab. 2. Recording time represents the subject's total data acquisition time and NoS sensor represents the number of measured samples from the FBG sensor. The maximum relative error was 5.86 %. For the entire data set, 95.49 % (95.61 % for males and 95.37 % for females) of the values lied within the ±1.96 SD range for the heart rate determination. Figure 8 shows the Bland-Altman Plots of the heart rate measurements for two male (left) and for two female (right) subjects. Based on Bland-Altman statistical analysis, we can state with confidence that no basic systematic errors occurred in our measurements. The sensor showed satisfactory results in measuring both the respiratory and the heart rates with acceptable accuracy without systematic errors.

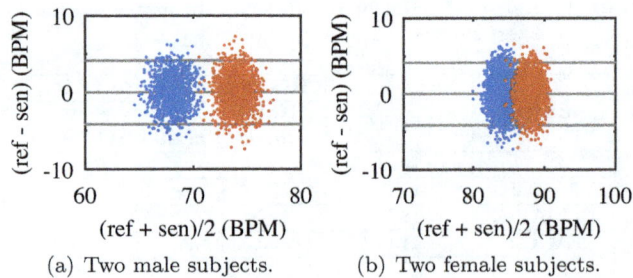

(a) Two male subjects.   (b) Two female subjects.

Fig. 8: Statistical analysis using the Bland-Altman Plots for heart rate measurements.

Tab. 2: Statistical data for heart rate measurements in 4 test subjects.

Subject	Rec. time (s)	Heart Rate (RR)	
		NoS sensor	Samples in ± 1.96 SD (%)
MAN1	1032	1118	95.64
MAN2	1157	1363	95.58
FEMALE1	1876	2542	95.13
FEMALE2	1198	1680	95.61

4.   Conclusion

Here we reported on the validation of a novel fiber-optic sensor system suitable for simultaneous cardiac and respiration monitoring during Magnetic Resonance Imaging (MRI) examinations. The sensor's functionality was verified by performing a series of real measurements carried out in a clinical setup (Prostejov Private Clinic) on four test subjects after obtaining their written informed consents. During data collection, the subjects were asked to express their personal feeling of comfort level. None of the subjects experienced any feeling of discomfort. The Bland-Altman statistical analysis of the acquired data demonstrated that there were no basic systematic errors in the measurement data. The sensor showed satisfactory results in accurately measuring both respiratory and heart rates. For the entire data set 95.49 % of the values lied within the ± 1.96 SD range for the heart rate determination and 96.10 % for the respiratory rate determination. The results of heart rate measurements were characterized by a maximum relative error of 5.86 % while the respiratory rate measurements were characterized by a maximum relative error of 4.41 %.

Our satisfactory results provide promising means to leverage the advancement of research in the field of noninvasive vital sign monitoring in MRI environments. Furthermore, our method and system enable the clinical staff to predict and detect impending hyperventilation and panic attacks in patients while undergoing an MRI examination.

For our future research, we are very excited to report that the clinicians at the Prostejov Private Clinic in the Czech Republic have agreed to deploy our novel sensor system to investigate its utility and evaluate its efficacy in predicting and detecting impending hyperventilation and panic attacks in their patients during MRI examinations.

Acknowledgment

This article was supported by the project of the Technology Agency of the Czech Republic TA04021263 and by Ministry of Education of the Czech Republic within the projects Nos. SP2017/128 and SP2017/79. The research has been partially supported by the Ministry of Education, Youth and Sports of the Czech Republic through the grant project no. CZ.1.07/2.3.00/20.0217 within the frame of the operation programme Education for competitiveness financed by the European Structural Funds and from the state budget of the Czech Republic. This article was also supported by the Ministry of the Interior of the Czech Republic within the project No. VI20152020008.

References

[1] PITTING, A., J. J. ARCH, C. W. R. LAM and M. G. CRASKE. Heart rate and heart rate variability in panic, social anxiety, obsessive-compulsive, and generalized anxiety disorders at baseline and in response to relaxation and hyperventilation. *International Journal of Psychophysiology*. 2013, vol. 87, iss. 1, pp. 19–27. ISSN 0167-8760. DOI: 10.1016/j.ijpsycho.2012.10.012.

[2] MEURET, A. E., T. RITZ, F. H. WILHELM and W. T. ROTH. Voluntary hyperventilation in the treatment of panic disorder - Functions of hyperventilation, their implications for breathing training, and recommendations for standardization. *Clinical Psychology Review*. 2005, vol. 25, iss. 3, pp. 285–306. ISSN 0272-7358. DOI: 10.1016/j.cpr.2005.01.002.

[3] HARRIS, L. M., J. ROBINSON and R. G. MENZIES. Predictors of panic symptoms during magnetic resonance imaging scans. *International Journal of Behavioral Medicine*. 2001, vol. 8, iss. 1, pp. 80–87. ISSN 1070-5503. DOI: 10.1207/S15327558IJBM0801_06.

[4] DZIUDA, L. Fiber-optic sensors for monitoring patient physiological parameters: a review of applicable technologies and relevance to use during MRI procedures. *Journal of Biomedical Optics*. 2015, vol. 20, iss. 1, pp. 1–23. ISSN 1083-3668. DOI: 10.1117/1.JBO.20.1.010901.

[5] CHETHANA, K., A. S. GURU PRASAD, S. N. OMKAR and S. ASOKAN. Fiber bragg grating sensor based device for simultaneous measurement of respiratory and cardiac activities. *Journal of Biophotonics*. 2016, vol. 10, iss. 2, pp. 278–285. ISSN 1864-063X. DOI: 10.1002/jbio.201500268.

[6] DZIUDA, L., M. KREJ and F. W. SKIBNIEWSKI. Fiber bragg grating strain sensor incorporated to monitor patient vital signs during MRI. *IEEE Sensors Journal*. 2013, vol. 13, iss. 12, pp. 4986–4991. ISSN 1530-437X. DOI: 10.1109/JSEN.2013.2279160.

[7] DZIUDA, L., F. W. SKIBNIEWSKI, M. KREJ and J. LEWANDOWSKI. Monitoring respiration and cardiac activity using fiber Bragg grating-based sensor. *IEEE Transactions on Biomedical Engineering*. 2012, vol. 59, iss. 7, pp. 1934–1942. ISSN 0018-9294. DOI: 10.1109/TBME.2012.2194145.

[8] FAJKUS, M., J. NEDOMA, R. MARTINEK, V. VASINEK, H. NAZERAN and P. SISKA. A Non-invasive Multichannel Hybrid Fiber-optic Sensor System for Vital Sign Monitoring. *Sensors*. 2016, vol. 17, iss. 1, pp. 1–17. ISSN 1424-8220. DOI: 10.3390/s17010111.

[9] FAJKUS, M., J. NEDOMA, P. SISKA and V. VASINEK. FBG sensor of breathing encapsulated into polydimethylsiloxane. In: *Proceedings of SPIE: Optical Materials and Biomaterials in Security and Defence Systems Technology XIII*. Bellingham: SPIE, 2016, pp. 1–6. ISBN 978-151060392-9. DOI: 10.1117/12.2241663.

[10] NEDOMA, J., M. FAJKUS, P. SISKA, R. MARTINEK and V. VASINEK. Non-invasive fiber optical probe encapsulated into polydimethylsiloxane for measuring respiratory and heart rate of the human body. *Advances in Electrical and Electronic Engineering*. 2017, vol. 15, no. 1, pp. 93–100. ISSN 1336-1376. DOI: 10.15598/aeee.v15i1.1923.

[11] LIU, Z. and H. Y. TAM. Industrial and medical applications of fiber Bragg gratings. *Chinese Optics Letters*. 2016, vol. 14, iss. 12, pp. 1–5. ISSN 1671-7694. DOI: 10.3788/COL201614.120007.

[12] BHOWMIK, K., E. AMBIKAIRAJAH, G. D. PENG, Y. LUO and G. RAJAN. High-sensitivity polymer fibre Bragg grating sensor for biomedical applications. *IEEE Sensors Applications Symposium*. Catania: IEEE, 2016, pp. 1–5. ISBN 978-1-4799-7250-0. DOI: 10.1109/SAS.2016.7479822.

[13] NEDOMA, J., M. FAJKUS, R. MARTINEK and V. VASINEK. Non-invasive Fiber-Optic Biomedical Sensor for the Monitoring of Vital Signs of Human Body. *Advances in Electrical and Electronic Engineering*. 2017, vol. 15, no. 2, pp. 336–342. ISSN 1336-1376. DOI: 10.15598/aeee.v15i2.2131.

[14] KERSEY, A. D., M. A. DAVIS, H. J. PATRICK, M. LEBLANC, K. P. KOO, C. G. ASKINS, M. A. PUTHAM and E. J. FRIEBELE. Fiber grating sensors. *Journal of Lightwave Technology*. 1997, vol. 15, iss. 8, pp. 1442–1463. ISSN 0733-8724. DOI: 10.1109/50.618377.

[15] FAJKUS, M., I. NAVRUZ, S. KEPAK, A. DAVIDSON, P. SISKA, J. CUBIK and V. VASINEK. Capacity of wavelength and time division multiplexing for quasi-distributed measurement using fiber bragg gratings. *Advances in Electrical and Electronic Engineering*. 2015, vol. 13, no. 5, pp. 575–582. ISSN 1336-1376. DOI: 10.15598/aeee.v13i5.1508.

[16] FAJKUS, M., J. NEDOMA, S. KEPAK, L. RAPANT, R. MARTINEK, L. BEDNAREK, M. NOVAK and V. VASINEK. Mathematical model of optimized design of multi-point sensoric measurement with Bragg gratings using wavelength divison multiplex. In: *Proceedings of SPIE: Optical Modelling and Design IV*. Brussels: SPIE, 2016, pp. 1–8. ISBN 978-151060134-5. DOI: 10.1117/12.2239551.

[17] NEDOMA, J., M. FAJKUS, L. BEDNAREK, J. FRNDA, J. ZAVADIL and V. VASINEK. Encapsulation of FBG sensor into the PDMS and its effect on spectral and temperature characteristics. *Advances in Electrical and Electronic Engineering*. 2016, vol. 14, no. 4, pp. 460–466. ISSN 1336-1376. DOI: 10.15598/aeee.v14i4.1786.

[18] NEDOMA, J., M. FAJKUS and V. VASINEK. Influence of PDMS encapsulation on the sensitivity and frequency range of fiber-optic interferometer. In: *Proceedings of SPIE: Optical Materials and Biomaterials in Security and Defence Systems Technology XIII*. Edinburg: SPIE, 2016, pp. 1–10. ISBN 978-151060392-9. DOI: 10.1117/12.2243170.

[19] FENDINGER, N. J. Polydimethylsiloxane (PDMS): Environmental Fate and Effects. In: *Organosilicon Chemistry IV: From Molecules to Material*. Weinheim: Wiley-VCH Verlag, 2005, pp. 626–638. ISBN 978-3-527-62077-7.

[20] MARTINEK, R., R. KAHANKOVA, H. NAZERAN, J. KONECNY, J. JEZEWSKI, P. JANKU, P. BILIK, J. ZIDEK, J. NEDOMA and M. FAJKUS. Non-invasive Fetal Monitoring: A Maternal Surface ECG Electrode Placement-based Novel Approach for Optimization of Adaptive Filter Control Parameters

Using the LMS and RLS Algorithms. *Sensors.* 2017, vol. 17, iss. 5, pp. 1–31. ISSN 1424-8220. DOI: 10.3390/s17051154.

[21] MARTINEK, R., J. NEDOMA, M. FAJKUS, R. KAHANKOVA, R. KONECNY, P. JANKU, S. KEPAK, P. BILIK and H. NAZERAN. A Phonocardiographic-Based Fiber-Optic Sensor and Adaptive Filtering System for Noninvasive Continuous Fetal Heart Rate Monitoring. *Sensors.* 2017, vol. 17, iss. 4, pp. 1–26. ISSN 1424-8220. DOI: 10.3390/s17040890.

[22] Signa HDxt 1.5T magnetic resonance scanner. *GE Healthcare* [online]. 2017. Available at: http://www3.gehealthcare.in/en/products/categories/magnetic-resonance-imaging/signa-hdxt-1-5t.

[23] BLAND, J. M. and D.G. ALTMAN. Measuring agreement in method comparison studies. *Statistical methods in laboratory medicine.* 1999, vol. 8, iss. 2, pp. 135–160. ISSN 1050-1647.

About Authors

Jan NEDOMA was born in 1988 in Prostejov. In 2012 he received a Bachelor's degree from VSB–Technical University of Ostrava, Faculty of Electrical Engineering and Computer Science, Department of Telecommunications. Two years later, he received his Master's degree in the field of Telecommunications in the same workplace. He is currently employee and a Ph.D. student of Department of Telecommunications at VSB–Technical University of Ostrava. He works in the field of fiber-optic sensor systems.

Marcel FAJKUS was born in 1987 in Ostrava. In 2009 he received a Bachelor's degree from VSB–Technical University of Ostrava, Faculty of Electrical Engineering and Computer Science, Department of Telecommunications. Two years later, he received a Master's degree in the field of Telecommunications in the same workplace. He is currently employee and a Ph.D. student of Department of Telecommunications at VSB–Technical University of Ostrava. He works in the field of optical communications and fiber-optic sensor systems.

Martin NOVAK was born in 1989 in Prostejov. In 2012 he received a Bachelor's degree from VSB–Technical University of Ostrava, Faculty of Electrical Engineering and Computer Science, Department of Telecommunications. Three years later, he received his Master's degree in the field of Telecommunications in the same workplace. He is currently Ph.D. student of Department of Telecommunications at VSB–Technical University of Ostrava. He works in the field of optical communications and fiber-optic sensor systems.

Nela STRBIKOVA was born in 1993 in Sumperk. In 2017 she received a Master's degree from VSB–Technical University of Ostrava, Faculty of Electrical Engineering and Computer Science, Department of Cybernetics and Biomedical Engineering. She wants to continue with her studies in Ph.D. program in Department of Technical Cybernetics.

Vladimir VASINEK was born in Ostrava. In 1980 he graduated in Physics, specialization in Optoelectronics, from the Science Faculty of Palacky University. He was awarded the title of RNDr at the Science Faculty of Palacky University in the field of Applied Electronics. The scientific degree of Ph.D. was conferred upon him in the branch of Quantum Electronics and Optics in 1989. He became an associate professor in 1994 in the branch of Applied Physics. He has been a professor of Electronics and Communication Science since 2007. He pursues this branch at the Department of Telecommunications at VSB–Technical University of Ostrava. His research work is dedicated to optical communications, optical fibers, optoelectronics, optical measurements, optical networks projecting, fiber-optic sensors, MW access networks. He is a member of many societies: OSA, SPIE, EOS, Czech Photonics Society; he is a chairman of the Ph.D. board at the VSB-Technical University of Ostrava. He is also a member of habitation boards and the boards appointing to professorship.

Homer NAZERAN holds B.Sc., M.Sc. and Ph.D. degrees in Electrical (Honors), Clinical and Biomedical Engineering from UT Austin, Case Western Reserve and University of Texas Southwestern Medical Center (UTSWM) at Dallas/UTA, respectively. He has close to 3 decades of experience in industry and academia and has practiced and taught biomedical engineering in the Middle East, Europe, Australia and USA. In Australia, with Professor Andrew Downing he co-founded the School of Engineering at the Flinders University of South Australia, introduced and established the electrical and electronics and biomedical engineering degree programs (1991 to 2001). He returned to the University of Texas at Arlington as a visiting professor in 1997 and 2001. He joined UTEP in 2002 to create and establish biomedical engineering degree programs at the Department of Electrical and Computer Engineering. His research interests are in the areas of computer modeling of physiological systems, intelligent biomedical instrumentation and biomedical signal processing as applied to chronic health conditions and telemedicine. He has more than 150 journal and conference articles in his research areas published in IEEE Engineering in Medicine and Biology Society (EMBS) and other flagship international conference proceedings. He is a reviewer for several national and international journals in his related fields including IEEE Transactions on Biomedical Engineering, Medical and Biological Engineering and Computing, Biomedical Engineering Online and others. His teaching interests are in electronics, biomedical instrumentation, physiological systems, and biomedical signal processing. He is also interested in development of novel teaching methods, lifelong learning and critical thinking habits in the classroom and interdisciplinary education based on application of nonlinear dynamics systems (complexity) theory. His research, teaching and professional activities have been supported by NIH, NSF, and DOE among others.

Jan VANUS was born in 1972 in Czech Republic. He is currently Academic Staff Member, Department of Cybernetics and Biomedical Engineering of the Faculty of Electrical Engineering and Computer Science, VSB–Technical University of Ostrava. The research title is "Design and application Intelligent Buildings Control Systems". His research interests include Smart Home and Smart Home Care (remote comfort control, visualization, Data Mining, Big Data processing, energy management, optimization, recognition, classification, prediction), speech signal processing, adaptive filters, voice communication with control system.

Frantisek PERECAR was born in 1989 in Presov. In 2011 he received a Bachelor's degree from University of Zilina. Two years later, he received his Master's degree in the field of Telecommunications in the same workplace. He is currently a Ph.D. student of Department of Telecommunications at VSB–Technical University of Ostrava. He works in the field of fiber-optic sensor systems.

Radek MARTINEK was born in 1984 in Czech Republic. In 2009 he received Master's degree in Information and Communication Technology from VSB–Technical University of Ostrava. Since 2012 he worked here as a research fellow. In 2014 he successfully defended his dissertation thesis titled "The use of complex adaptive methods of signal processing for refining the diagnostic quality of the abdominal fetal electrocardiogram". He works as an associate professor at VSB–Technical University of Ostrava since 2017.

22

Noninvasive Fetal Heart Rate Monitoring: Validation of Phonocardiography-Based Fiber-Optic Sensing and Adaptive Filtering Using the NLMS Algorithm

Jan NEDOMA[1], Marcel FAJKUS[1], Stanislav KEPAK[1], Jakub CUBIK[1],
Radana KAHANKOVA[2], Petr JANKU[3], Vladimir VASINEK[1],
Homer NAZERAN[4], Radek MARTINEK[2]

[1]Department of Telecommunications, Faculty of Electrical Engineering and Computer Science,
VSB–Technical university of Ostrava, 17. listopadu 15, 708 33 Ostrava, Czech Republic
[2]Department of Cybernetics and Biomedical Engineering, Faculty of Electrical Engineering and Computer
Science, VSB–Technical University of Ostrava, 17. listopadu 15, 708 33 Ostrava, Czech Republic
[3]Department of Gynecology and Obstetrics, Faculty of Medicine, Masaryk University and
University Hospital Brno, Jihlavska 20, 625 00 Brno, Czech Republic
[4]Department of Electrical and Computer Engineering, College of Engineering,
University of Texas El Paso, 500 W University Ave, El Paso, TX 79968, United States of America

jan.nedoma@vsb.cz, marcel.fajkus@vsb.cz, stanislav.kepak@vsb.cz, jakub.cubik@vsb.cz,
radana.kahankova@vsb.cz, janku.petr@fnbrno.cz, vladimir.vasinek@vsb.cz, hnazeran@utep.edu,
radek.martinek@vsb.cz

Abstract. *Here we present the evaluation results of our novel noninvasive phonocardiographic-based fiber-optic sensor for fetal Heart Rate (fHR) detection using adaptive filtering and the NLMS Algorithm. The sensor uses two interferometric probes encapsulated inside a PolyDiMethylSiloxane (PDMS) polymer. Based on real data acquired from pregnant women in a suitable research laboratory environment, once they had given their written informed consents, we created a simplified dynamic signal model of the distribution of maternal and fetal heart sounds inside the maternal body. Building upon this signal model, we verified the functionality of our novel fiber-optic sensor and its associated adaptive filtering system using the NLMS Algorithm. The main reason why we chose this technology to develop our system was that it allows monitoring the fHR without exposing the fetus to any external energies or radiation (in contrast to the ultrasound-based Cardiotocography Method). We used objective criteria such as: Signal to Noise Ratios: SNR_{in}, SNR_{out} and Percentage Root-mean-square Difference (PRD) for our evaluations.*

Keywords

ElectroMagnetic Interference (EMI), fetal Heart Rate (fHR), fetal PhonoCardioGraphy (fPCG), Fiber-optic sensor, maternal PhonoCardioGraphy (mPCG), Normalized Least Mean Square (NLMS) algorithm, PolyDiMethylSiloxane (PDMS).

1. Introduction

In this article, we report on the evaluation of a noninvasive method for fetal Heart Rate (fHR) detection and monitoring during gestation, labor, and delivery based on fetal PhonoCardioGraphy (fPCG). Our proposed method relies on the combined capabilities of fiber-optic sensing and adaptive filtering (implementing the Normalized Least Mean Square - NLMS - Algorithm). In our recent work reported elsewhere [1] and [2], we developed an adaptive system, which enabled us to measure the fetal Heart Rate (fHR) by means of fPCG signal peak detection using the maternal abdominal PhonoCardioGrams (aPCGs). We observed that diagnostic-quality fPCG signals required for accurate fHR detection are contaminated by an unwanted maternal component (the mPCG signals) in addition to other technical and biological interferences. We showed that as the spectral contents of the fPCG and mPCG signals overlap in the frequency domain, common filtering methods such as signal subtraction, linear filtering, and others are ineffective in extracting reliable fHR information and therefore cannot be used.

Our recent research as well as others have also indicated that Fiber-optic technologies such as Fiber Bragg Gratings (FBGs) or interferometers are used increasingly in many biomedical applications; see articles [3], [4], [5], [6], [7], [8], [9], [10] and [11]. Building upon these advancements, we developed our novel sensor that uses two non-invasive interferometric probes encapsulated in a PolyDiMethylSiloxane (PDMS) polymer with the designation Sylgard 184.

The well-established conventional Phonocardiography is based on the scanning of acoustic signals by means of a microphone placed on the thorax. As for fetal Phonocardiography, the microphone is placed on the maternal abdomen [12], [13] and [14].

Our solution described here is based on the scanning of acoustic signals by means of two Mach-Zehnder interferometric fiber-optic probes. The advantages of these interferometers are their immunity to Electro-Magnetic Interferences (EMI), and their ability to measure any changes in the optical path length (such as the core refraction index, fiber length and the wavelength used). Therefore, the smallest measurable frequency due to any phenomena resulting in the change of the above-mentioned physical properties is theoretically unlimited [15] and [16].

To perform our system evaluations, we needed to use synthetic data. For generating suitable synthetic signals, we conducted a set of measurements on pregnant women in a suitable research laboratory environment after obtaining their written consents. We then created a simplified dynamic signal model for the distribution of maternal and fetal heart sounds inside the maternal body. Based upon this signal model, we generated synthetic data with properties as close as possible to the real data. The necessity to use synthetic data at this stage of our research was further justified by considering the fact that our patent-pending interferometric sensors have yet to be legislatively approved for clinical testing on pregnant women. It is important to emphasize that legislative regulations for use of new technology on pregnant women are extremely strict (as an unborn fetus is critically sensitive to external energies

such as mechanical pressure, electromagnetic radiation, change in temperature, and so on).

In current clinical practice, clinicians use either ultrasound-based methods such as CardioTocoGraphy (CTG), which measures the fetal heart rate along with maternal uterine contractions, or fetal Echocardiography (fECHO) to diagnose fetal congenital heart defects from the 20th to the 23rd week of pregnancy [17] and [18]. These sophisticated technologies are now integral parts of routine modern obstetrics. It is important to emphasize that the CTG technology has helped clinicians reduce the mortality rate of newborn babies during delivery. In spite of this considerable impact, it is generally recognized that this technology has some disadvantages such as high sensitivity to noise caused by maternal movements and the need to frequently reposition the ultrasound transducers. Also, this method is not suitable for long-term continuous fetal heart rate monitoring due to the potentially harmful influence of ultrasonic radiation on the fetus.

Our method and system, once statistically and clinically proven and validated, offer a number of advantages (in contrast to the currently used ultrasound-based CTG and other conventional methods), including their applicability to continuous long-term fHR monitoring without exposing the fetus to any radiation as well as their compatibility with Magnetic Resonance Imaging (MRI) environments. The continuous long-term monitoring capability of our system is highly desirable, especially in those cases in which the pregnant woman faces a dangerous situation (such as after an accident), and it becomes absolutely essential to perform a time consuming MRI examination to ensure that the unborn child is intact and safe. The other specific advantage of our technology is that it can be used in water deliveries.

2.　Methods

2.1.　Fetal Phonocardiography

Fetal PhonoCardioGraphy (fPCG) was discovered during the 17th century by Kergardec, Marsac, and Kennedy [19]. Although fPCG was discovered a very long time ago, interest in this research area has only grown over the past few years. This figure shows the number of peer-reviewed articles that appear in the Science Direct, the Institute of Electrical and Electronics Engineers (IEEE) and the National Institute of Health (PubMed) databases.

The PCG signal is composed of two main acoustic components (the first heart sound S1 and second heart sound S2), see Fig. 1, and two additional heart sounds (S3 and S4). S1 is systolic and is connected with the closure of bicuspid and tricuspid valves at the beginning of ventricular contraction. S2 is diastolic and is produced by the closure of semilunar valves. The third heart sound (S3) is pro-diastolic and appears when a valve muscle quivers during the fast phase of blood flow into the valve. The fourth heart sound (S4) is presystolic and is a sign of the quivering of valve muscle during systole in the atrium. The last two mentioned heart sounds (S3 and S4) are not common for adults, and their presence is a sign of cardiac insufficiency [20].

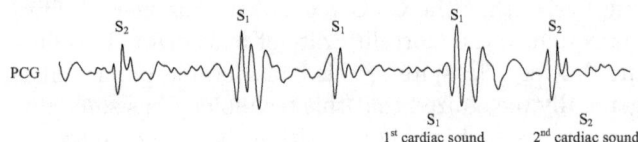

Fig. 1: Basic components of PCG signals.

2.2. PCG-Based Fibre-Optic Sensor

Our fiber-optic sensor is encapsulated inside polydimethylsiloxane [21], [22] and [23] and is comprised of two Mach-Zehnder interferometric components formed by 1×2 and 3×3 power couplers with an even split ratio; see Fig. 2.

Fig. 2: Our noninvasive fiber-optic measurement probe.

The reference fiber is stored in a stable environment. The output beams are recombined at a second 3×3 coupler. The output signal is detected by photodetectors. The resultant optical intensity after 3×3 coupler can be described by the following Eq. (1).

$$I_n = A_n + B_n \cos\left[\phi(t) + \phi_{drift}(t) + (n-1)\frac{2\pi}{3}\right], \quad (1)$$

where n represents the coupler output index with a value of 1, 2 or 3. The symbol A_n represents the mean value of optical intensity (DC component). Symbol B_n represents the optical intensity variation amplitude depending on fringe visibility, $\phi(t)$ represents the signal of interest, and $\phi_{drift}(t)$ is a quasi-static phase shift due to coupler properties. For the extraction of the

proper signal, it is necessary to use a demodulation algorithm [24].

2.3. Implementation of the Adaptive NLMS Algorithm

The measurands sensed by our interferometric sensors generated the fetal heart rate information, which was then fed into an adaptive stochastic system using the Root Mean Square Error (RMSE) criterion. This stochastic approach required a large number of measurements to produce powerful statistics. This consideration led to the utilization of the Normalized Least Mean Square (NLMS) Algorithm, which is a representative of basic stochastic gradient-based adaptation methods; see articles [25] and [26].

The Normalized Least Mean Square (NLMS) Algorithm is a variant of the Least Mean Square Algorithm. The former is able to accelerate the convergence speed with a reasonable computational cost and selects a normalized step-size μ_n, which results in both a stable and fast converging adaptation algorithm, see [27] and [28]. Implementation of the NLMS Algorithm can be summarized as follows:

```
BEGIN w⃗(n=0)=0⃗
FOR (n=1,2,...,N):
y(n)=w⃗ᵀ(n)·x⃗(n)
e(n)=d(n)-y(n)
w⃗(n+1)=w⃗(n)+μ(n)·e(n)·x⃗(n).
```

The step-size μ_n can be described as follows Eq. (2).

$$\mu(n) = \frac{\mu}{\delta + \vec{x}^T(n) \cdot \vec{x}(n)}. \quad (2)$$

Finally, we obtain the following Eq. (3).

$$\vec{w}(n) = \vec{w}(n-1) + \mu\frac{e(n) \cdot \vec{x}(n)}{\delta + \vec{x}^T(n) \cdot \vec{x}(n)}, \quad (3)$$

where $\mu \in (0,2]$ and $\delta > 0$. Parameter δ represents the regularization parameter (prevents the denominator of Eq. (4) becoming zero).

3.   Results

Our measurement system comprised of a novel fiber-optic sensor and its associated adaptive filtering system for fetal Heart Rate (fHR) monitoring is shown in Fig. 3. The adaptive system consists of two measurement sensors (FC/APC type) which were placed on the chest and abdomen, optical interrogator and DSP (Digital Signal Processing) unit for the recording, amplification, digitalization, demodulation and filtering the measured signals. Optical interrogator consists of DFB (Distributed Feedback Laser) laser with wavelength 1549.5 nm and output power of 3 mW and three InGaAs Amplified Photodetectors (Indium Gallium Arsenide). Signal was digitalized by National Instruments card NI-USB 6210 with the sampling frequency of 250 kHz and analyzed by software application written in the LabView (2015, National Instruments, Austin, Texas, USA) [29] and [30].

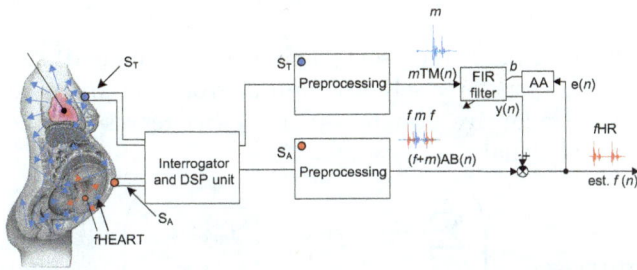

Fig. 3: Basic scheme of our sensor and its associated NLMS adaptive system for fHR monitoring.

Measurements (Fig. 4) were performed in a suitable research laboratory environment on 8 pregnant women (GA = 36–42 weeks) after obtaining their written informed consents. The test subjects were between the age of 21 and 37, their weight was between 57 kg and 103 kg, and their height was between 156 cm and 196 cm. Based on the obtained results we can state that no significant differences were found in the quality of the collected data based on the subjects' age, weight, and height.

Fig. 4: An example of real data acquisition from a volunteer subject.

Using real data, we created a simplified dynamic model of sound distribution in the human body to generate suitable synthetic signals such as: ST (signals from sensors placed on the chest) and SA signals (from sensors placed on the abdomen). Our PCG signal model was inspired by contributions made by AL-MASI et al. [31] and [32], who devoted considerable

efforts to generating synthetic PCG signals. In addition, we greatly benefited from our own research in generating realistic synthetic physiological and pathological fECG signals [33] and [34] in order to evaluate the performance of our system.

Figure 5 shows an ideal mPCG signal after removing the mother's breathing artifacts (using a Butterworth second-order band-pass filter with corner frequencies: $f_L = 10$ Hz, and $f_H = 400$ Hz, respectively). This signal served as a reference input for our adaptive system running the NLMS Algorithm. The filtered results enabled us to determine the mHR (by performing mPCG signal peak detection). Maternal first and second heart sounds are denoted as mS_1 and mS_2, respectively, in Fig. 5.

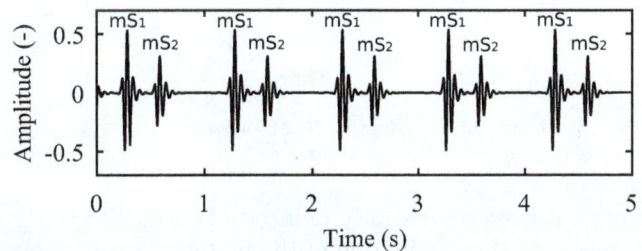

Fig. 5: The reference synthetic mPCG signal based on real measurements made from thoracic (S_T) sensors.

Figure 6 shows an ideal fPCG waveform after preprocessing the maternal signal. We need to emphasize here that the first fetal heart sounds (fS_1) result from the closing of the fetal tricuspid and mitral valves and the second fetal heart sounds (fS_2) are produced by the closure of the fetal pulmonic and aortic valves.

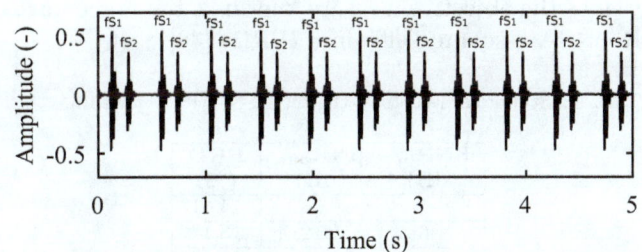

Fig. 6: The reference ideal synthetic fPCG signals based on real measurements from abdominal (S_A) sensors.

Figure 7 shows an example of the primary abdominal PCG (aPCG) synthetic input signal measured by the abdominal sensor. The aPCG signal (made up of the fPCG and mPCG components) is applied to the adaptive NLMS Algorithm. For determination of the fetal Heart Rate (fHR), it is necessary to detect fS1 components in the composite aPCG signals, which is a difficult task without advanced signal processing.

Figure 8 shows an example of the output from our adaptive system using the NLMS Algorithm. Based on these results we can observe that: the mPCG compo-

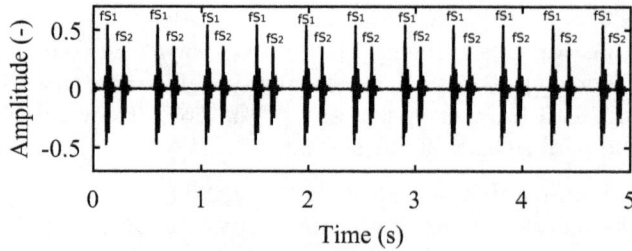

Fig. 7: The reference ideal synthetic fPCG signals based on real measurements from abdominal (S_A) sensors.

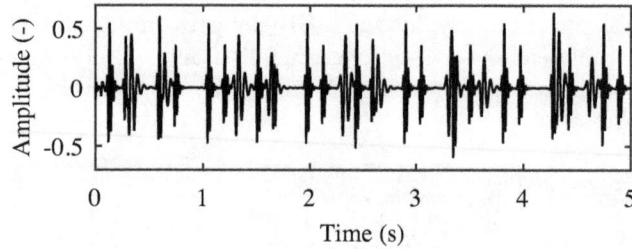

Fig. 8: Output of the adaptive NLMS system.

nent has been significantly reduced. This figure clearly shows that the elimination of the maternal component is not ideal; nevertheless, this component is reduced well under the level of fPCG signals. Using the filtered signal, we can use conventional techniques [35], [36] and [37] to determine the fHR information from the fPCG signals.

Table 1 summarizes our experimental results. The performance of our adaptive system using the NLMS Algorithm was evaluated by finding the differences between input (SNR_{in}) and output (SNR_{out}) values as well as the objective measure known as the Percentage Root-mean-square Difference (PRD) [38].

Tab. 1: Statistical results of the tested NLMS Algorithm.

SNR_{in} (dB)	SNR_{out} (dB)	PRD (%)
−7	0.98	14.61
−6	1.12	13.14
−5	1.33	10.74
−4	1.43	9.98
−3	1.48	9.35
−2	1.57	7.9
−1	1.65	6.74
0	1.71	5.69
1	1.70	5.74

The SNR_{in} value can be calculated by using the following Eq. (4):

$$SNR_{in} = 10\log\left(\frac{\sum_{n=1}^{N-1}[sig_{usef}(n)]^2}{\sum_{n=1}^{N-1}[sig_{noise}(n) - sig_{usef}(n)]^2}\right), \quad (4)$$

where $sig_{usef}(n)$ is a desired signal (modelled reference course of S_T) and $sig_{noise}(n)$ is a noise signal (mPCG is measured up in the abdominal part - S_A).

The SNR_{OUT} value can be calculated by using the following equation Eq. (5):

$$SNR_{out} = 10\log\left(\frac{\sum_{n=1}^{N-1}[sig_{des}(n)]^2}{\sum_{n=1}^{N-1}[sig_{pre}(n) - sig_{usef}(n)]^2}\right), \quad (5)$$

where $sig_{pre}(n)$ represents a predicted (estimated) signal, or more precisely, the output from the proposed NLMS adaptive system and $sig_{des}(n)$ represents the desired signal.

$$PRD\,(\%) = \left(\frac{\sum_{n=1}^{N}[sig_{usef}(n) - sig_{pre}(n)]^2}{\sum_{n=1}^{N}sig_{usef}(n)}\right) \cdot 100. \quad (6)$$

One way to quantify the difference between the reference and the output signal: $sig_{pre}(n)$ is by using the PRD as given by equation Eq. (6) below:

4. Conclusion

In this article we focused on the validation of our novel patent-pending interferometric PPG-based sensor and its associated adaptive filtering system using the NLMS Algorithm for effective processing of aPCG signals to extract fPCG signals and fHR information. In the evaluations of the signal filtering quality of our system, we used objective parameters such as SNR and PRD.

The main reason why we chose the fiber-optic technology to develop our system was that it enables fHR monitoring without exposing the fetus to any radiation (in contrast to the ultrasound-based CTG method). Our innovative system offers a number of advantages including applicability to continuous long-term fHR monitoring without exposing the fetus to any radiation as well as compatibility with Magnetic Resonance Imaging (MRI) environments. The long-term monitoring capacity of our system is highly desirable, especially in those cases when the pregnant woman faces a dangerous situation (such as after an accident), and it becomes absolutely necessary to perform a time consuming MRI examination to ensure that the unborn fetus is intact and safe. The other specific advantage of our technology is that it can be used in water deliveries.

In our future research, we intend to use data from clinical practice to investigate a variety of challenging research topics such as the influence of sensor placement, fetal position and gestational age on aPCG signal filtering, fPCG signal extraction, and fHR monitoring.

Acknowledgment

This article was supported by the project of the Technology Agency of the Czech Republic TA04021263 and by Ministry of Education of the Czech Republic within the projects Nos. SP2017/128 and SP2017/79. The research has been partially supported by the Ministry of Education, Youth and Sports of the Czech Republic through the grant project no. CZ.1.07/2.3.00/20.0217 within the frame of the operation programme Education for competitiveness financed by the European Structural Funds and from the state budget of the Czech Republic. This article was also supported by the Ministry of the Interior of the Czech Republic within the projects Nos. VI20152020008.

References

[1] MARTINEK, R., J. NEDOMA, M. FAJKUS, R. KAHANKOVA, J. KONECNY, P. JANKU, S. KEPAK, P. BILIK and H. NAZERAN. A Phonocardiographic-Based Fiber-Optic Sensor and Adaptive Filtering System for Noninvasive Continuous Fetal Heart Rate Monitoring. *Sensors*. 2017, vol. 17, iss. 4, pp. 1–26. ISSN 1424-8220. DOI: 10.3390/s17040890.

[2] MARTINEK, R., R. KAHANKOVA, H. NAZERAN, J. KONECNY, J. JEZEWSKI, P. JANKU, P. BILIK, J. ZIDEK, J. NEDOMA and M. FAJKUS. Non-Invasive Fetal Monitoring: A Maternal Surface ECG Electrode Placement-Based Novel Approach for Optimization of Adaptive Filter Control Parameters Using the LMS and RLS Algorithms. *Sensors*. 2017, vol. 17, iss. 5, pp. 1–31. ISSN 1424-8220. DOI: 10.3390/s17051154.

[3] RORIZ, P., L. CARVALHO, O. FRAZAO, J. L. SANTOS and J. A. SIMOES. From conventional sensors to fibre optic sensors for strain and force measurements in biomechanics applications: A review. *Journal of Biomechanics*. 2014. vol. 47, iss. 6, pp. 1251–1261. ISSN 0021-9290. DOI: 10.1016/j.jbiomech.2014.01.054.

[4] DZIUDA, L. Fiber-optic sensors for monitoring patient physiological parameters: a review of applicable technics and relevance to use during MRI procedures. *Journal of Biomedical Optics*. 2015, vol. 20, iss. 1, pp. 1–23. ISSN 1560-2281. DOI: 10.1117/1.JBO.20.1.010901.

[5] CHETHANA, K., A. S. GURU PRASAD, S. N. OMKAR and S. ASOKAN. Fiber bragg grating sensor based device for simultaneous measurement of respiratory and cardiac activities. *Journal of Biophotonics*. 2016, vol. 10, iss. 2, pp. 278–285. ISSN 1864-063X. DOI: 10.1002/jbio.201500268.

[6] DZIUDA, L., M. KREJ and F. W. SKIBNIEWSKI. Fiber bragg grating strain sensor incorporated to monitor patient vital signs during MRI. *IEEE Sensors Journal*. 2013, vol. 13, iss. 12. pp. 4986–4991. ISSN 1530-437X. DOI: 10.1109/JSEN.2013.2279160.

[7] DZIUDA, L., F. W. SKIBNIEWSKI, M. KREJ and J. LEWANDOWSKI. Monitoring respiration and cardiac activity using fiber Bragg grating-based sensor. *IEEE Transactions on Biomedical Engineering*. 2012, vol. 59, iss. 7, pp. 1934–1942. ISSN 0018-9294. DOI: 10.1109/TBME.2012.2194145.

[8] FAJKUS, M., J. NEDOMA, R. MARTINEK, V. VASINEK, H. NAZERAN and P. SISKA. A Non-invasive Multichannel Hybrid Fiber-optic Sensor System for Vital Sign Monitoring. *Sensors*. 2017, vol. 17, iss. 1, pp. 1–17. ISSN 1424-8220. DOI: 10.3390/s17010111.

[9] NEDOMA, J., M. FAJKUS, P. SISKA, R. MARTINEK and V. VASINEK. Non-invasive fiber optical probe encapsulated into polydimethylsiloxane for measuring respiratory and heart rate of the human body. *Advances in Electrical and Electronic Engineering*. 2017, vol. 15, no. 1, pp. 93–100. ISSN 1336-1376. DOI: 10.15598/aeee.v15i1.1923.

[10] FAJKUS, M., J. NEDOMA, P. SISKA and V. VASINEK. FBG sensor of breathing encapsulated into polydimethylsiloxane. In: *Proceedings of SPIE: Optical Materials and Biomaterials in Security and Defence Systems Technology XIII*. Edinburg: SPIE, 2016, pp. 1–6. ISBN 0277-786X. DOI: 10.1117/12.2241663.

[11] FAJKUS, M., J. NEDOMA, S. KEPAK, L. RAPANT, R. MARTINEK, L. BEDNAREK, M. NOVAK and V. VASINEK. Mathematical model of optimized design of multi-point sensoric measurement with Bragg gratings using wavelength divison multiplex. In: *Optical Modelling and Design IV*. Brussels: Proceedings of SPIE, 2016, pp. 1–7. ISBN 978-1-5106-0134-5. DOI: 10.1117/12.2239551.

[12] TAN, B. H. and M. MOGHAVVEMI. Real time analysis of fetal phonocardiography. In: *IEEE Region 10 Annual International Conference*. Kuala Lumpur: IEEE, 2000, pp. 135–140. ISBN 0-7803-6355-8. DOI: 10.1109/TENCON.2000.888405.

[13] MOGHAVVEMI, M., B. H. TAN and S. Y. TAN. A non-invasive PC-based measurement of fetal phonocardiography. *Sensors and Actuators A: Physical*. 2003, vol. 107, iss. 1, pp. 96–103. ISSN 0924-4247. DOI: 10.1016/S0924-4247(03)00254-1.

[14] VARADY, P., L. WILDT, Z. BENYO and A. HEIN. An advanced method in fetal phonocardiography. *Computer Methods and programs in Biomedicine*. 2003, vol. 71, iss. 3, pp. 283–296. ISSN 0169-2607. DOI: 10.1016/S0169-2607(02)00111-6.

[15] LOPEZ-HIGUERA, J. M. *Handbook of optical fibre sensing technology*. 1st ed. New York: Wiley, 2002. ISBN 978-0-471-82053-6.

[16] GOODWIN, E. P. and J. C. WYANT. *Field guide to interferometric optical testing*. 1st ed. Bellingham: SPIE, 2006. ISBN 978-0819465108.

[17] DEVANE, D., J. G. LALOR, S. DALY, W. MCGUIRE, A. CUTHBERT and V. SMITH. Cardiotocography versus intermittent auscultation of fetal heart on admission to labour ward for assessment of fetal wellbeing. *Cochrane Database of Systematic Reviews*. 2017, vol. 2017, iss. 1, pp. 1–46. ISSN 1469-493X. DOI: 10.1002/14651858.CD005122.pub5.

[18] GRIVELL, R. M., Z. ALFIREVIC, G. M. GYTE and D. DEVANE. Antenatal cardiotocography for fetal assessment. *The Cochrane database of systematic reviews*. 2015, vol. 123, iss. 4, pp. 1–30. ISSN 1469-493X.

[19] KENNEDY, E. *Observations on Obstetric Auscultation: With an Analysis of the Evidences of Pregnancy, and an Inquiry Into the Proofs of the Life and Death of the Fetus in Utero (Classic Reprint)*. 2nd ed. New York: Forgotten Books, 2015. ISBN 978-1332272440.

[20] AYRES-DE-CAMPOS, D., C. Y. SPONG and E. CHANDRAHARAN. FIGO consensus guidelines on intrapartum fetal monitoring: Cardiotocography. *International Journal of Gynecology & Obstetrics*. 2016, vol. 133, iss. 1, pp. 13–24. ISSN 0020-7292. DOI: 10.1016/j.ijgo.2016.02.005.

[21] NEDOMA, J., M. FAJKUS, L. BEDNAREK, J. FRNDA, J. ZAVADIL and V. VASINEK. Encapsulation of FBG sensor into the PDMS and its effect on spectral and temperature characteristics. *Advances in Electrical and Electronic Engineering*. 2016, vol. 14, no. 4, pp. 460–466. ISSN 1336-1376. DOI: 10.15598/aeee.v14i4.1786.

[22] NEDOMA, J., M. FAJKUS and V. VASINEK. Influence of PDMS encapsulation on the sensitivity and frequency range of fiber-optic interferometer. In: *Proceedings of SPIE: Optical Materials and Biomaterials in Security and Defence Systems Technology XIII*. Edinburg: SPIE, 2016, pp. 1–7. ISBN 978-1-5106-0392-9. DOI: 10.1117/12.2243170.

[23] FENDINGER, N. J. *Polydimethylsiloxane (PDMS): Environmental Fate and Effects*. 1st ed. Weinheim: Wiley-VCH Verlag GmbH, 2005. ISBN 978-94-009-1507-7.

[24] TODD, M. D., M. SEAVER and F. BUCHOLTZ. Improved, operationally-passive interferometric demodulation method using 3×3 coupler. *Electronics Letters*. 2002, vol. 38, iss. 15, pp. 784–786. ISSN 0013-5194. DOI: 10.1049/el:20020569.

[25] HAYKIN, S. S. *Adaptive filter theory*. 5th ed. New Jersey: Pearson Education, 2008. ISBN 978-0132671453.

[26] VASEGHI, S. V. and V. SAEED. *Advanced signal processing and digital noise reduction*. 1st ed. Berlin: Springer-Verlag, 2013. ISBN 978-3-322-92773-6.

[27] UNCINI, A. *Fundamentals of adaptive signal processing*. Cham: Springer International Publishing, 2015. ISBN 978-3-319-02807-1.

[28] FARHANG-BOROUJENY, B. *Adaptive Filters: Theory and Applications*. 2nd ed. Chichester: John Wiley & Sons, 2013. ISBN 978-1-119-97954-8.

[29] ZAZULA, D., D. DONLAGIC and S. SPRAGER. Application of Fibre-Optic Interferometry to Detection of Human Vital Sign. *Journal of the Laser and Health Academy*. 2012, vol. 2012, no. 1, pp. 27–32. ISSN 1855-9921. DOI: 10.13140/2.1.2033.4726.

[30] BRANDSTETTER, P. and L. KLEIN. Second Order Low-Pass and High-Pass Filter Designs using Method of Synthetic Immittance Elements. *Advances in Electrical and Electronic Engineering*. 2013, vol. 11, no. 1, pp. 16–21. ISSN 1336-1376. DOI: 10.15598/aeee.v11i1.800.

[31] ZHANG, D. Wavelet Approach for ECG Baseline Wander Correction and Noise Reduction. In: *IEEE Engineering in Medicine and Biology 27th Annual Conference*. Shanghai: IEEE, 2005, pp. 1212–1215. ISBN 0-7803-8741-4. DOI: 10.1109/IEMBS.2005.1616642.

[32] ALMASI, A., M. B. SHAMSOLLAHI and L. SEDHADJI. Bayesian denoising framework of phonocardiogram based on a new

dynamical model. *IRBM*. 2013, vol. 34, iss. 3, pp. 214–225. ISBN 978-1-4244-4121-1. DOI: 10.1016/j.irbm.2013.01.017.

[33] ALMASI, A., M. B. SHAMSOLLAHI and L. SEDHADJI. A dynamical model for generating synthetic Phonocardiogram signals. In: *Proceedings of the Annual International Conference of the IEEE Engineering in Medicine and Biology Society, EMBS*. Boston: IEEE, 2011, pp. 5686–5689. ISBN 978-1-4244-4121-1. DOI: 10.1109/IEMBS.2011.6091376.

[34] MARTINEK, R., A. SINCL, J. VANUS, M. KELNAR, P. BILIK, Z. MACHACEK and J. ZIDEK. Modelling of fetal hypoxic conditions based on virtual instrumentation. *Advances in Intelligent Systems and Computing*. Paris: Springer, 2016, pp. 249–259. ISBN 978-3-319-29504-6. DOI: 10.1007/978-3-319-29504-6_25.

[35] MARTINEK, R., M. KELNAR, P. KOUDELKA, J. VANUS, P. BIKIK, P. JANKU, H. NAZERAN and J. ZIDEK. A novel LabVIEW-based multi-channel non-invasive abdominal maternal-fetal electrocardiogram signal generator. *Physiological Measurement*. 2016, vol. 37, iss. 2, pp. 238–256. ISSN 0967-3334. DOI: 10.1088/0967-3334/37/2/238.

[36] TANG, H., T. LI, T. QIU and Y. PARK. Fetal Heart Rate Monitoring from Phonocardiograph Signal Using Repetition Frequency of Heart Sounds. *Journal of Electrical and Computer Engineering*. 2016, vol. 2016, no. 2404267, pp. 1–6. ISSN 2090-0147. DOI: 10.1155/2016/2404267.

[37] CHETLUR ADITHYA, P., R. SANKAR, W. A. MORENO and S. HART. Trends in fetal monitoring through phonocardiography: Challenges and future directions. *Biomedical Signal Processing and Control*. 2017, vol. 33, iss. 1, pp. 289–305. ISSN 1746-8094. DOI: 10.1016/j.bspc.2016.11.007.

[38] BLANCO-VELASCO, M., F. ROLDAN, J. GODINO, J. BLANCO, C. ARMIENS and F. LOPEZ-FERRERAS. On the use of PRD and CR parameters for ECG compression. *Medical Engineering and Physics*. 2005, vol. 27, iss. 9, pp. 798–802. ISSN 1350-4533. DOI: 10.1016/j.medengphy.2005.02.007.

About Authors

Jan NEDOMA was born in 1988 in Prostejov. In 2012 he received a Bachelor's degree from VSB–Technical University of Ostrava, Faculty of Electrical Engineering and Computer Science, Department of Telecommunications. Two years later, he received his Master's degree in the field of Telecommunications in the same workplace. He is currently employee (science and research assistant) and a Ph.D. student of Department of Telecommunications at VSB–Technical University of Ostrava. He works in the field of fiber optic sensor systems.

Marcel FAJKUS was born in 1987 in Ostrava. In 2009 he received a Bachelor's degree from VSB–Technical University of Ostrava, Faculty of Electrical Engineering and Computer Science, Department of Telecommunications. Two years later, he received a Master's degree in the field of Telecommunications in the same workplace. He is currently employee and a Ph.D. student of Department of Telecommunications at VSB–Technical University of Ostrava. He works in the field of optical communications and fiber optic sensor systems.

Stanislav KEPAK was born in 1987 in Ostrava. In 2011 he received Master's degree in the feld of Telecommunications. He is currently Ph.D. student and he works in the feld of optical communications and fiber optic sensor systems.

Jakub CUBIK was born in 1986 in Olomouc. In 2009 received Bachelor's degree on VSB–Technical University of Ostrava, Faculty of Electrical Engineering and Computer Science, Department of Telecommunications. Two years later he received on the same workplace his Master's degree in the field of Telecommunications. He is currently Ph.D. student, and he works in the field of optical communications and fiber optic sensor systems.

Radana KAHANKOVA was born in 1991 in Opava, Czech Republic. She received her Bachelors's degree at the VSB–Technical University of Ostrava, the Department of Cybernetics and Biomedical Engineering in 2014. Two years later at the same department, she received her Master's degree in the field of Biomedical Engineering. She is currently pursuing her Ph.D in Technical Cybernetics. Her current research is focused on improving the quality of electronic fetal monitoring.

Petr JANKU was born on 30th May 1969 in Brno, Czech Republic. He graduated from the Medical Faculty Masaryk University in Brno. He works in the Department of Obstetrics and Gynecology of the University Hospital Brno as a Deputy Head responsible for Obstetrics. His research work concentrates on monitoring of fetuses.

Vladimir VASINEK was born in Ostrava. In 1980 he graduated in Physics, specialization in Optoelectronics, from the Science Faculty of Palacky

University. He was awarded the title of RNDr. at the Science Faculty of Palacky University in the field of Applied Electronics. The scientific degree of Ph.D. was conferred upon him in the branch of Quantum Electronics and Optics in 1989. He became an associate professor in 1994 in the branch of Applied Physics. He has been a professor of Electronics and Communication Science since 2007. He pursues this branch at the Department of Telecommunications at VSB–Technical University of Ostrava. His research work is dedicated to optical communications, optical fibers, optoelectronics, optical measurements, optical networks projecting, fiber optic sensors, MW access networks. He is a member of many societies: OSA, SPIE, EOS, Czech Photonics Society; he is a chairman of the Ph.D. board at the VSB–Technical University of Ostrava. He is also a member of habitation boards and the boards appointing to professorship.

Homer NAZERAN holds B.Sc., M.Sc. and Ph.D. degrees in Electrical (Honors), Clinical and Biomedical Engineering from UT Austin, Case Western Reserve and University of Texas Southwestern Medical Center (UTSWM) at Dallas/UTA, respectively. He has close to 3 decades of experience in industry and academia and has practiced and taught biomedical engineering in the Middle East, Europe, Australia and USA. In Australia, with Professor Andrew Downing he co-founded the School of Engineering at the Flinders University of South Australia, introduced and established the electrical and electronics and biomedical engineering degree programs (1991 to 2001). He returned to the University of Texas at Arlington as a visiting professor in 1997 and 2001. He joined UTEP in 2002 to create and establish biomedical engineering

degree programs at the Department of Electrical and Computer Engineering. His research interests are in the areas of computer modeling of physiological systems, intelligent biomedical instrumentation and biomedical signal processing as applied to chronic health conditions and telemedicine. He has more than 150 journal and conference articles in his research areas published in IEEE Engineering in Medicine and Biology Society (EMBS) and other flagship international conference proceedings. He is a reviewer for several national and international journals in his related fields including IEEE Transactions on Biomedical Engineering, Medical and Biological Engineering and Computing, Biomedical Engineering Online and others. His teaching interests are in electronics, biomedical instrumentation, physiological systems, and biomedical signal processing. He is also interested in development of novel teaching methods, lifelong learning and critical thinking habits in the classroom and interdisciplinary education based on application of nonlinear dynamics systems (complexity) theory. His research, teaching and professional activities have been supported by NIH, NSF, and DOE among others.

Radek MARTINEK was born in 1984 in Czech Republic. In 2009 he received Master's degree in Information and Communication Technology from VSB–Technical University of Ostrava. Since 2012 he worked here as a research fellow. In 2014 he successfully defended his dissertation thesis titled "The use of complex adaptive methods of signal processing for refining the diagnostic quality of the abdominal fetal electrocardiogram". He works as an associate professor at VSB–Technical University of Ostrava since 2017.

23

Active Pre-Equalizer for Broadband over Visible Light

Tomas STRATIL[1], Petr KOUDELKA[1], Radek MARTINEK[2], Tomas NOVAK[3]

[1]Department of Telecommunications, Faculty of Electrical Engineering and Computer Science, VSB–Technical University of Ostrava, 17. listopadu 15, 708 33 Ostrava, Czech Republic
[2]Department of Cybernetics and Biomedical Engineering, Faculty of Electrical Engineering and Computer Science, VSB–Technical University of Ostrava, 17. listopadu 15, 708 33 Ostrava, Czech Republic
[3]Department of General Electrical Engineering, Faculty of Electrical Engineering and Computer Science, VSB–Technical University of Ostrava, 17. listopadu 15, 708 33 Ostrava, Czech Republic

tomas.stratil@vsb.cz, petr.koudelka@vsb.cz, radek.martinek@vsb.cz, tomas.novak1@vsb.cz

Abstract. *This paper introduces a new technology called Broadband over Visible Light (BVL) which combines two technology solutions like Visible Light Communication(VLC) and Broadband over Power Line (BPL). This new technology is suitable for converting modern LED lighting systems into communication systems. However, there are some deficiencies in BVL technology such as the low bandwidth of LED optical transmitters. Pre-equalization may be solution of this problem. This paper proposes higher bandwidth using the pre-equalization circuit. Also, it shows real experimental results demonstrating an improvement of bandwidth and transmission rate.*

Keywords

Equalization, Labview, pre-equalizaton, Software-defined radio, Visible Light Communication, VLC.

1. Introduction

The recent development in the area of white LEDs caused their common use as a highly efficient alternative to the conventional sources of optical radiation in the visible range. This development brought progressive changes in the lighting technology. The physical principle of white LEDs allows their use for communication purposes. The physical principles, including changes in trends of the lighting technology, caused the emergence of a new research direction generally called Visible Light Communication (VLC), which is a deriva-tive of original research direction generally known as indoor Optical Wireless Communication (indoor OWC), operating exclusively in the infra-red spectrum of optical radiation. The objective of this research direction is merging lighting and communications [1], [2] and [?].

The Broadband over Visible Light (BVL) is a new research direction based essentially on VLC technology. Again, it is intended to utilize the visible spectrum of optical radiation as a communication direction to the end user (downlink) and to utilize the infra-red spectrum of optical radiation (940 nm) in the reverse communication direction (uplink). Moreover, compared to the VLC concept, in the case of BVL, it is intended to use the chipset of the Broadband over Power Line (BPL) technology, which, inter alia, allows the use of the OFDM MQAM modulation format at the number of 1155 sub-carriers in the frequency range from 2 MHz to 32 MHz (for example HomePlug AV). The BVL technology should, by its nature, enable transmission speed of 100 Mb·s^{-1}. Additionally, the BVL technology provides connectivity over power conductors in an efficient way. It gives us the opportunity to transmit the modulated signal to the optical transmitter by its power lines, and use visible light as a wireless data transmitter.

White light LEDs as transmitters for communication link have the big disadvantage as low bandwidth. Low bandwidth is caused by optoelectronic response of the LED and due to physical principles of fluorescence in a thin layer of phosphor which is responsible for creating white light from blue light. Fluorescence inserts some delay to the optical signal and thus influences the maximum bit rate. White power LEDs achieve several MHz of bandwidth [8]. Some researchers achieve

10 MHz of bandwidth due to optical band pass filter on detection side, where only blue part of wavelength processes pass on photodetector without delayed yellow part from fluorescence. Equalization techniques are used for elimination of LEDs optoelectronic response to achieve higher bandwidth [6] and [7]. This article deals with this actual issue and brings new unpublished results, which can help to develop BVL technology.

2. Bandwidth of LEDs

Broadband Power Line technology operates in the frequency range from 2 MHz to 32 MHz at HomePlug AV specification and provides a 200 Mb·s^{-1} PHY channel rate and 150 Mb·s^{-1} information rate. Suitable bandwidth of optical transmitter has to be reached for cooperating HomePlug AV specification with VLC. The frequency response of high power light LED Philips Fortimo LED DLM 3000 44 W/830 was measured. Network analyzer Rhode-Schwarz ZVB 4 (3 kHz to 4 GHz) [9] was used with our own designed circuit Bias-T [12]. PIN photodetector Thorlabs PDA10A-EC was applied on detection side. Philips Fortimo LED DLM 3000 44 W/830 has system efficiency 62 lm/W. This LED light source offers an advantage for VLC measuring and testing, because of its concept of construction. There were used blue LED chips, which are directly placed on the aluminum block for effective cooling. External diffuser in front of blue LED chips converts part of blue (lower) wavelengths range into higher wavelengths due to phosphor layer as seen in Fig. 1. This white light LED has patented remote phosphor technology. This LED light source meets requirements for future duplex communication. Photodetector operating in the infrared wavelength range could be used behind diffuser with the phosphor layer. Measured values of frequency response were fitted by cubic function to

eliminate roughness caused by noise and other signal distortion. The smoother curve of frequency response provides ideal conditions to design the pre-equalization circuit. -3 dBm bandwidth was achieved at 2 MHz with a high power white light LED as shown in Fig. 6, and at 7 MHz with blue part of radiation without phosphor effect.

2.1. Design of Pre-Equalizer

The purpose of equalization is to compensate signal distortion in a communication channel. The main distortion of the signal was produced by the optical transmitter at VLC channel due to an optoelectronic response of LED and delay of the phosphor [6]. We used well-known equalization techniques and techniques for designing filters to reach suitable bandwidth. Appropriate bandwidth was reached with a really simple electronic circuit, because the simplicity of designed circuit was the important requirement.

Pre-equalization circuit was designed according to the reversal of measured frequency response, which can be seen in Fig. 6. The circuit is composed of an active and mainly passive part, which determines shape of frequency response. The passive part of the circuit is high pass filter, which causes 35 dBm attenuation, as it can be seen in Fig. 3. The active part of the circuit contains operational amplifier OPA847 eliminating attenuation of the passive part. The complete circuit provides frequency response closed to the reversal of measured frequency response. Circuit diagram of pre-equalizer can be seen in Fig. 2. The transfer function of the pre-equalization circuit was expressed as:

$$H(j\omega) = \left(1 + \frac{R_5}{R_4}\right) \cdot \frac{R_3}{\frac{R_2}{j\omega R_2 C + 1} + R_3}, \quad (1)$$

where ω is defined as $2\pi f$.

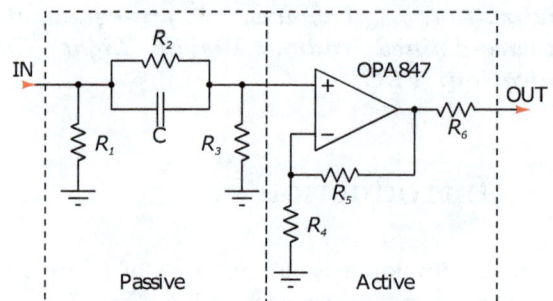

Fig. 2: Circuit diagram of active pre-equalizer with OPA847.

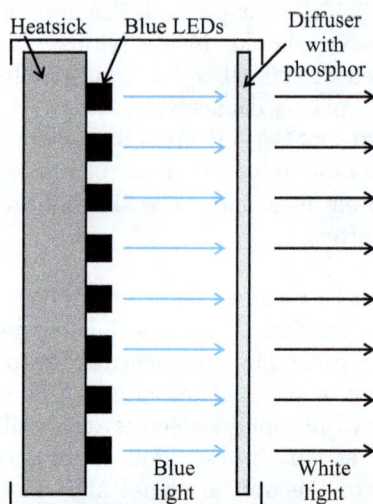

Fig. 1: Concept of Philips FORTIMO LED DLM 3000.

2.2. Simulations

Passive part of the circuit shapes the curve of frequency response and contains parallel connection of resistor R_2

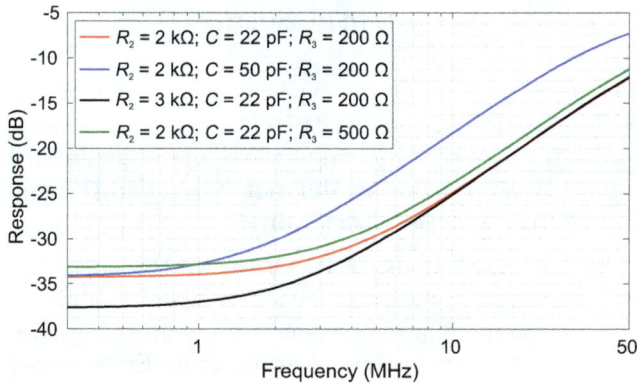

Fig. 3: Simulation results of pasive part of pre-equalizer and effect components values on frequency response.

and capacitor C and resistor R_3, parallelly connected to them. Figure 3 shows the curves from simulations, where the effect of components values on the shape of frequency response can be noticed. The best result of reversal frequency response was reached by the components $R_2 = 2$ kΩ, $C = 22$ pF and $R_3 = 200$ Ω.

Operational amplifier OPA847 was used in active part of the pre-equalization circuit. The OPA847 provides a unique combination of a very low input voltage noise, along with a very low distortion output stage to give one of the highest dynamic range of op amps available. Voltage-feedback of op amps, unlike current-feedback designs, can use a wide range of resistor's values to set up their gain. R_4 was set to 39.2 Ω and R_5 was optimized according to desired gain. Using this guideline ensures that the noise added at the output due to the Johnson noise of the resistors does not significantly increase the total noise over the 0.85 V/$\sqrt{\text{Hz}}$ input voltage noise for the op amp itself. This R_4 value is suggested as a good starting point for the design of the circuit.

Curves gained by simulations of different adjusted values of feedback resistor R_5 are shown in Fig. 4,

where open-source simulation software Qucs was used. R_4 value was 39.2 Ω whereas we were trying to achieve ideal amplification for the pre-equalization circuit by adjusting value R_5. $R_5 = 750$ Ω provided the best result of reversal frequency response and achieved gain +20 V/V of operation amplifier. Values of R_4 and R_5 affected complete amplification of active part of the circuit. On the other hand, they also affected shape of the frequency response of overall pre-equalization circuit as shown in Fig. 4.

2.3. Measurement

The frequency response of constructed pre-equalizer was measured and the results were compared with simulations. Network analyzer Rhode-Schwarz ZVB 4 (3 kHz to 4 GHz) was used. Results from simulations, measurements and reverse are shown in the Fig. 5. Frequency response from simulations and measurements of the pre-equalization circuit are almost same to the reverse frequency response, which is desirable. Operational amplifier OPA847 operates up to 40 MHz as seen in Fig. 5. This is due to high amplification of OPA847, however 40 MHz is sufficient for BVL technology solution. The designed pre-equalizer circuit attenuates the input signal by about 1 dBm in the frequency range 0.2 to 2 MHz. Designed and constructed circuit achieves power level 15 dBm at 40 MHz. Bandwidth from 2 MHz to 40 MHz of designed circuit is compliant according to HomePlug AV technology solution of Broadband Power Line communication.

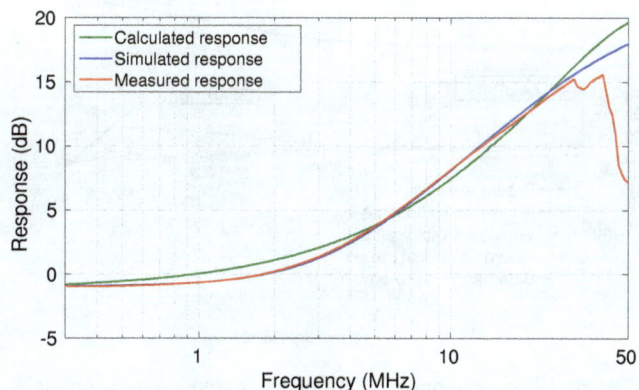

Fig. 5: Frequency response of designed circuit given by reverse, simulation and measurement.

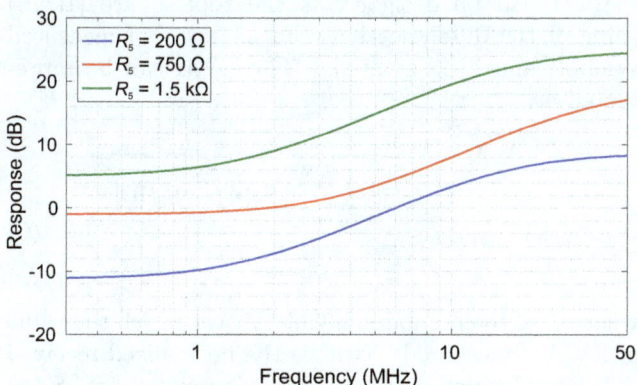

Fig. 4: Simulation results of active part of pre-equaliser and different feedback resistor values effect on frequency response.

3. Testing Effect of Pre-Equalizer on Bandwith and Modulation

To verify designed and constructed pre-equalizer effect, a new measurement was done on Philips Fortimo LED

DLM 3000 44 W/830 by aforementioned vector network analyzer. The frequency response of LED light
source without designed circuit was measured at first,
then with pre-equalizer. To allow determining influence of fluorescence in the phosphor, measurement was
repeated without diffuser with the phosphor layer on
the mentioned white light LED source. Uniform distance 40 cm between the optical transmitter and photodetector was set.

The pre-equalizer measurement results are shown
in Fig. 6. It is evident how designed circuit of pre-
equalizer influences frequency response of phosphor
based white light and also blue part of radiation without influencing the delay due to fluorescence at phosphor layer. A bandwidth of -3 dBm was achieved at
2 MHz with white light LED without pre-equalization,
whilst, a bandwidth of -10 dBm was achieved at
6 MHz. In the case of white light LED without
pre-equalization, -3 dBm bandwidth was achieved at
2 MHz and -10 dBm bandwidth at 6 MHz. When with
the pre-equalization circuit connected between the network analyzer and Bias-T, -3 dBm bandwidth was obtained at 3 MHz and -10 dBm bandwidth at 40 MHz.
It proves how pre-equalization mitigates natural inclinations to low bandwidth of semiconductor phosphor
based LED light transmitters. The bandwidth of blue
part of radiation without the effect of luminescence
achieved amplification by 5 dBm at 20 MHz frequency
due to pre-equalization.

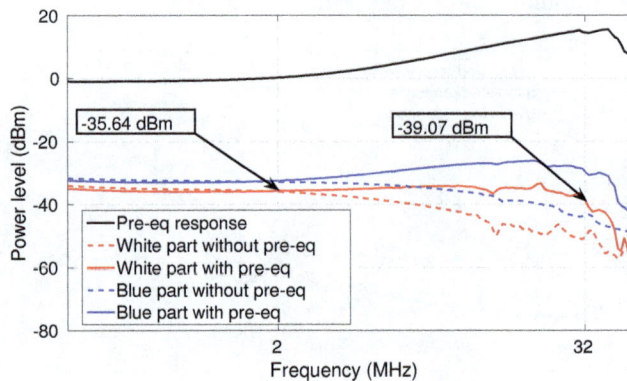

Fig. 6: Effect pre-equalizer on frequency response of white light
LED and blue light LED.

The power level of the signal has higher inclination to drop toward increasing frequency, due to influence of phosphor layer. Diffuser with phosphor layer
decrease power level. BVL technology with respect
to HomePlug AV technology solution operates from
2 MHz to 32 MHz, hence BVL solution needs to achieve
this bandwidth. An attenuation of 3.43 dBm was
achieved with mentioned bandwidth due to designed
pre-equalization circuit.

3.1. Experimental Setup

The block diagram for the experimental setup is shown
in Fig. 7. RF VSG NI PXI-5670 (Vector signal generator) [11] was used to generate the digitally modulated
signal. MQAM digital modulation was tested [10] and
[14], 4QAM was used specifically.

Vector signal analyzer RF VSA NI PXI-5661 [11]
was used on the receiver side. The signal modulated
by a digital modulation scheme was monitored by con-
stellation diagram and simultaneously an Error Vector
Magnitude (EVM) was measured. The EVM provides
a comprehensive measure of the quality of the digitally
modulated signal. We used it to verify pre-equalization
circuit effect on the transmitted digital signal, depending on the symbol rate (used bandwidth).

SI PIN photodetector ThorLabs PDA10A-EC is
operating in the wavelength range from 200 nm to
1100 nm. Photodetector PDA10A-EC has an effective
area $A_{eff} = 0.8$ mm^2 only, therefore N-BK7 Plano-
Convex Lens with a focal length of 25.4 mm was used.
Thanks to the lens, an adequate signal output was obtained to verify the functionality at a realistic distance
of 3 m between transmitter and receiver. Center frequencies 5 MHz, 10 MHz, 15 MHz and 20 MHz were
used in digital modulation scheme.

The detected signal can be represented by:

$$y(n) = g(n)x(n) + \eta(n), \qquad (2)$$

where $g(n)$ and $\eta(n)$ represent the multiplicative and
additive impairments to the detected signal. The multiplicative impairments can be a result of channel estimation errors or IQ imbalances, for example. The additive impairments are usually caused by thermal noise
and are modeled as an *i.i.d.* (Independent and Identi-
cally Distributed random variables) complex AWGN
samples with Power Spectral Density (PSD) of $N_0/2$.

EVM can be designed as the root-square (RMS)
value of the difference between an array of measured
symbols and ideal symbols. The EVM can be represented as:

$$EVM_{RMS} = \sqrt{\frac{\frac{1}{N}\sum_{n=1}^{N}|S_r(n) - S_t(n)|^2}{P_0}}, \qquad (3)$$

where N is the number of symbols over which the value
of EVM is measured. $S_r(n)$ is the normalized received
n^{th} symbol which is disrupted by Gaussian noise. $S_t(n)$
is the ideal transmitted value of the nth symbol $x(n)$,
and P_0 is either the maximum normalized ideal symbol power or the average power of all symbols for the

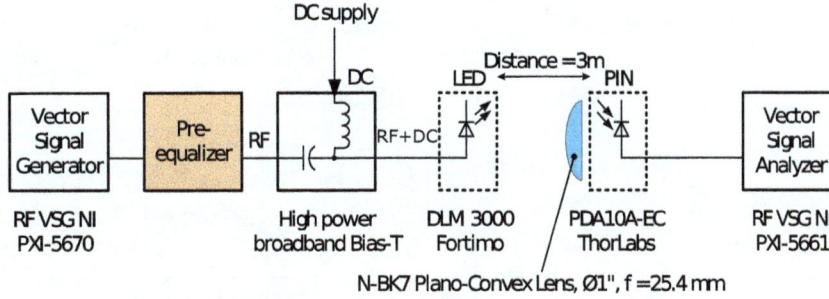

Fig. 7: Block diagram of experimental measurement.

chosen modulation. P_0 can be represented by:

$$P_0 = \frac{1}{M} \sum_{m=1}^{M} |S_m|^2. \tag{4}$$

The EVM value is normalized with average symbol energy to remove the dependency of EVM on the modulation order. Consider the detected signal in Eq. (2), where $g(n) \approx 1$. For non-data-aided receivers, the EVM is:

$$EVM_{RMS} = \sqrt{\frac{\frac{1}{N} \sum_{n=1}^{N} |y(n) - \tilde{x}(n)|^2}{P_0}}, \tag{5}$$

where $\tilde{x}(n)$ are transmitted symbols, which are estimated and used to measure the EVM value.

According to [13], the EVM for QAM signals is:

$$EVM_{QAM} =$$
$$= \left[\frac{1}{SNR} - 8\sqrt{\frac{3}{2\pi(M-1)SNR}} \sum_{i=1}^{\sqrt{M}-1} \gamma_i e^{\frac{3\beta_i^2 SNR}{2(M-1)}} \right.$$
$$\left. + \frac{12}{M-1} \sum_{i=1}^{\sqrt{M}-1} \gamma_i \beta_i \mathrm{erfc}\left(\sqrt{\frac{3\beta_i^2 SNR}{2(M-1)}} \right) \right]^{1/2}, \tag{6}$$

where

$$\gamma_i = 1 - \frac{i}{\sqrt{M}}, \text{ and } \beta_i = 2i - 1. \tag{7}$$

The EVM of a QAM signal in Eq. (6) can be divided into two parts. The first part is $1/SNR$, which represents the ideal EVM when no errors are introduced to the symbol detection. The second part is QAM signal, which is the sum of the exponential and error function, representing the reduction in measured EVM due to the error detection.

3.2. Results and Discussions

The results were measured by the block diagram shown in Fig. 7. 4QAM digital modulation scheme was transmitted via white part of radiation with effect of the luminescence and results are shown in Fig. 8. Small differences were measured between EVM values of VLC system without the pre-equalizaion circuit and with pre-equalization for center frequency 5 MHz. The pre-equalization circuit had significant influence at higher frequencies.

Significant improvement of EVM values was verified with VLC system with pre-equalization. Most significant difference of EVM values was for center frequency of 20 MHz due to frequency response of pre-equalization circuit seen in Fig. 6. EVM value increased when symbol rate was increased, due to non-linearity of frequency spectrum.

EVM values were increased due to low signal power level and natural inclinations of the frequency response as shown in Fig. 6. Higher bandwidth was used with higher symbol rate and it increased EVM and decreased communication possibility because of inadequate frequency response. The VLC system was not suitable for use in higher central frequencies and higher bandwidth for digital modulations without pre-equalization. The VLC system with the pre-equalization circuit provided compliant conditions, thus higher central frequencies and bandwidth could be used.

4. Conclusion

In this paper pre-equalization of VLC transmitter has been presented. In order to get suitable frequency response of VLC transmitter based on phosphor white LED light source, we have been proposed equalization circuit used in our VLC system. The aforementioned HomePlug AV technology bandwidth from 2 MHz to 32 MHz was achieved with 3.43 dBm attenuation by commercial phosphorescent white light LED and proposed equalizer circuit. The objectives of this paper were to achieve suitable frequency response for the mentioned BVL technology solution. The proposed system demonstrably improves operational bandwidth in VLC system and could be considered as suitable system improvement for future Broadband over Visible Light (BVL) technology deploy.

(a) 5 MHz.

(b) 10 MHz.

(c) 15 MHz.

(d) 20 MHz.

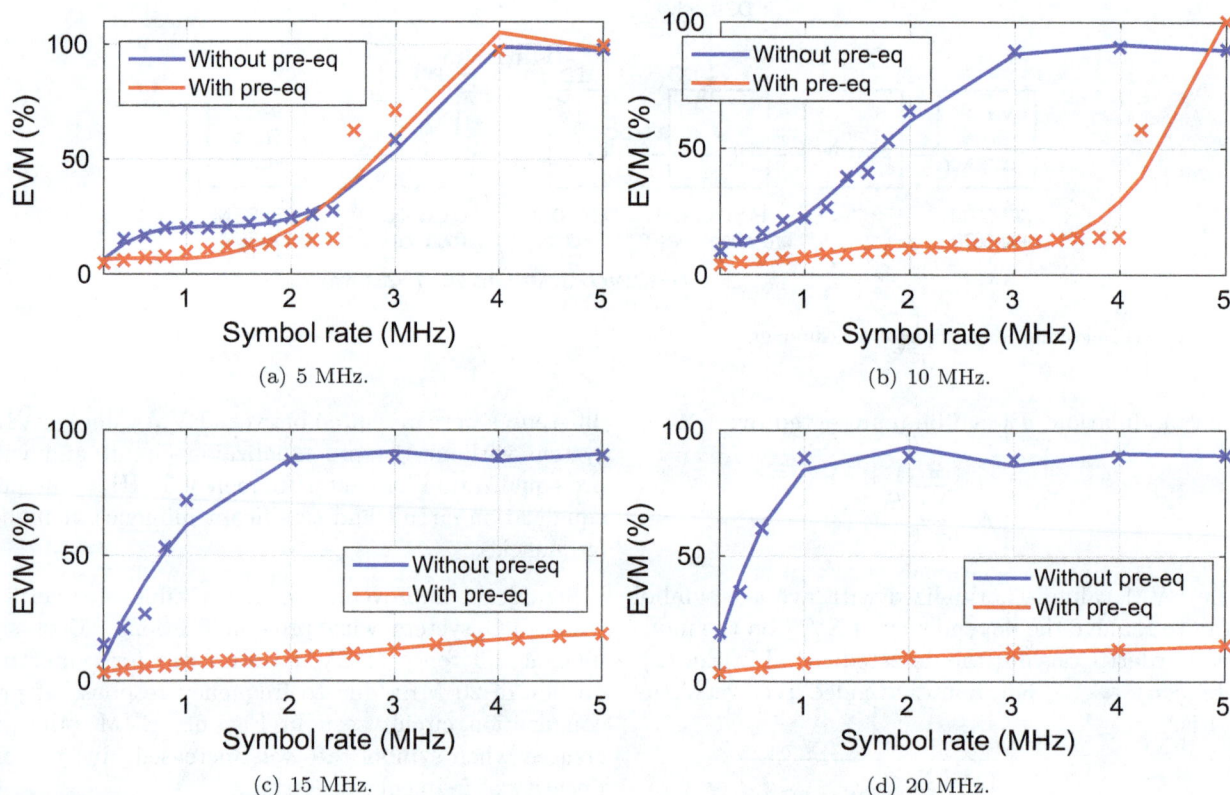

Fig. 8: Resulst of measurement EVM depending on symbol rate and carry frequency transmit 4QAM modulation via white light LED.

Acknowledgment

The research described in this article could be carried out thanks to the active support of the Ministry of Education of the Czech Republic within the projects no. SP2017/97: Remote Control of Public Lighting Luminaires via the Smart Technology Support and SP2017/128: Virtual instrumentation for the measurement and testing IV. This article was supported by projects Technology Agency of the Czech Republic TG01010137: BroadbandLIGHT. The research has been partially supported by the Ministry of Interior of the Czech Republic through grant project MVCR No. VI20172019071: Analysis of visibility of transport infrastructure for safety increasing during night, sunrise and sunset.

References

[1] MCCULLAGH, M. J. and D. R. WISELY. 155 Mbit/s optical wireless link using a bootstrapped silicon APD receiver. *Electronics letters.* 1994, vol. 30, no. 5, pp. 430–432. ISSN 0013-5194.

[2] CARRUTHERS, J. B. and J. M. KAHN. Angle Diversity for Nondirected Wireless Infrared Communication. *IEEE International Conference on Communications.* Atlanta: IEEE, 1998, pp. 1665–1670. ISBN 0-7803-4788-9. DOI: 10.1109/ICC.1998.683113.

[3] TANAKA, Y., T. KOMINE, S. HARUYAMA and M. NAKAGAWA. Indoor Visible Light Data Transmission System Utilizing White LED Lights. *IEICE transactions on communications.* 2003, vol. E86-B, no. 8, pp. 2440–2454. ISSN 0916-8516.

[4] DIMITROV, S. and H. HAAS. Information Rate of OFDM-Based Optical Wireless Communication Systems With Nonlinear Distortion. *Journal of Lightwave Technology.* 2013, vol. 31, no. 6, pp. 918–929. ISSN 2160-8881. DOI: 10.1109/jlt.2012.2236642.

[5] HUANG, X., S. CHEN, Z. WANG, J. SHI, Y. WANG, J. XIAO and N. CHI. 2.0-Gb/s Visible Light Link Based on Adaptive Bit Allocation OFDM of a Single Phosphorescent White LED. *IEEE Photonics Journal.* 2015, vol. 7, no. 5, pp. 1–8. ISSN 1943-0655. DOI: 10.1109/JPHOT.2015.2480541.

[6] HUANG, X., Z. WANG, J. SHI, Y. WANG and N. CHI. 1.6 Gbit/s phosphorescent white

LED based VLC transmission using a cascaded pre-equalization circuit and a differential outputs PIN receiver. *Optics Express*. 2015, vol. 23, no. 17, pp. 22034–22042. ISSN 1094-4087. DOI: 10.1364/OE.23.022034.

[7] MINH, H. L., D. O'BRIEN, G. FAULKNER, L. ZENG, K. LEE, D. JUNG, Y. J. OH and E. T. WON. 100-Mb/s NRZ Visible Light Communications Using a Postequalized White LED *IEEE Photonics Technology Letters*, 2009, vol. 21, iss. 15, pp. 1063–1065. ISSN 1041-1135. DOI: 10.1109/LPT.2009.2022413.

[8] O'BRIEN, D. C., H. LE MINH, G. FAULKNER, M. WOLF, L. GROBE, J. LI, and O. BOUCHET, Indoor Gigabit optical wireless communications: Challenges and possibilities. In: *International Conference on Transparent Optical Networks*. Munich: IEEE, 2010, pp. 1–6. ISBN 978-1-4244-7799-9. DOI: 10.1109/ICTON.2010.5549136.

[9] KIM, N.-T. Ultra-wideband bias-tee design using distributed network synthesis. *IEICE Electronics Express*. 2013, vol. 10, no. 15, pp. 1–8. ISSN 1349-2543. DOI: 10.1587/elex.10.20130472.

[10] KOUDELKA, P., J. LATAL, P. SISKA, J. VITASEK, A. LINER, R. MARTINEK and V. VASINEK. Indoor visible light communication: modeling and analysis of multi-state modulation. In: *Proceedings of Laser Communication and Propagation through the Atmosphere and Oceans, 9224*. San Diego: SPIE, 2015, pp. 1I-1–1I-8. ISBN 978-162841251-2. DOI: 10.1117/12.2063090.

[11] MARTINEK, R., J. ZIDEK and K. TOMALA. BER measurement in software defined radio systems. *Przeglad Elektrotechniczny*. 2013, vol. 89, iss. 2B, pp. 205–210. ISSN 0033-2097.

[12] STRATIL, T., P. KOUDELKA, J. JANKOVYCH, V. VASINEK, R. MARTINEK and T. PAVELEK. Broadband over Visible Light: High power wideband bias-T solution. In: *10th International Symposium on Communication Systems, Networks and Digital Signal Processing (CSNDSP)*. Prague: IEEE, 2016, pp. 1–5. DOI: 10.1109/CSNDSP.2016.7574002.

[13] MAHMOUD, H. A. and H. ARSLAN. Error vector magnitude to SNR conversion for nondata-aided receivers. *IEEE Transactions on Wireless Communications*. 2009, vol. 8, no. 5, pp. 2694–2704. ISSN 1536-1276. DOI: 10.1109/TWC.2009.080862.

[14] KOUDELKA, P., P. SOLTYS, R. MARTINEK, J. LATAL, P. SISKA, S. KEPAK and V. VASINEK. Utilization of M-QAM modulation during optical wireless Car to Car communication. In: *OptoElectronics and Communication Conference and Australian Conference on Optical Fibre Technology*. Melbourne: IEEE, 2014. pp. 452-454. ISBN 978-1-922107-21-3.

About Authors

Tomas STRATIL was born in 1990 in Olomouc, Czech republic. In 2015 He received Master's degree in optical communication from VSB–Technical University of Ostrava. His research interests include Visible light communication and Smart technologies.

Petr KOUDELKA was born in 1984 in Prostejov, Czech Republic. In 2006 received Bachelor's degree on VSB–Technical University of Ostrava, Faculty of Electrical Engineering and Computer Science, Department of telecommunications. Two years later he received on the same workplace his Master's degree in the field of Optoelectronics. In 2015 he successfully defended his dissertation thesis titled "Study of the Infoor Optical Wireless Network in the Visible Optical Radiation". He works as an Asistant Professor at VSB–Technical University of Ostrava since 2016. His current research interests include Wireless Optical Communications, Optical Access Networks and Smart City technologies.

Radek MARTINEK was born in 1984 in Czech Republic. In 2009 he received Master's degree in Information and Communication Technology from VSB–Technical University of Ostrava. Since 2012 he worked here as a Research Fellow. In 2014 he successfully defended his dissertation thesis titled „The Use of Complex Adaptive Methods of Signal Processing for Refining the Diagnostic Quality of the Abdominal Fetal Electrocardiogram". He became an Associate Professor in Technical Cybernetics in 2017 after defending the habilitation thesis titled "Design and Optimization of Adaptive Systems for Applications of Technical Cybernetics and Biomedical Engineering Based on Virtual Instrumentation". He works as an Associate Professor at VSB–Technical University of Ostrava since 2017. His current research interests include: Digital Signal Processing (Linear and Adaptive Filtering, Soft Computing - Artificial Intelligence and Adaptive Fuzzy Systems, Non-Adaptive Methods, Biological Signal Processing, Digital Processing of Speech Signals); Wireless Communications (Software-Defined Radio); Power Quality Improvement. He has more than 70 journal and conference articles in his research areas.

Tomas NOVAK was born in 1972 in Pribram, Czech Republic. In 1996 received Master's degree on VSB–Technical University of Ostrava, Faculty of Electrical Engineering and Computer Science, Department of Electrical Engineering. Seven years later he received on the same workplace his Ph.D. degree in the field of Electric Light and Diagnostic. Now he works as an Associate Professor at the same university. His current research interests include Public Lighting, Lighting Pollutions, Interior Light Controlling and Smart City technologies.

24

APPLICATION OF AFM MEASUREMENT AND FRACTAL ANALYSIS TO STUDY THE SURFACE OF NATURAL OPTICAL STRUCTURES

Dinara SOBOLA[1], Stefan TALU[2], Petr SADOVSKY[1],
Nikola PAPEZ[1], Lubomir GRMELA[1]

[1]Physics Department, Faculty of Electrical Engineering and Communication,
Brno University of Technology, Technicka 8, 616 00 Brno, Czech Republic
[2]Department of Automotive Engineering and Transports, Discipline of Descriptive Geometry and Engineering
Graphics, Faculty of Mechanical Engineering, Technical University of Cluj-Napoca, 103-105 B-dul Muncii
Street, 400641 Cluj-Napoca, Romania

sobola@vutbr.cz, stefan_ta@yahoo.com, petrsad@feec.vutbr.cz, xpapez04@stud.feec.vutbr.cz,
grmela@feec.vutbr.cz

Abstract. *The wings scales of the butterflies were studied by Atomic Force Microscopy (AFM) in the air. Measurements were done without special preparation of species in order to observe the surface in real conditions. The data of probe microscopy (figures) confirm AFM to be a powerful technique for determining features of the insects' wings. These features play a key role in optical phenomena which makes fascinating wings coloration. The structure determines light reflection, propagation, and diffraction. AFM imaging was done at the areas of specific colors without scale separation.*

Keywords

Atomic force microscopy, diffraction grating, fractal analysis, structural coloration, wing surface.

1.   Introduction

The natural structures are sources of inspiration for design of artificial devices for many years. One of such structures could be found at the wings of the butterflies and moths [1]. The exceptional optical and mechanical properties make them important for a number of applications [2], [3] and [4].

The fractal concept is widely used in biological sciences to characterize the irregular complex structures [5], [6] and [7]. On the other hand, fractal geometry offers new and valuable opportunities to describe and compare complex individual or species-specific patterns. It provides an integrative measure that captures the complexity of a whole pattern when explored at different scales, which would be a great help to study their variability and functionality [8] and [9]. Fractal analysis describes the geometrical complexity in the wings of several, taxonomically different butterflies, in terms of their fractal dimension. It was used in several studies from biological literature [10]. Two groups of butterflies were chosen for this study. Morphology and surface structure of the wings scales were investigated. First is Euploea mulciber, known as "Striped Blue Crow", and the second is Morpho didius, also named as "Giant Blue Morpho". Both species exhibit strongly angle dependent coloration of wings. Our analysis was carried out using 10 specimens for each species. The darkest and the brightest areas of wings were studied. Even ordinary optical microscopy can show that the scales are different along the wing surface; it depends on the distance from the body. Some of them are even modified into the tiny tubes. Forewings of the butterflies were studied at discal and postdicsal areas. The choice of areas could be explained by differences in color and consequently the surface structure (Fig. 1): the one that reflects light and looks shiny, and the other looking dark and lusterless.

Fig. 1: Rows of scales, the iridescence is created by physical phenomena.

2. Experimental

2.1. Atomic Force Microscopy

AFM (Atomic Force Microscope) NTEGRA (NT-MDT production) was used to study the surface topography in semi-contact mode. No special preparation of the sample was done: just cutting of a piece of wing by scissors and fixing it by a tape on the substrate from the bottom side of the wing. All measurements were performed in the same laboratory, at room temperature (296 ± 1 K) and 50 ± 1 % relative humidity. The measurements were repeated three times for each sample on different reference areas, to validate the reproducibility of the data. Statistical analyses were performed using the GraphPad InStat version 3.20 computer software package (GraphPad, San Diego, CA, USA) [11]. The data of AFM represent only the surface appearance, without explanation of inner structure of the wings. The scale surface ridges are responsible for direction of light wave's propagation. These structures define the part of spectrum which is absorbed or penetrates to the next bottom layers. The upper topography texture of the wings represents diffraction grating (Fig. 2).

Smaller scanning area allows detailed observation of the surface topography (Fig. 3).

The native software of the microscope provides processing of the results. Figure 4 demonstrates High-High correlation graphs of both species. It characterizes lateral distribution of surface features - the distance of features' high correlation. The distance between surface features as well as their shape forms the diffraction grating on the surface and contributes the scale color.

The characteristic wavelengths of the spatial periodicity are measured using a spatial power spectrum (Tab. 1). The wavelength periodicity also shows the difference between species and different colors on the same wing. There is one large peak (sharp at a particular wavelength) on each graph and smaller peaks surrounding (Fig. 5).

2.2. Fractal Analysis

Cube counting method, based on the linear interpolation, applied for AFM data, was used for fractal analysis of the butterflies' wings, which is described in detail in [12]. Cube counting method [12] is derived directly from a definition of box-counting fractal dimension. The algorithm is based on the following steps: a cubic lattice with lattice constant l is superimposed on the z-expanded surface. Initially l is set at $X = 2$ (where X is length of edge of the surface), resulting in a lattice of $2 \times 2 \times 2 = 8$ cubes. Then $N(l)$ is the number of all cubes that contain at least one pixel of the image. The lattice constant l is then reduced stepwise by factor of 2 and the process is repeated until l equals to the distance between two adjacent pixels. The slope of a plot of log $N(l)$ versus log$1 = l$ gives the fractal dimension D directly. The results of the fractal dimensions (D) for AFM images of wings areas of all samples are shown in Fig. 6.

The results of the fractal dimensions (D) with coefficients of correlation (R_2) are given in Tab. 2. For all analyzed cases (Tab. 2), the coefficients of correlation (R_2) associated with fractal dimensions D were greater than 0.99 representing a good linear correlation. An (R_2) of 1.0 indicates that the regression line perfectly fits the data.

The texture of wings scales is semiregular. It is also found that the geometrical complexity of the butterflies' wings shows clear distinctions for the two groups of species in terms of their fractal dimension. These results provide us additional methods for distinguishing species and their distinctive colors (differences in intensity and tonality).

2.3. Thermocamera Imaging

Since AFM is limited by height of the sample, the imaging was carried out on flat scales of the wings (not on veins of wing construction). But these construction elements are well seen on thermocamera images. Thermocamera imaging mostly depends on the material nature and its inner structure. It partly represents macrostructure of the wing.

The thermocamera was used to follow the behavior of wings in relation to heated objects: the butterfly

(a) Giant blue morpho samples (black color).

(b) Striped blue crow samples (brown color).

(c) Giant blue morpho samples (blue color).

(d) Striped blue crow samples (blue color).

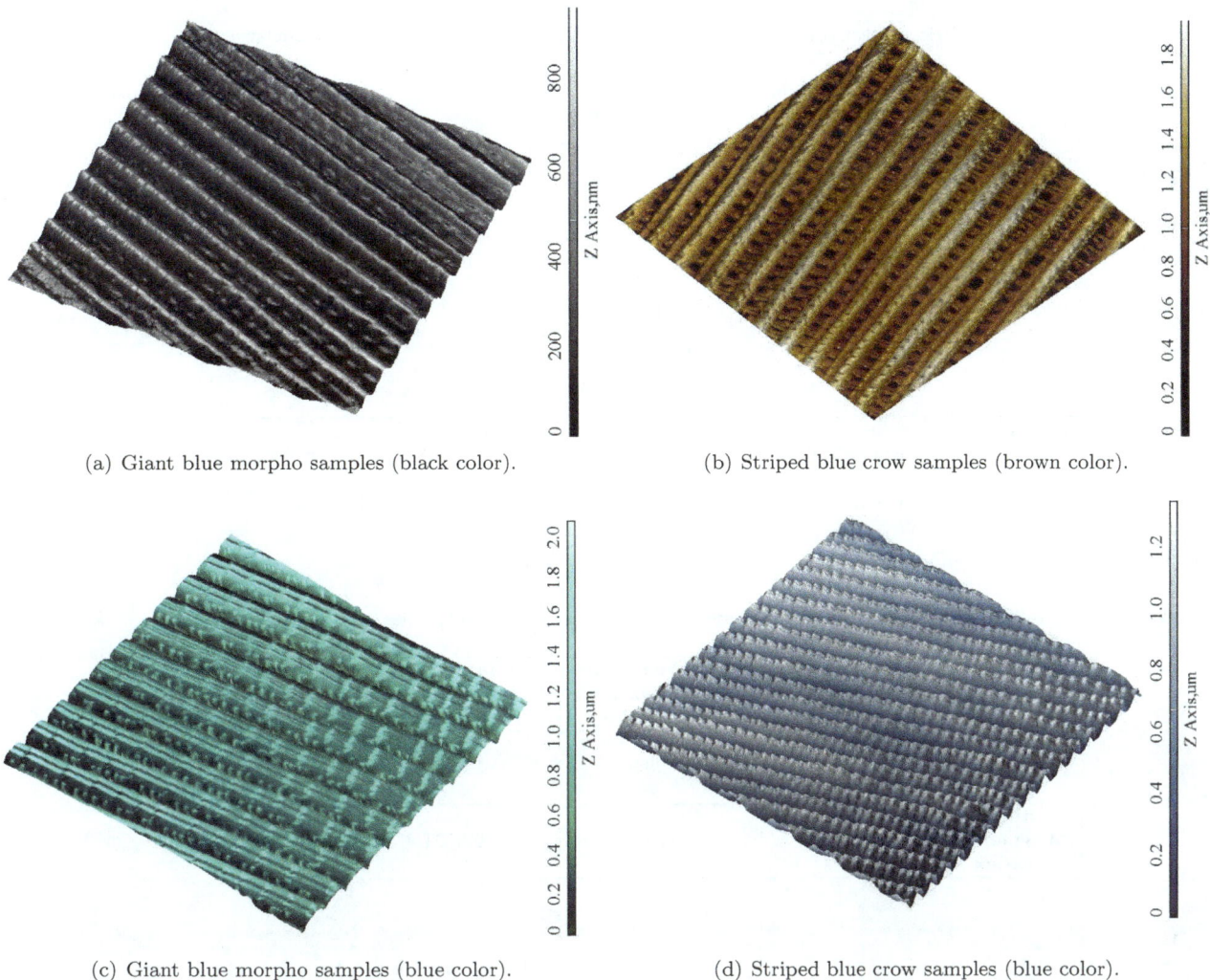

Fig. 2: AFM images of wing area (area scan 20×20 μm).

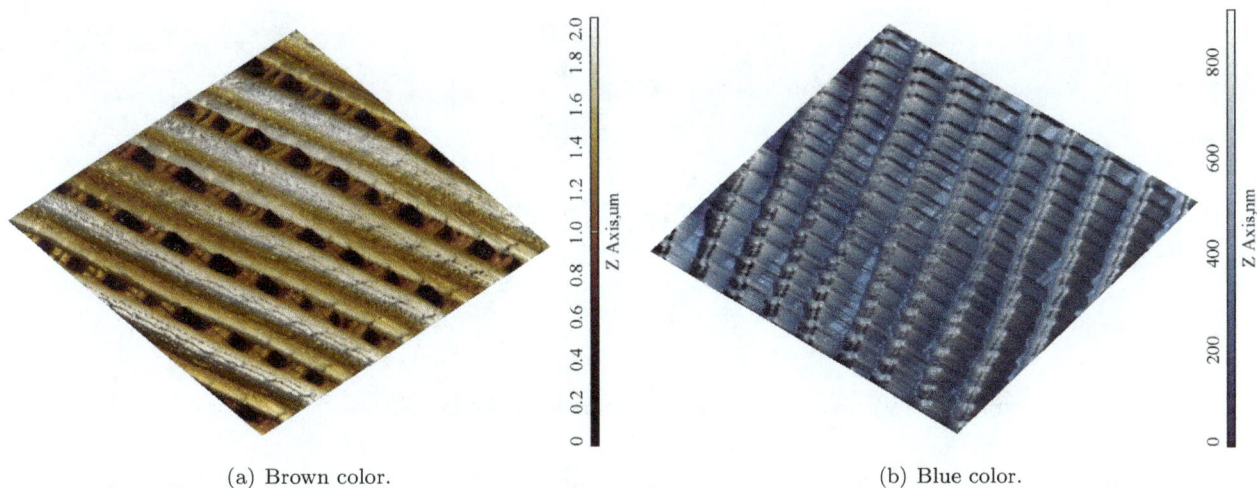

(a) Brown color.

(b) Blue color.

Fig. 3: AFM images of wings area of striped blue crow samples (area scan 10×10 μm).

wings seem to be transparent. The heated pattern is well observed through the wing in the image from camera (Fig. 7). The transparence helps butterflies to be invisible for most of predators. It is well known that a number of predators have very good vision in IR spectrum. Only the veins show the

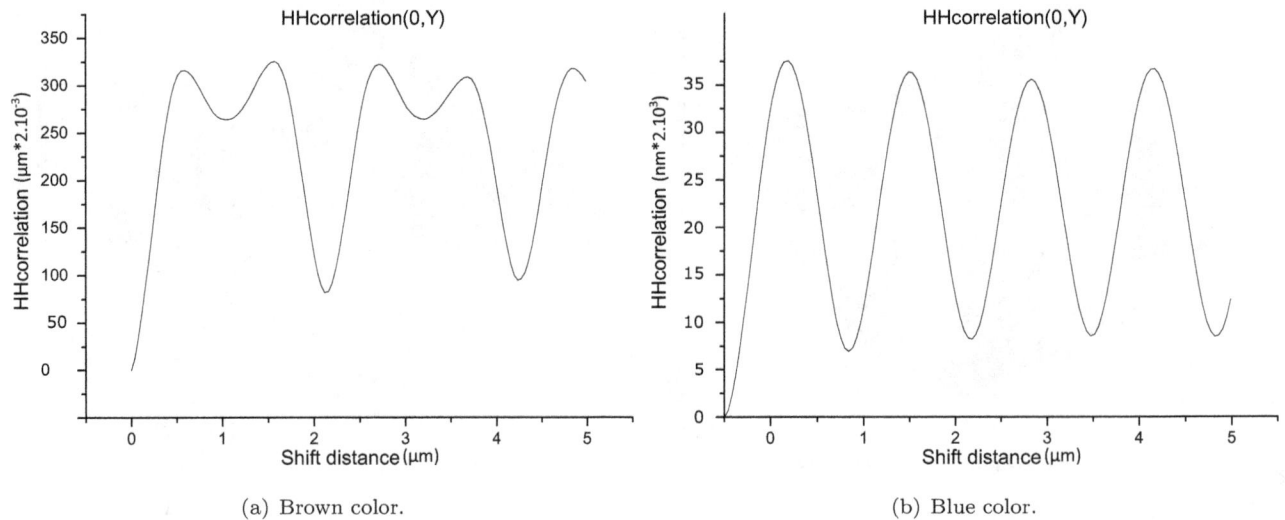

(a) Brown color.

(b) Blue color.

Fig. 4: High-High correlation graphs that correspond to Striped Blue Crow samples.

Tab. 1: Radial wavelength and radial wavelength index for the two groups of butterflies.

Parameters	Striped Blue Crow samples (brown color)	Striped Blue Crow samples (blue color)	Giant Blue Morpho samples (black color)	Giant Blue Morpho samples (blue color)
Radial Wavelength (μm)	10.0060	9.9990	1.6660	19.9990
Radial Wavelength Index	0.0616	0.0995	0.0794	0.0686

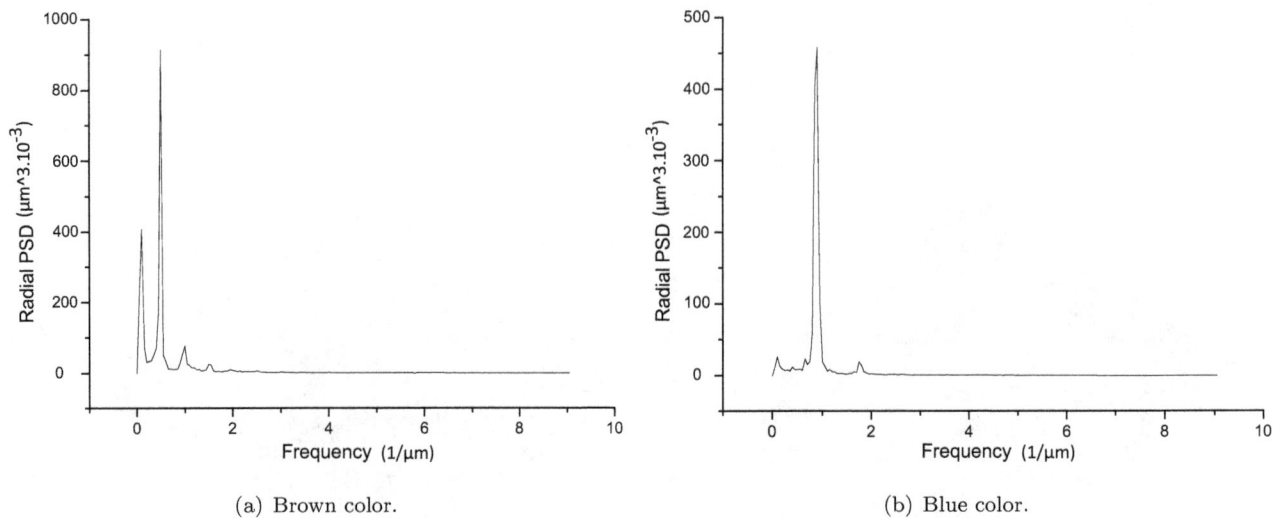

(a) Brown color.

(b) Blue color.

Fig. 5: Radial power spectrum density graphs that correspond to Striped Blue Crow samples.

presence of the wings between thermocamera and heated object. This could be considered as one more protected mechanism for surviving (besides imitation of tree and grass leaves or eye spots on the wings).

The basic properties of the height values distribution of the surface samples (including its variance, skewness, and kurtosis), computed according to [12] are shown in Tab. 3.

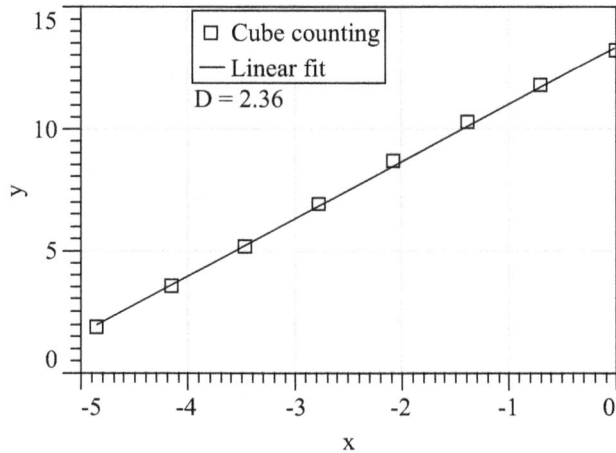

(a) Striped Blue Crow samples (blue color).

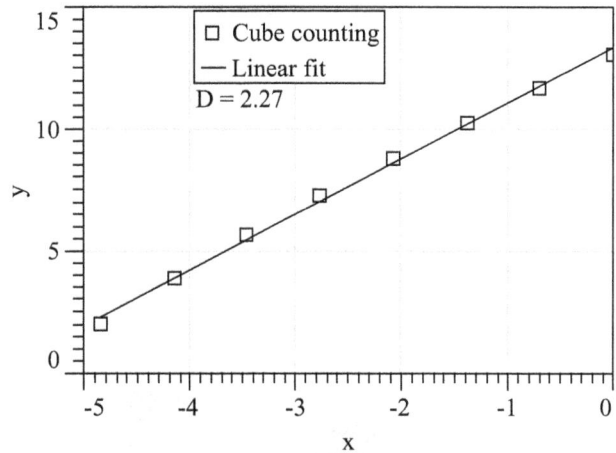

(b) Striped Blue Crow samples (brown color).

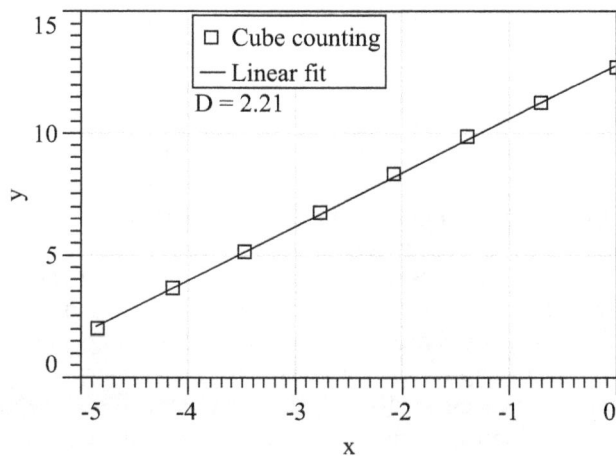

(c) Giant Blue Morpho samples (black color).

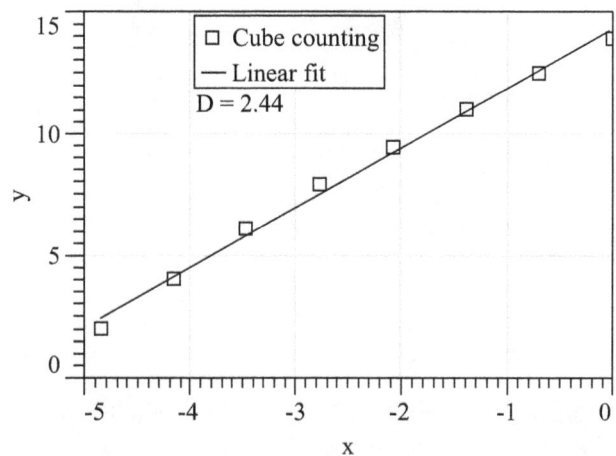

(d) Giant Blue Morpho samples (blue color).

Fig. 6: Fractal dimension for AFM images of wings areas (area scan 20×20 µm).

Tab. 2: The fractal dimensions (D) with coefficients of correlation (R^2) determined by the cube counting method, based on the linear interpolation type, of the two groups of butterflies: a) Striped Blue Crow sample; b) Giant Blue Morpho sample. Scanning square areas of 20×20 µm.

Parameters	Striped blue crow samples (brown color)	Striped blue crow samples (blue color)	Giant blue morpho samples (black color)	Giant blue morpho samples (blue color)
D	2.27 ± 0.02	2.36 ± 0.02	2.21 ± 0.018	2.44 ± 0.018
R^2	0.995	0.995	0.996	0.996

3.   Discussion

A lot of natural objects have a fractal structure. The structure is repeated at the lower metric range. The wing scales are fractal photonic structures which are developed by nature. It is one of the ways to manipulate the energy of sun. In order to analyze the influence of a structure on coloration, we scanned different colors area of two species. Branches of micro and nano-sized features form the wing scale [13]. Although the diffraction grating is observed at all cases,

there are differences in correlation lengths and fractal data of the surfaces. Diffraction grating of E. mulciber specie was studied by F. Mika in [14] by SEM. The advantages of AFM in this case before SEM are measurements in real conditions (air, humidity) and obtain real 3D data about surfaces topography. Correlation lengths ($Lx = 0.882$ µm and $Ly = 0.352$ µm for blue color, and $Lx = 0.800$ µm $Ly = 299$ µm for brown color of Striped Blue Crow wing sample) are the characteristic lengths over the scanned surface. Correlation length depends of roughness of the surface: the higher roughness, the smaller correlation length [15]. Consid-

Tab. 3: The basic properties of the height values distribution (including its variance, skewness and kurtosis) of the two groups of butterflies: a) Striped Blue Crow sample; b) Giant Blue Morpho sample. Scanning square areas of 20×20 μm.

The basic properties of the height values distribution of the surface samples	Striped blue crow samples (brown color)	Striped blue crow samples (blue color)	Giant blue morpho samples (black color)	Giant blue morpho samples (blue color)
	Values	Values	Values	Values
Ra (Sa) (μm)	0.439	0.66	0.816	0.351
Rms (Sq) (μm)	0.543	0.828	0.933	0.456
Skew (Ssk) (-)	-0.279	0.755	-0.106	-1.35
Kurtosis (Sku) (-)	-0.07	0.248	-1.14	0.981
Inclination θ (°)	3.2	5.7	4.7	5.2
Inclination φ (°)	-0.2	7.4	32.7	-173.6

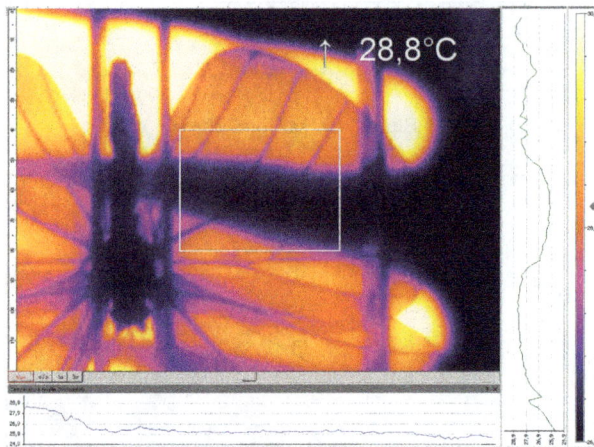

Fig. 7: Thermocamera image of fingers behind the wings.

ering this, the attention is paid to the quality of the AFM measurements, since the noise is close to zero correlation length [15]. The data for fractal analysis were measured with the same magnification. Fractal dimensions correlate with the scales morphology. So, black and brown color surfaces have lower fractal dimension than the blue areas. And consequently, these values relate to reflectance of the surface [16], [17], [18], [19] and [20]. Multilayered structure of the scales and their arrangement contribute the colors of wings. However, surface features also play important role (which are diffraction gratings for visible light). They are well seen in AFM images [21], [22] and [23]. The chitin elements have different shapes: the oriented layers on the ridges and well visible notches constriction. The cross-ribs between ridges form a complex three-dimensional structure. The variation of surface structure provides obtaining the colors from iridescent to antiglare.

complicated semi-ordered combination of surface features. The sizes of these features are comparable to visible light wavelengths. The studied 3D topography is a first step in color producing: the diffraction grating of the scale surface defines further propagation, transmission and reflection of light. The wings seem to be transparent at IR radiation since it is possible to see heated objects through the wings by thermo-camera. Here, by combining correlational and experimental evidence, we also describe the surface of scales as physical structures, which have a fractal nature and can reveal biologically meaningful information. The ridges of the scale surface have fractal properties and are oriented in one direction. The morphology of the analyzed samples provides additional description about the structural features of the butterflies' wings 3D surface topography. Both theory analysis and experimental results suggest that AFM, the statistical and fractal analysis can provide additional insight into the wings 3D morphology and can be included in an algorithmic mathematical model.

Acknowledgment

Research described in the paper was financially supported by the Ministry of Education, Youth and Sports of the Czech Republic under the project CEITEC 2020 (LQ1601), by the National Sustainability Program under grant LO1401, by the Grant Agency of the Czech Republic under no. GACR 15-05259S and by Internal Grant Agency of Brno University of Technology, grant No. FEKT-S-17-4626. For the research, infrastructure of the SIX Center was used.

4.   Conclusion

There are many optical systems created by nature. One of them is the wing scale which has superior optical and hydrophobic surface properties. It was found that different color areas have different three-dimensional (3D) structure of the surface. The 3D structure represents

References

[1] DALLAEVA, D. and P. TOMANEK. AFM study of structure influence on butterfly wings coloration. *Advances in Electrical and Electronic Engineering*. 2012, vol. 10, no. 2, pp. 120–124. ISSN 1804-3119. DOI: 10.15598/aeee.v10i2.616.

[2] KINGSOLVER, J. G. Butterfly Engineering. *Scientific American*. 1985, vol. 253, iss. 2, pp. 106–113. ISSN 0036-8733.

[3] GRUVERMAN, A., B. J. RODRIGUEZ and S. V. KALININ. Nanoscale electromechanical and mechanical imaging of butterfly wings by Scanning Probe Microscopy. *Journal of Scanning Probe Microscopy*. 2006, vol. 1, iss. 2, pp. 74–78. ISSN 1557-7937. DOI: 10.1166/jspm.2006.008.

[4] BIRO, L. P., Z. BALINT, K. KERTESZ, Z. VERTESY, G. I. MARK, Z. E. HORVATH, J. BALAZS, D. MEHN, I. KIRICSI, V. LOUSSE and J.-P. VIGNERON. Role of photonic-crystal-type structures in the thermal regulation of a Lycaenid butterfly sister species pair. *Physical Review*. 2003, vol. 67, iss. 2, pp. 1–7. ISSN 2469-9926. DOI: 10.1103/PhysRevE.67.021907.

[5] TALU, S. *Micro and nanoscale characterization of three dimensional surfaces. Basics and applications*. 1st ed. Cluj-Napoca: Napoca Star Publishing House, 2015. ISBN 978-606-690-349-3.

[6] TALU, S. Mathematical methods used in monofractal and multifractal analysis for the processing of biological and medical data and images. *Animal Biology and Animal Husbandry. International Journal of the Bioflux Society*. 2012, vol. 4, iss. 1, pp. 1–4. ISSN 2067-6344.

[7] TALU, S. Texture analysis methods for the characterisation of biological and medical images. *Extreme Life, Biospeology and Astrobiology*. 2012, vol. 4, iss. 1, pp. 8–12. ISSN 2067-6360. DOI: 10.5772/8912.

[8] JOVANI, R., L. PEREZ-RODRIGUEZ and F. MOUGEOT. Fractal geometry for animal biometrics: a response to Kuhl and Burghardt. *Trends in Ecology and Evolution*. 2013, vol. 28, no. 9, pp. 499–500. ISSN 0169-5347. DOI: 10.1016/j.tree.2013.06.004.

[9] PEREZ-RODRIGUEZ, L., R. JOVANI and F. MOUGEOT. Fractal geometry of a complex plumage trait reveals bird's quality. *Proceedings of the Royal Society B: Biological Sciences*. 2013, vol. 280, iss. 1755, pp. 1–6. ISSN 0962-8452. DOI: 10.1098/rspb.2012.2783.

[10] CASTREJON-PITA, A. A., A. SARMIENTO-GALAN, J. R. CASTREJON-PITA and R. CASTREJON-GARCIA. Fractal Dimension in Butterflies' Wings: a novel approach to understanding wing patterns? *Journal of Mathematical Biology*. 2005, vol. 50, iss. 5, pp. 584–594. ISSN 0303-6812. DOI: 10.1007/s00285-004-0302-6.

[11] GraphPad InStat software, version 3.20. *GraphPad Software* [online]. 2016. Available at: http://www.graphpad.com/instat/instat.htm.

[12] KLAPETEK, P., D. NECAS and C. ANDERSON. Gwyddion software user guide, version 2.28. *Czech Metrology Institute* [online]. 2012. Available at: http://gwyddion.net/.

[13] KERTESZ, K., G. PISZTER, E. JAKAB, Z. BALINT, Z. VERTESY and L. P. BIRO. Color change of Blue butterfly wing scales in an air - Vapor ambient. *Applied Surface Science*. 2013, vol. 281, iss. 1, pp. 49–53. ISSN 0169-4332. DOI: 10.1016/j.apsusc.2013.01.037.

[14] MIKA, F., J. MATEJKOVA-PLSKOVA, S. JIWAJINDA, P. DECHKRONG and M. SHIOJIRI. Photonic Crystal Structure and Coloration of Wing Scales of Butterflies Exhibiting Selective Wavelength Iridescence. *Materials*. 2012, vol. 5, iss. 5, pp. 754–771. ISSN 1996-1944. DOI: 10.3390/ma5050754.

[15] GUPTA, V. K. and R. A. JANGID. Microwave response of rough surface with auto-correlation functions, RMS heights and correlation lengths using active remote sensing. *Indian Journal of Radio and Space Physics*. 2011, vol. 40, iss. 3, pp. 137–146. ISSN 0975-105X.

[16] PIECHACZEK, M., L. SMEDOWSKI and S. PUSZ. Evaluation of the possibilities of applying fractal analysis for the characterization of molecular arrangement of carbon deposits in comparison to conventional instrumental methods. *International Journal of Coal Geology*. 2015, vol. 139, iss. 1, pp. 40–48. ISSN 0166-5162. DOI: 10.1016/j.coal.2014.06.026.

[17] TALU, S., M. BRAMOWICZ, S. KULESZA, A. GHADERI, V. DALOUJI, S. SOLAYMANI, M. FATHI KENARI and M. GHORANNEVISS. Fractal features and surface micromorphology of diamond nanocrystals. *Journal of Microscopy*. 2016, vol. 264, iss. 2, pp. 143–152. ISSN 1365-2818. DOI: 10.1111/jmi.12422.

[18] TALU, S., M. BRAMOWICZ, S. KULESZA, A. SHAFIEKHANI, A. GHADERI, F. MASHAYEKHI and S. H. SOLAYMANI. Microstructure and Tribological Properties of FeNPs@a-C:H Films by Micromorphology Analysis and Fractal Geometry. *Industrial and Engineering Chemistry Research*. 2015, vol. 54, iss. 33, pp. 8212–8218. ISSN 1520-5045. DOI: 10.1021/acs.iecr.5b02449.

[19] STACH, S., Z. GARCZYK, S. TALU, S. SO-
LAYMANI, A. GHADERI, R. MORADIAN, N.
BERYANI NEZAFAT, S. M. ELAHI and H. GHO-
LAMALI. Stereometric parameters of the Cu/Fe
NPs thin films. *The Journal of Physical Chem-
istry C.* 2015, vol. 119, iss. 31, pp. 17887–17898.
ISSN 1932-7455. DOI: 10.1021/acs.jpcc.5b04676.

[20] TALU, S., S. SOLAYMANI, M. BRAMOWICZ,
N. NASERI, S. KULESZA and A. GHADERI.
Surface micromorphology and fractal geometry of Co/CP/X (X = Cu, Ti, SM and Ni)
nanoflake electrocatalysts. *RSC Advances.* 2016,
vol. 6, iss. 32, pp. 27228–27234. ISSN 2046-2069.
DOI: 10.1039/C6RA01791F.

[21] TALU, S., S. SOLAYMANI, M. BRAMOWICZ,
S. KULESZA, A. GHADERI, S. SHAHPOURI
and S. M. ELAHI. Effect of electric field direction
and substrate roughness on three-dimensional self-
assembly growth of copper oxide nanowires. *Journal of Materials Science: Materials in Electronics.*
2016, vol. 27, iss. 9, pp. 9272–9277. ISSN 1573-
482X. DOI: 10.1007/s10854-016-4965-8.

[22] NASERI, N., S. SOLAYMANI, A. GHADERI,
M. BRAMOWICZ, S. KULESZA, S. TALU,
M. POURREZA and S. GHASEMI. Microstructure, morphology and electrochemical properties
of Co nanoflake water oxidation electrocatalyst
at micro-and nanoscale. *RSC Advances.* 2017,
vol. 7, iss. 21, pp. 12923–12930. ISSN 2046-2069.
DOI: 10.1039/C6RA28795F.

[23] RAMAZANOV, S., S. TALU, D. SOBOLA,
S. STACH, G. RAMAZANOV. Epitaxy of
silicon carbide on silicon: Micromorpholog-
ical analysis of growth surface evolution.
Superlattices and Microstructures. 2015,
vol. 86, iss. 1, pp. 395–402. ISSN 0749-6036.
DOI: 10.1016/j.spmi.2015.08.007.

About Authors

Dinara SOBOLA was born in 1988 in Kaspiysk,
Russian Federation. She received her M.Sc. degree from Dagestan State University in 2010. She
finnished Ph.D. study in Physical Electronics and
Nanotechnology at Faculty of Electrical Engineering
and Communication, Brno University of Technology
in 2015 and continues to work as researcher in Optoelectronic Characterisation of Nanostructures. Her
research interests include scanning probe microscopy,
wide band gap semiconductors.

Stefan TALU was born in Floresti, county Vaslui,
Romania, on 31st July 1964. He graduated as a mechanical engineer from University of Craiova, Faculty
of Mechanics, Romania in 1988. He received his Ph.D.
degree (in technical sciences with specialty in technology of machine building) and Associate Professor from
the Technical University of Cluj-Napoca, Romania,
in 1998 and 2002 respectively. His research interests
include fractal/multifractal geometry, descriptive
geometry, mathematical algorithms for solving the
optimum problems and computer aided design.

Petr SADOVSKY is a researcher at the Department
of Physics at Faculty of Electrical Engineering and
Communication, Brno University of Technology. His
scientific interests include processing of signals.

Nikola PAPEZ was born in 1989 in Brno, Czech
Republic. He received the M.Sc. degree in Communications and Informatics from Brno University of
Technology, Czech Republic, in 2016 and currently
continues with postgraduate study in Physical Electronics and Nanotechnology at the same University.
His scientific activity is directed toward the areas of
solar cells, its stability and surface morphology.

Lubomir GRMELA is a Professor, Head of
Department of Physics, a researcher in the group of
Optoelectronic Characterisation of Nanostructures.
His research interests include preparation and charac-
terization of structures for electronics.

Appendix

The basic properties of the height values distribution,
including its variance, skewness and kurtosis, computed according the Ref. [12] is defined as follows:

- RMS value of the height irregularities: this quantity is computed from data variance.

- *Ra* value of the height irregularities: this quantity
is similar to RMS value with the only difference in
exponent (power) within the data variance sum.
As for the RMS this exponent is $q = 2$, the *Ra*
value is computed with exponent $q = 1$ and absolute values of the data (zero mean).

- Height distribution skewness: computed from 3rd
central moment of data values.

- Height distribution kurtosis: computed from 4th
central moment of data values.

- Mean inclination of facets in area: computed by
averaging normalized facet direction vectors.

- Variation, which is calculated as the integral of the
absolute value of the local gradient.

25

Formation Process and Properties of Ohmic Contacts Containing Molybdenum to AlGaN/GaN Heterostructures

Wojciech MACHERZYNSKI, Jacek GRYGLEWICZ, Andrzej STAFINIAK,
Joanna PRAZMOWSKA, Regina PASZKIEWICZ

Department of Microelectronics and Nanotechnology, Faculty of Microsystem Electronics and Photonics,
Wroclaw University of Technology, Janiszewskiego 11/17, 50-370 Wroclaw, Poland

wojciech.macherzynski@pwr.edu.pl, jacek.gryglewicz@pwr.edu.pl, andrzej.stafiniak@pwr.edu.pl,
joanna.prazmowska@pwr.edu.pl, regina.paszkiewicz@pwr.edu.pl

Abstract. *Properties of wide bandgap semiconductors as chemical inertness to harsh conditions and possibility of working at high temperature ensure possible applications in the field as military, aerospace, automotive, engine monitoring, flame detection and solar UV detection. Requirements for ohmic contacts in semiconductor devices are determined by the proposed application. These contacts to AlGaN/GaN heterostructure for application as high temperature, high frequency and high power devices have to exhibit good surface morphology and low contact resistance. The latter is a crucial factor in limiting the development of high performance AlGaN/GaN devices. Lowering of the resistance is assured by rapid thermal annealing process. The paper present studies of Ti/Al/Mo/Au ohmic contacst annealed at temperature range from 825 °C to 885 °C in N_2 atmosphere. The electrical parameters of examined samples as a function of the annealing process condition have been studied. Initially the annealing temperature increase caused lowering of the contacts resistance. The lowest value was noticed for the temperature of annealing equal to 885 °C. Further increase of annealing temperature led to deterioration of contact resistance of investigated ohmic contacts.*

Keywords

AlGaN/GaN, ohmic contacts, RTA, RTP, surface morphology, Ti/Al/Mo/Au.

1. Introduction

The ohmic contacts in AlGaN/GaN semiconductor devices have crucial influence on device performance [1], [2], and [3]. At the high electron mobility transistor (HEMT) ohmic contacts govern transconductance and saturation current. The AlGaN/GaN HEMTs are capable of handling higher current densities than other III-V high electron mobility transistors due to higher two-dimensional electron gas (2DEG) density (10^{13} cm^{-2} or higher) accumulated on the AlGaN/GaN interface [4] and [5].

The thermal stability of AlGaN/GaN heterostructures and their chemical inertness engender difficulties in ohmic contact formation. Smooth surface morphology for high edge definition and minimal contact resistance are essential for desirable device behavior. To achieve a change from Schottky contact after deposition metallization to ohmic contact, samples were annealed at different temperatures. High annealing temperatures, usually over 800 °C, are required to establish good ohmic contact performance [6], [7], [8], [9], [10], [11], [12], and [13]. On the other hand, so high annealing temperature causes changes on the heterostructure and metal-semiconductor interface, which in turn leads to alteration of 2DEG parameter - carrier mobility [14].

At this stage of investigation we have to seek for compromise between appropriate ohmic contact performance and 2DEG parameters. In our studies, for Ti/Al/Mo/Au ohmic contacts, the temperature had to be above 800 degrees to reach this compromise. To minimize the deterioration of 2DEG parameters, the thermal annealing was led in time as short as possible to reach good ohmic performance.

Also, the high temperature annealing has strong influence on the microstructure and the surface morphology of the ohmic contact. The reasons of the impact of annealing temperature on the microstructure of Ti/Al/Ni/Au are the low melting temperature of aluminum (660 °C) and migration followed by coalescence of agglomerates [15].

In this study Ti/Al/Mo/Au metallization scheme have been used. The Ti/Al based ohmic contact is one of the most prevalent metallization schemes of ohmic contact to AlGaN/GaN heterostructures [6], [7], [8], [9], [10], [11], [12], and [13]. A titanium layer is essential as, at elevated temperatures, the Ti participates in the reaction with nitrides on the interface and forms TiN [8]. This reaction extracts nitrogen and generates N-vacancies. N-vacancies act as n-type dopants and create a highly doped layer underneath the metallization, leading to low-contact resistance of the Ti/Al based ohmic contact. The aluminum is the layer which is responsible for the formation of the ohmic contact to AlGaN/GaN heterostructures. In general, there is no standard annealing temperature that leads to low resistance ohmic contact. Research studies indicated on different temperature that exhibited successful ohmic contact formation [6], [7], [8], [9], [10], [11], [12], and [13].

Low resistance ohmic contacts to AlGaN/GaN are of great importance because an improvement of their electrical properties would lead to enhancement of the device performance. In this paper we report the influence of annealing temperature on the current-voltage characteristics and contact resistance R_c.

2.   Experimental Details

The AlGaN/GaN heterostructure applied in this study consisted of AlGaN/GaN grown by metalorganic vapor phase epitaxy (MOVPE) on sapphire substrate. Prior to metal deposition, the native oxide (Ga_2O_3) was removed from all samples surfaces by etching in HCl:H_2O (1:1) solution, followed by a deionised water rinsing and drying in N_2 flow. Then, the samples were immediately loaded into the vacuum chamber of an evaporation system. The metallic contact consisting of Ti/Al/Mo/Au (23/100/40/190 nm) was deposited on the substrate under vacuum conditions with a base pressure lower than 10^{-6} mbar. The metal layers were deposited by using an electron beam evaporator (Ti, Al, Mo) and resistance heater (Au). The transfer length method (TLM) mesa isolation was achieved by means of a 80 nm deep mesa etch performed by Cl_2/BCl_3/Ar reactive ion etching. The Ti/Al/Mo/Au ohmic metallizations were annealed at various temperatures in rapid thermal annealing (RTA) system. The temperature of each annealing process was changed over the range

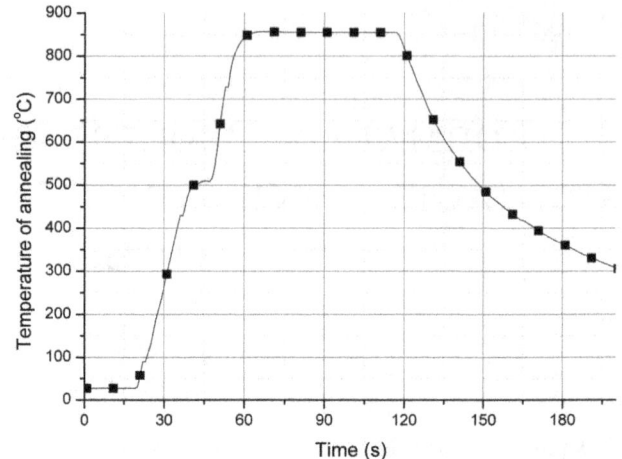

Fig. 1: Temperature characteristic of thermal annealing at 855 °C.

from 825 °C to 855 °C and the annealing time of 60 seconds was kept for all samples (Fig. 1).

To study the influence of the annealing process parameters on the properties of the Ti/Al/Mo/Au metallization, the electrical parameters (I-V characteristic and contact resistance Rc) were measured. For contact resistance Rc measuring we adopted the TLM (four probes mode) test structure. The distance between contact were 31, 20, 10 and 6 μm. The current-voltage (I-V) characteristics were measured on the two neighboring contacts from TLM test structure.

3.   Result and Discussion

Figure 2 shows the current-voltage characteristics of Ti/Al/Mo/Au metallization as a function of annealing temperature. When the distance between measured neighboring pads is lower (Fig. 2(a) - 6 μm, Fig. 2(b) - 10 μm) the current at a given voltage as expected increases. However, the decreasing of distance between pads is reflected at more visible non-linearity of I-V characteristics (Fig. 2). It could mean, that on the m-s interface remains a barrier. The influence of temperature of RTA annealing process shows, that the smallest total resistance RT at given voltage was achieved at 855 °C (Fig. 2). First, at given voltage with increase of annealing temperature up to 855 °C the total resistance RT decreased.

However, increase of annealing temperature above the temperature of 855 °C caused the increase of total resistance RT. But the shape of I-V characteristic remains slightly non-linear. Figure 3 shows the contact resistance Rc of Ti/Al/Mo/Au metallization as a function of annealing temperature. For calculation of contact resistance, we adopted the TLM method. The resistance for given distance between pads and anneal-

(a) distance between measured metal pads: 5 μm

(b) distance between measured metal pads: 10 μm

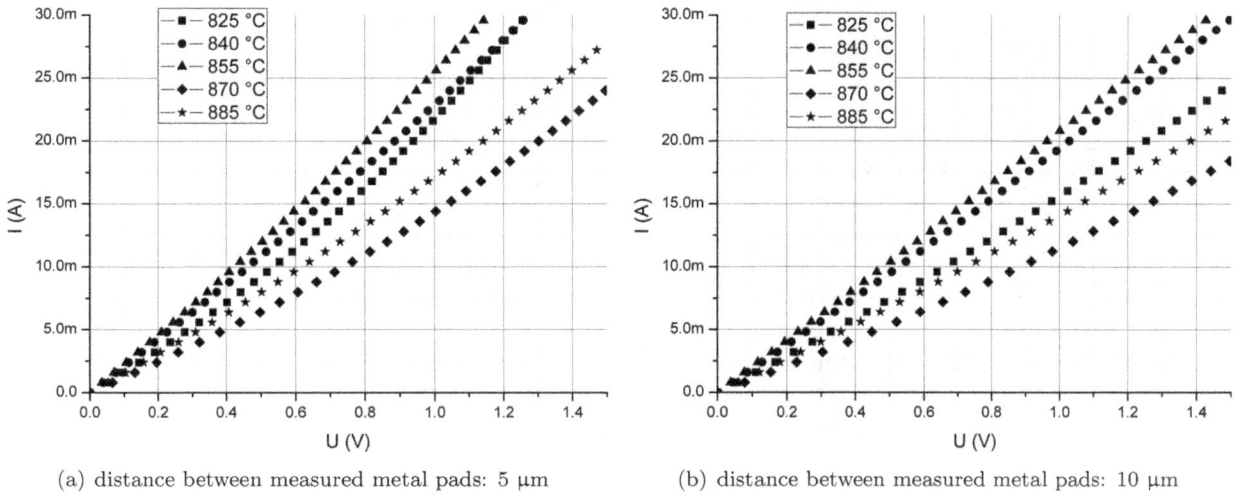

Fig. 2: Current-Voltage characteristic of Ti/Al/Mo/Au contact after annealing at various temperatures.

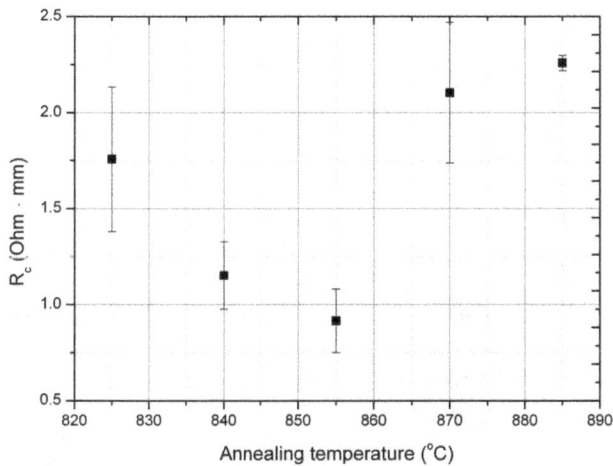

Fig. 3: Contact resistance Rc as a function of annealing temperature of Ti/Al/Mo/Au contacts for AlGaN/GaN heterostructures.

phology and the heterogeneous chemical composition of the ohmic contact is the formation of Al droplets

(a) before annealing

ing temperature was calculated from I-V characteristics at given voltage (0.2 V). Also for those contacts the resistance RC have the smallest value at 855 °C.

SEM was used to characterize the film smoothness and edge acuity. As shown in Fig. 4(b), the surface of ohmic contacts changed after annealing at high temperature (855 °C) but is still smooth enough and the edge acuity is proper for a variety of applications in semiconductor devices. Rough surface of ohmic contacts is disadvantageous for reliability and stability [11].

The roughness appeared due to the Al in the ohmic contact scheme. Impact of annealing temperature on the microstructure of studied Ti/Al/Mo/Au was not so large as on the Ti/Al/Ni/Au ohmic contacts (Fig. 5) examined earlier [15], [17], but it is still easily observed. At the ohmic contact with Ni barrier (Ti/Al/Ni/Au) the primary mechanism responsible for the poor mor-

(b) after annealing at 855 °C RTA 60 s

Fig. 4: SEM micrographs of ohmic contacts.

Fig. 5: SEM micrographs of ohmic contacts containing Ni barrier (Ti/Al/Ni/Au) instead of Mo.

above 660 °C (melting point of Al), [17]. On the final roughness and morphology of ohmic contacts influence not only presence of Al layer but also the thicknesses and composition of the rest layers of ohmic metallization. In particular the type of barrier layer for gold [17]. Comparison of two technologies, with Ni (Fig. 5) and Mo (Fig. 4) layers, shown the better properties of the molybdenum layer.

Because Al plays an essential role in the ohmic contact formation, it would be a challenge to avoid of its application. SEM micrographs of Ti/Al/Mo/Au ohmic contacts annealed at various temperature (not shown) did not show large differences in the topography, only sometimes some cracks of Mo layer have been observed (Fig. 4). It was observed that a smooth surface, superior edge acuity and lowest contact resistance of 0.92 $\Omega \cdot$mm were obtained for the sample annealed at 855 °C.

4. Conclusion

It has been demonstrated the influence of temperature of rapid thermal annealing process on the ohmic contact performance of Ti/Al/Mo/Au metallization to AlGaN/GaN heterostructures. For all studied samples, a lower contact resistance of 0.92 $\Omega \cdot$mm, a good surface morphology and edge acuity were achieved when annealing the samples at 855 °C for 60 s. However, the I-V characteristics still remain slightly non-linear. It means, that the metal-semiconductor Ti/Al/Mo/Au-AlGaN/GaN contacts still have a barrier. Further temperature increase of thermal annealing process did not influence on the shape of I-V characteristic. What is more, it caused the increase of contact resistance RC. Our results indicated, that further optimization of this Ti/Al/Mo/Au contact has to be made.

Acknowledgment

This work was co-financed by the European Union within European Regional Development Fund, through grant Innovative Economy (POIG.01.01.02-00-008/08-05), National Science Centre under the grant no. DEC-2012/07/D/ST7/02583, by National Centre for Research and Development through Applied Research Program grant no. 178782, program LIDER no. 027/533/L-5/13/NCBR/2014, by Wroclaw University of Technology statutory grants and Slovak-Polish International Cooperation Program no. SK-PL-2015-0028.

References

[1] QUIAO, D., Z. F. GUAN, J. CARLTON, S. S. LAU, G. S. SULLIVAN, L. JIA, L. S. YU, P. M. ASBECK, S. S. LAU, S. H. LIM, Z. L. WEBER, T. E. HAYNES and J. B. BARNER. Ta-based interface ohmic contacts to AlGaN/GaN heterostructures. *Journal of Applied Physics*. 2001, vol. 89, iss. 10, pp. 5543–5546. ISSN 0021-8979. DOI: 10.1063/1.1365431.

[2] JACOBS, B., M. C. J. C. M. KRAMER, E. J. GELUK and F. KAROUTA. Optimisation of the Ti/Al/Ni/Au ohmic contact on AlGaN/GaN FET structures. *Journal of Crystal Growth*. 2002, vol. 241, iss. 1-2, pp. 15–18. ISSN 0022-0248. DOI: 10.1016/S0022-0248(02)00920-X.

[3] MAHAJAN, S. S., A. DHAUL, R. LAISHRAM, S. KAPOOR, S. VINAYAK and B. K. SEHGAL. Micro-structural evaluation of Ti/Al/Ni/Au ohmic contacts with different Ti/Al thicknesses in AlGaN/GaN HEMTs. *Materials Science and Engineering B*. 2014, vol. 183, iss. 1, pp. 47–53. ISSN 0921-5107. DOI: 10.1016/j.mseb.2013.12.005

[4] AMBACHER, O., J. SMART, J. R. SHEALY, N. G. WEIMANN, K. CHU, M. MURPHY, W. J. SCHAFF, L. F. EASTMAN, R. DIMITROV, L. WITTMER, M. STUTZMANN, W. RIEGER and J. HILSENBECK. Two-dimensional electron gases induced by spontaneous and piezoelectric polarization charges in N- and Ga-face AlGaN/GaN heterostructures. *Journal of Applied Physics*. 1999, vol. 85, iss. 6, pp. 3222–3233. ISSN 0021-8979. DOI: 10.1063/1.369664.

[5] SMORCHKOVA, I. P., C. R. ELSASS, J. P. IBBETSON, R. VETURY, B. HEYING, P. FINI, E. HAUS, S. P. DENBAARS, J. S. SPECK and U. K. MISHRA. Polarization-induced charge and electron mobility in AlGaN/GaN heterostructures grown by plasma-assisted molecular-beam epitaxy. *Journal of Applied Physics*. 1999,

vol. 86, iss. 8, pp. 4520–4526. ISSN 0021-8979. DOI: 10.1063/1.371396.

[6] RUVIMOV, S., Z. LILIENTAL-WEBER, J. WASHBURN, D. QIAO, S. S. LAU and P. K. CHU. Microstructure of Ti/Al ohmic contacts for n-AlGaN. *Applied Physics Letters*. 1998, vol. 73, iss. 18, pp. 2582–2584. ISSN 0003-6951. DOI: 10.1063/1.122512.

[7] WANG, L., F. M. MOHAMMED, B. OFUONYE and I. ADESIDA. Ohmic contacts to n/*pm* GaN capped AlGaN/AlN/GaN high electron mobility transistors. *Applied Physics Letters*. 2007, vol. 91, iss. 1, pp. 012113-1–012113-3. ISSN 0003-6951. DOI: 10.1063/1.2754371.

[8] RUVIMOV, S., Z. LILIENTAL WEBER, J. WASHBURN, K. J. DUXSTAD, E. E. HALLER, Z. F. FAN, S. N. MOHAMMAD, W. KIM, A. E. BOTCHKAREV and H. MORKOC. Microstructure of Ti/Al and Ti/Al/Ni/Au Ohmic contacts for nGaN. *Applied Physics Letters*. 1996, vol. 69, iss. 11, pp. 1556–1558. ISSN 0003-6951. DOI: 10.1063/1.117060.

[9] WANG, L., F. M. MOHAMMED and I. ADESIDA. Dislocation-induced nonuniform interfacial reactions of Ti/Al/Mo/Au ohmic contacts on AlGaN/GaN heterostructure. *Applied Physics Letters*. 2005, vol. 87, iss. 14, pp. 141915-1–141915-3. ISSN 0003-6951. DOI: 10.1063/1.2081136.

[10] FAY, M. W., G. MOLDOVAN, N. J. WESTON, P. D. BROWN, I. HARRISON, K. P. HILTON, A. MASTERTON, D. WALLIS, R. S. BALMER, M. J. UREN and T. MARTIN. *Journal of Applied Physics*. 2004, vol. 96, iss. 10, pp. 5588–5595. ISSN 0021-8979. DOI: 10.1063/1.1796514.

[11] YANXU, Z., C. WEIWEI, F. YUYU, D. YE and X. CHEN. Effects of rapid thermal annealing on ohmic contact of AlGaN/GaN HEMTs. *Journal of Semiconductors*. 2014, vol. 35, iss. 2, pp. 026004-1–026004-4. ISSN 1674-4926. DOI: 10.1088/1674-4926/35/2/026004.

[12] LALINSKY, T., G. VANKOL, Z. MOZOLOVA, J. LIDAY, P. VOGRINCIC, A. VINCZE, F. UHEREK, S. HASCIKL and I. KOSTIC. Nb-Ti/Al/Ni/Au Ohmic Metallic System to AlGaN/GaN. In: *International Conference on Advanced Semiconductor Devices and Microsystems*. Smolenice Castle: IEEE, 2006, pp. 151–154. ISBN 1-4244-0369-0. DOI: 10.1109/ASDAM.2006.331176.

[13] FLOROVIC, M., J. KOVAC, P. KORDOS, J. SKRINIAROVA, T. LALINSKY, S. HASCIK, M. MICHALKA, D. DONOVAL and F. UHEREK. Electrical properties of ohmic contacts for Al0.3Ga0.7N/GaN semiconductor devices. In: *International Conference on Advanced Semiconductor Devices and Microsystems*. Smolenice: IEEE, 2008, pp. 103–106. ISBN 978-1-4244-2326-2. DOI: 10.1109/ASDAM.2008.4743291.

[14] MACHERZYNSKI, W. and B. PASZKIEWICZ. Study of interface reactions between Ti/Al/Ni/Au metallization and AlGaN/GaN heterostructures. *Central European Journal of Physics*. 2013, vol. 11, iss. 2, pp. 258–263. ISSN 1895-1082. DOI: 10.2478/s11534-012-0158-0.

[15] MACHERZYNSKI, W., A. STAFINIAK, A. SZYSZKA, J. GRYGLEWICZ, B. PASZKIEWICZ, R. PASZKIEWICZ and M. TLACZALA. Effect of annealing temperature on the morphology of ohmic contact Ti/Al/Ni/Au to n-AlGaN/GaN heterostructures. *Optica Applicata*. 2009, vol. 39, iss. 4, pp. 673–679. ISSN 1899-7015.

[16] MACHERZYNSKI, W., K. INDYKIEWICZ and B. PASZKIEWICZ. Chemical analysis of Ti/Al/Ni/Au ohmic contacts to AlGaN/GaN heterostructures. *Optica Applicata*. vol. 43, iss. 1, pp. 67–72. ISSN 1899-7015.

[17] MACHERZYNSKI, W. and B. PASZKIEWICZ. Development of diffusion barriers for Ti/Al based ohmic contact to AlGaN/GaN heterostructures. In: *The Ninth International Conference on Advanced Semiconductor Devices and Microsystems*. Smolenice: IEEE, 2012, pp. 203–206. ISBN 978-1-4673-1195-3. DOI: 10.1109/ASDAM.2012.6418532.

About Authors

Wojciech MACHERZYNSKI received his M.Sc. degree in Electronic from Wroclaw University of Technology, Poland in 2005 and Ph.D. degree from the Wroclaw University of Technology in 2011. Now he is assistant professor at Wroclaw University of Technology. His research is focused on the technology of semiconductors devices in particular on development of the metal-semiconductor junction.

Jacek GRYGLEWICZ received his M.Sc. degree in Electrical Engineering from Wroclaw University of Technology, Poland in 2009 and Ph.D. degree from the Wroclaw University of Technology in 2015. Now he is assistant professor at Wroclaw University of Technology. His research is focused on device processing and parameter evaluation of nitrides-based devices: HEMTs and sensors. He is co-author of 16 scientific publications.

Andrzej STAFINIAK received M.Sc. degree (2008) and Ph.D. degree (2015) in electronics from Wroclaw University of Technology. Since then, he has been assistant professor in Division of Microelectronics and Nanotechnology, Wroclaw University of TechnologyWroclaw University of Technology. His current research has focused on development of process technology and measurements of nanostructures based devices.

Joanna PRAZMOWSKA received her M.Sc. degree in Electronic from Wroclaw University of Technology, Poland in 2005 and Ph.D. degree from Wroclaw University of Technology in 2011. Now she is assistant professor at Wroclaw University of Technology. Her research interest embraces technology of semiconductor devices i.e. lithography process development of electronic, optoelectronic devices as well as gas sensors.

Regina PASZKIEWICZ received her M.Sc. degree in Electrical Engineering from St. Petersburg Electrotechnical University, St. Petersburg, Russia in 1982 and Ph.D. degree from the Wroclaw University of Technology in 1997. Now she is full professor at Wroclaw University of Technology. Her research is focused on the technology of (Ga, Al, In)N semiconductors, microwave and optoelectronic devices technological processes development.

26

STUDY OF INTERFACE OF OHMIC CONTACTS TO AlGaN/GaN HETEROSTRUCTURE

Joanna PRAZMOWSKA, Wojciech MACHERZYNSKI, Regina PASZKIEWCZ

Department of Microelectronics and Nanotechnology, Faculty of Microsystem Electronics and Photonics, Wroclaw University of Technology, Janiszewskiego 11/17, 50-370 Wroclaw, Poland

joanna.prazmowska@pwr.edu.pl, wojciech.macherzynski@pwr.edu.pl, regina.paszkiewicz@pwr.edu.pl

Abstract. *The paper embraces studies of the interface of ohmic contacts and AIIIBV-N heterostructure. The TiAl based metallization stack was investigated. The Ti/Al/Ni/Au contact to AlGaN/GaN heterostructures fabricated by metal-organic vapour phase epitaxy was examined using three methods i.e. etching of annealed contact metallization, fractures (prepared at room temperature and after a bath in liquid nitrogen) and microsections imaging. The main focus was on the estimation of reaction range on the metal-semiconductor interface of samples. In the first method, the surface of AlGaN/GaN heterostructure after etching of metallization was studied by an optical microscope, scanning electron microscope and atomic force microscope. The changes of surface morphology of heterostructure directly reflect solid state reactions range between metallization and semiconductor. The range of reactions was also observed using the small-angle microsections method while the fractures analysis did not bring valuable information.*

Keywords

AIIIBV-N heterostructures, metal-semiconductor interface, Ti/Al/Ni/Au metallization.

1. Introduction

AlGaN/GaN heterostructure based High Electron Mobility Transistors (HEMTs) are good candidates for high-power and high-temperature application. This area of applications enforces thermal stability of applied materials, thus also ohmic/Schottky contacts have to be of good quality, appropriate morphology, low-resistance and thermal stability [1], [2]. Param-eters of ohmic contacts depend on various factors as e.g.: used metal stack, thicknesses of metal layers, heterostructure properties, and thermal annealing process parameters [3]. The standard scheme of metallization consists of Ti/Al/metal/Au, where metal could be one of Ni, Pd, Pt, Mo [4]. The metallization is processed in thermal annealing system after deposition. Depending on applied temperature of annealing, improvement or deterioration of electrical parameters of metal-semiconductor system occurs.

While the mechanism of contact formation to GaN is already understood, the ohmic contact formation to AlGaN/GaN needs further studies [5]. The most common method of investigation of metal-semiconductor (m-s) interface phenomena is TEM (transmission electron microscope) analysis [4], [5], [6] and [7]. In the paper, the m-s interface of Ti/Al/Ni/Au and AlGaN/GaN was studied using three methods. First technique included deposition of metallization on the surface of AlGaN/GaN, thermal annealing, etching of annealed ohmic contact and further observation of the etched surface of semiconductor using microscope methods as Scanning Electron Microscope (SEM) and Atomic Force Microscope (AFM). That gave an information on the m-s interface and permitted for estimation of the reaction range. Second one relied on fractures preparation and observation of the interface by application of SEM (Scanning Electron Microscope). The third method required polishing of the metallization near the edge of the sample. Obtained small-angle microsection permitted for observation of semiconductor surface.

2. Technology

The investigated samples contained metallization stacks deposited on AlGaN/GaN heterostructures fabricated using metalorganic vapour phase epitaxy method on Al_2O_3 substrates. Mesa structures were formed in chloride plasma of Reactive Ion Etching (RIE). The etching process was carried out through the oxide mask (SiO_2) deposited in Plasma-Enhanced Chemical Vapour Deposition (PECVD) system. The pattern was achieved in standard lithography process. The mask layer thickness was of about 300 nm.

Samples with Ti/Al/Ni/Au (20/100/40/150 nm) layers deposited in UHV system by electron beam (Ti, Ni) and resistance heater (Al, Au) evaporation were investigated. The test structures were fabricated using photolithography process in lift-off technique by application of LOR and SPR 700 bi-layer. The samples were coated by the layers using spin-coating method. Samples with Ti/Al/Ni/Au multilayer metallization were annealed in Rapid Thermal Annealing (RTA) system at a temperature of 820 °C.

To study the influence of annealing process on the properties of heterostructure the metallization layers after annealing were selectively etched layer by layer. The etching process was carried out in:

- iodine-potassium iodide (Au),

- H_3PO_4:H_2O (1:3) at 70 °C (Ni),

- H_3PO_4:HNO_3:CH_3COOH:H_2O (85:5:5:5) at 45 °C (Al),

- hydrogen peroxide at 65 °C (Ti).

Etching stages duration was 10 minutes each. After etching of metal layers optical microscope and SEM images of surface morphology of chosen area were recorded. This permitted to estimate the range of metal-semiconductor solid state reactions, follow the morphology changes and its correlation with m-s interface topography.

The study of application of fractures technique for the solid state reaction range on the interface of annealed metallization and AlGaN/GaN was carried out. The fractures technique was adopted from AIIIBV technology [8]. The mechanical incision of samples for this investigation was made using diamond blade. Samples were incised from the Al_2O_3 side for the depth of about 180 μm. Then samples were fractured at room temperature or immediately after bath in liquid nitrogen. Also samples fractured during the technological process because of the large stress were examined.

The last group were samples grinded and polished under small-angle near the edge of the sample. This method permitted for observation of the reactions range occurring on the interface in only little invasive way. In contrary to two above-mentioned methods, the sample after microsection preparation and observation can be used for further processing.

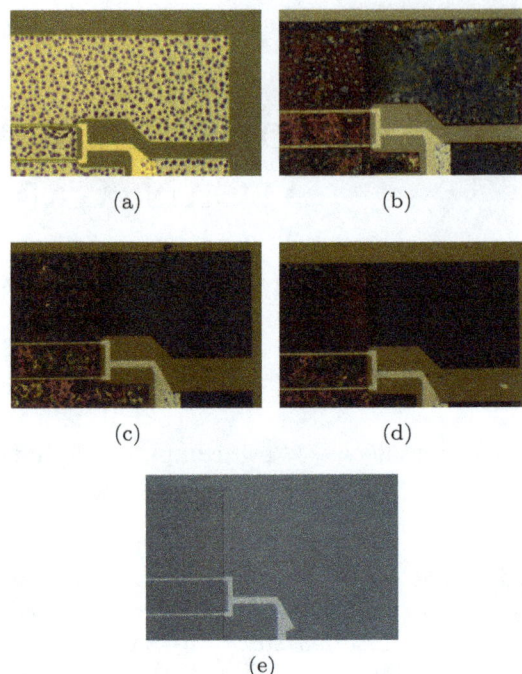

Fig. 1: Optical images of metallization surface (a) before etching and after etching of (b) Au, (c) Ni, (d) Al and (e) Ti.

3.   Results

The rapid thermal annealing process of ohmic contacts to AlGaN/GaN heterostructure based devices affects parameters of the contact. The selection of appropriate temperature value of annealing temperature was made based on presented and discussed earlier paper [9]. Application of annealing at temperature above 835 °C led to the degradation of AlGaN/GaN heterostructure surface beneath the contact. Temperature equal to 805 °C started reactions at the level of 2DEG position i.e. of about 25 nm from the surface. Therefore, the intermediate value of 820 °C was chosen for annealing of investigated samples. During thermal annealing of the Ni-based metallization the agglomerates appear on the surface (black spots in Fig 1(a) and Fig 3(a)). They are a consequence of migration and coalescence of melted Ni. The mechanism of agglomeration formation was already studied and described [10]. Figure 1 presents optical images of metallization surface Fig. 1(a) before etching, and remained metallizations layers after etching of Fig. 1(b) Au, Fig. 1(c) Ni, Fig. 1(d) Al and Fig. 1(e) Ti.

Each step of etching changed the surface of ohmic contact. The remained metal films (Fig. 1(b), Fig. 1(c), Fig. 1(d) and Fig. 1(e)) had slightly different topography. After first etching mostly the largest agglomerates of the contact (remarkable as black spots in Fig. 1(a)) were still observed. That could indicate lack of gold in the agglomerates volume, which corresponded to

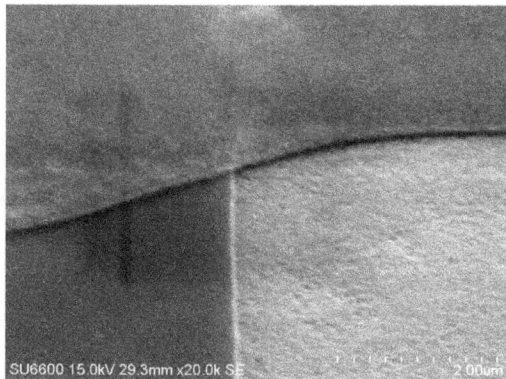

Fig. 2: SEM image embracing three types of areas: metallization on etched GaN (A) and mesa (A'), mesa surface (B), etched GaN (C).

[11]. It was observed that the remained metallization (Fig. 1(b), Fig. 1(c) and Fig. 1(d)) topographies did not reflect the m-s interface topography (Fig. 1(e)). Nevertheless, there were significant differences in topographies of surfaces beyond the ohmic contact on the mesa (Fig. 2, A) and outside of it (Fig. 2, A'). That may be explained by various topographies before metallization deposition (Fig. 2, B and C) and various materials (i.e. AlGaN and GaN).

Figure 3 presents SEM images of metallizations layers, Fig. 3(a) before etching and remained after etching

of Fig. 3(b) Au, Fig. 3(c) Ni, Fig. 3(d) Al and Fig. 3(e), Fig. 3(f) Ti.

SEM images confirmed graduate removal of the metallization layers in subsequent stages except of etching the Al layer. It could be explained by occurring of a metal compound that could not be etched in used solution in contrary to H_2O_2.

Step of Au etching caused a delamination near the agglomerates (Fig. 3(b)). The surface beneath the ohmic contact after etching of all layers of metallization changed significantly. Topography of this area is a reflection of the m-s interface. The topographies under agglomerates and outside of this area were nearly identical. The difference between surfaces of a sample before and after Al etching (Fig. 3(c) and Fig. 3(d) was nearly unnoticeable, which corresponded to optical microscope images (Fig. 1(c) and Fig. 1(d). It could be caused by etching of Al layer (or its compound) by solution of $H_3PO_4:H_2O$, used for Ni layer removal but it requires further analysis e.g. by EDX (Energy Dispersive X-ray Spectroscopy).

The consumption of the AlGaN layer was significant (Fig. 3(f)) – represented as dark spots on the picture. Figure 4(a) presented SEM image of two areas of mesa – bare surface (A) and surface with deposited and etched metallization (B). The AFM images were taken

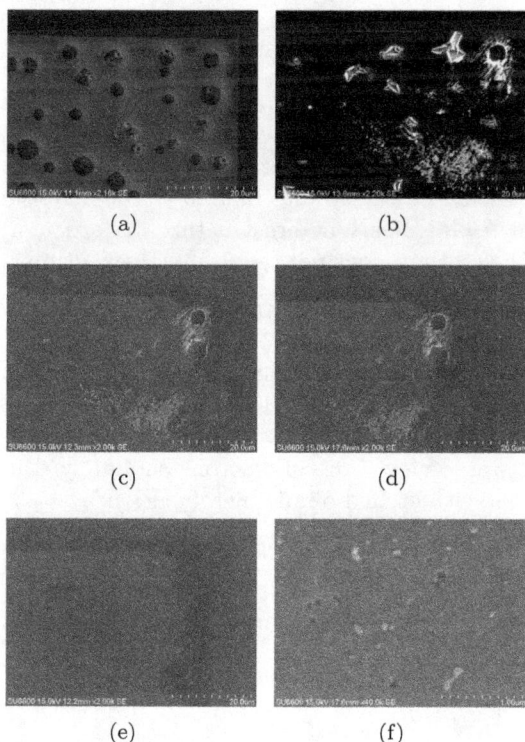

Fig. 3: SEM images of metallization surface before etching (a), and remained metallizations layers after etching of (b) Au, (c) Ni, (d) Al, (e) and (f) Ti.

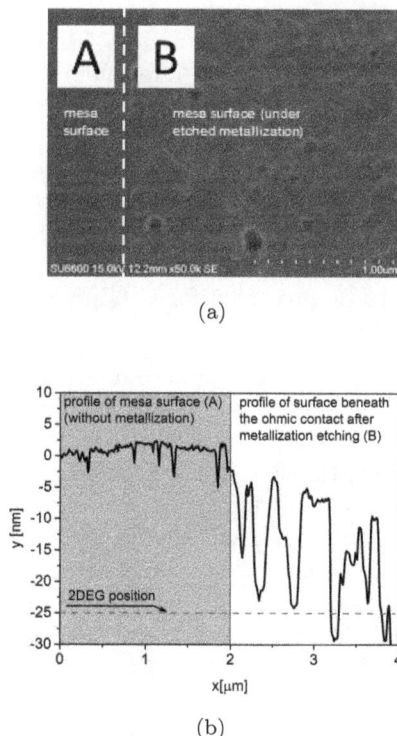

Fig. 4: (a) SEM image of semiconductor layer; mesa (A) and mesa surface after etching of ohmic contact (B), (b) profile of the surface after metallization etching (extracted from AFM image).

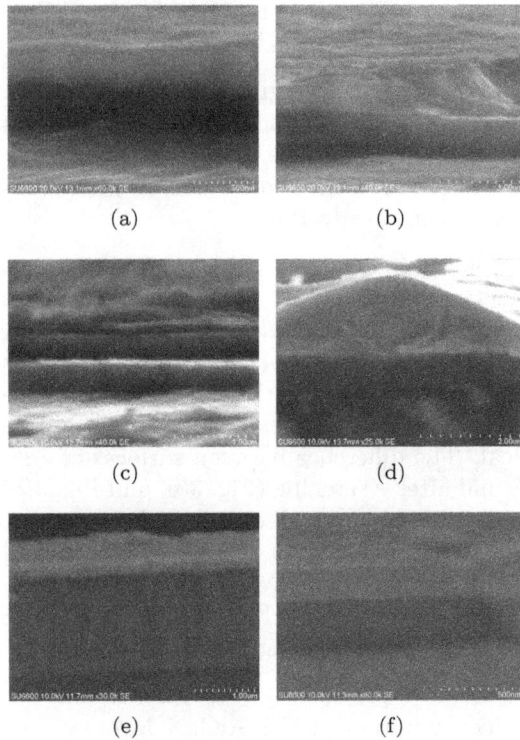

Fig. 5: Example SEM images of fractures of annealed Ti/Al/Ni/Au metallization and AlGaN/GaN heterostructure made at room temperature (a), (b), immediately after bath in LN_2 (c), (d) and fractured during technological process (e), (f).

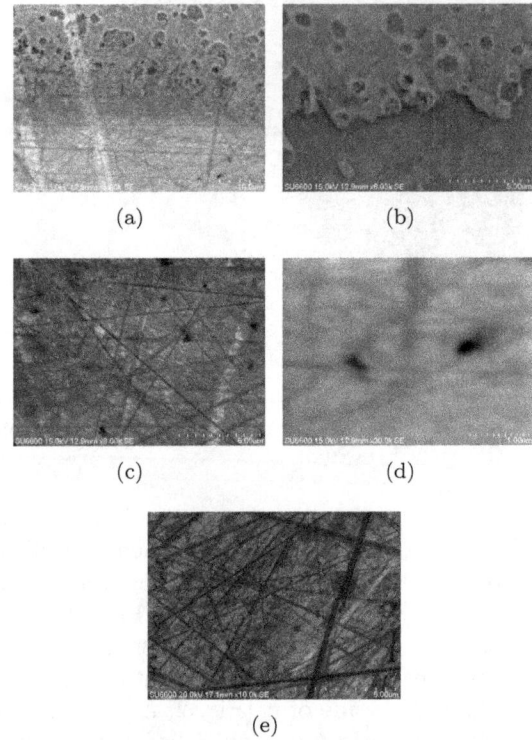

Fig. 6: Example SEM images of microsections of annealed Ti/Al/Ni/Au metallization and AlGaN/GaN heterostructure.

from the region embracing A and B. Figure 4(b) consisted of profile of the surface after metallization etching (extracted from AFM image). It confirmed phenomenon of AlGaN surface consumption. The depth of solid state reaction on the m-s interface reached even 30 nm within studied area. It seemed to penetrate the 2DEG region, but in this stage of research we are not able to estimate if heterostructure beneath metallization has to be preserved for obtaining the optimal parameters of m-s contact.

The applied fracture technique did not permit for observation of the penetration of ohmic contact into AlGaN/GaN heterostructure (Fig. 5). Despite obtained sharp fractures, the m-s interface on SEM images did not bring any valuable information. Bath in liquid nitrogen improved sharpness of the fractures, but it was still insufficient to investigate the changes occurring in the range of depth equal to 30 nm. This technique permitted only for distinguishing of ohmic contact and thickening metallization layer Fig. 5(d), Fig. 5(e) and Fig. 5(f).

The samples with annealed ohmic contacts were also examined using small-angle microsections. Figure 6(a) presents the microsection embracing two areas – metallization (upper part of the SEM image) and AlGaN/GaN heterostructure (bottom of the SEM im-

age). The spots on the surface of metallization could be agglomerates which are composed of materials in other propotions than metallization stack thus having other mechanical properties that led to more efficient polishing in those areas.

One of the disadvantages of this method is a tear off of the metallization in some areas of microsections. Nevertheless in the studies limited to the investigation of the AlGaN/GaN heterostructure surface morphology, the method does not require further optimization.

The study of the heterostructure surface beneath the ohmic contact (Fig. 6(c) and Fig. 6(d)) allows to observe the black spots among the polishing grooves that may indicate the interfacial reaction between the semiconductor and metallization. The size of the observed spots was of few tens of micron, similar as that for samples with etched ohmic metallization.

The spots were not observed on the microsection in the area outside the ohmic metallization Fig. 6(e).

4. Conclusion

The Ti/Al/Ni/Au metallizations to AlGaN/GaN heterostructures fabricated by metal-organic vapour phase epitaxy were studied. The range of reactions on the metal-semiconductor interface of samples was examined after etching of annealed ohmic contacts and using fractures.

At any stage of etching, topography of remained metallization did not reflect the topography of the m-s interface. Nevertheless, solid state reactions on the m-s interface were significant. The depth of reactions estimated on the bases of AFM images reached even 30 nm within studied area. SEM images of prepared fractures did not allow to estimate the depth of solid state reactions on the m-s interface. Application of LN_2 for fractures influenced slightly the sharpness of the fractures but did not enable analysis of solid state reactions on the m-s interface beneath annealed ohmic contact.

The applied method of small-angle microsections study permits for observation of range of reactions on the interface of Ti/Al/Ni/Au and AlGaN/GaN heterostructure. The reactions consequences are observable as black spots on the surface of semiconductor in the area beneath the ohmic contact.

Acknowledgment

This work was co-financed by the European Union within European Regional Development Fund, through grant Innovative Economy (POIG.01.01.02-00-008/08-05), National Science Center Poland under the grant no. DEC-2012/07/D/ST7/02583, by National Centre for Research and Development through Applied Research Program grant no. 178782, program LIDER no. 027/533/L-5/13/NCBR/2014, by Wroclaw University of Technology statutory grants and Slovak-Polish International Cooperation Program no. SK-PL-2015-0028.

References

[1] WANG, C., S.-J. CHO and N.-Y. KIM. Optimization of Ohmic Contact Metallization Process for AlGaN/GaN High Electron Mobility Transistor. *Transactions on Electrical and Electronic Materials*. 2013, vol. 14, iss. 1, pp. 32–35. ISSN 1229-7607. DOI: 10.4313/TEEM.2013.14.1.32.

[2] MOKRZYCKI, K. M. *Optimization of photolithography process with application of modern resists of Shipley Company*. Wroclaw. 2012. Master Thesis. Wroclaw University of Technology. Supervisor: Bogdan Jankowski.

[3] MACHERZYNSKI, W. and B. PASZKIEWICZ. Study of interface reactions between Ti/Al/Ni/Au metallization and AlGaN/GaN heterostructures. *Central European Journal of Physics*. 2013,

vol. 11, iss. 2, pp. 258–263. ISSN 1644-3608. DOI: 10.2478/s11534-012-0158-0.

[4] MACHERZYNSKI, W., K. INDYKIEWICZ and B. PASZKIEWICZ. Chemical analysis of Ti/Al/Ni/Au ohmic contacts to AlGaN/GaN heterostructures. *Optica Applicata*. 2013, vol. 43, iss. 1, pp. 67–72. ISSN 0078-5466. DOI: 10.5277/oa130109.

[5] VAN DAELEA, B., G. VAN TENDELOO, J. DERLUYN, P. SHRIVASTAVA, A. LORENZ, M. R. LEYS and M. GERMAIN. Mechanism for Ohmic contact formation on Si_3N_4 passivated AlGaN/GaN high-electron-mobility transistors. *Applied Physics Letters*. 2006, vol. 89, iss. 20, pp. 201908-1–201908-3. ISSN 0003-6951. DOI: 10.1063/1.2388889.

[6] TAKEI, Y., M. OKAMOTO, W. SAITO, K. TSUTSUI, K. KUKUSHIMA, H. WAKABAYASHI, Y. KATAOKA and H. IWAI. Ohmic Contact Properties Depending on AlGaN Layer Thickness for AlGaN/GaN High Electron Mobility Transistor Structures. *The Electrochemical Society Transactions*. 2014, vol. 61, iss. 4, pp. 265–270. ISSN 1938-5862. DOI: 10.1149/06104.0265ecst.

[7] WANG, C. and N.-Y. KIM. Electrical characterization and nanoscale surface morphology of optimized Ti/Al/Ta/Au ohmic contact for AlGaN/GaN HEMT. *Nanoscale Research Letter*. 2012, vol. 7, iss. 1, pp. 107-1–107-8. ISSN 1556-276X. DOI: 10.1186/1556-276X-7-107.

[8] JUNG, S. M., C. T. LEE and M. W. SHIN. Investigation of V-Ti/Al/Ni/Au Ohmic contact to AlGaN/GaN heterostructures with a thin GaN cap layer. *Semiconductor Science and Technology*. 2015, vol. 30, no. 7, pp. 075012–075012. ISSN 1361-6641. DOI: 10.1088/0268-1242/30/7/075012.

[9] CHIU, Y.-S., T.-M. LIN, H.-Q. NGUYEN, Y.-C. WENG, C.-L. NGUYEN, Y.-C. LIN, H.-W. YU, E. Y. CHANG and C.-T. LEE. Ti/Al/Ti/Ni/Au ohmic contacts on AlGaN/GaN high electron mobility transistors with improved surface morphology and low contact resistance. *Journal of Vacuum Science & Technology B*. 2014, vol. 32, iss. 1, pp. 011216–011216. ISSN 2166-2754. DOI: 10.1116/1.4862165.

[10] MACHERZYNSKI, W., A. STAFINIAK, A. SZYSZKA, J. GRYGLEWICZ, B. PASZKIEWICZ, R. PASZKIEWICZ and M. TLACZALA. Effect of annealing temperature on the morphology of ohmic contact Ti/Al/Ni/Au to n-AlGaN/GaN heterostructures. *Optica Applicata*. 2009, vol. 39, iss. 3, pp. 673–679. ISSN 0078-5466.

[11] SCHMID, A., C. SCHROETER, R. OTTO, M. SCHUSTER, V. KLEMM, D. RAFAJA and J. HEITMANN. Microstructure of V-based ohmic contacts to AlGaN/GaN heterostructures at a reduced annealing temperature. *Applied Physics Letters*. 2015, vol. 106, iss. 20, pp. 201908-1–201908-3. ISSN 0003-6951. DOI: 10.1063/1.4907735.

About Authors

Joanna PRAZMOWSKA received her M.Sc. degree in Electronics from Wroclaw University of Technology, Poland in 2005 and Ph.D. degree from Wroclaw University of Technology (WrUT) in 2011. Now she is assistant professor at WrUT. Her research interest embraces technology of semiconductor devices i.e. lithography process development of electronic, optoelectronic devices as well as gas sensors.

Wojciech MACHERZYNSKI received his M.Sc. degree in Electronics from Wroclaw University of Technology, Poland in 2005 and Ph.D. degree from the Wroclaw University of Technology in 2011. Now he is assistant professor at WrUT. His research is focused on the technology of semiconductors devices in particular on development of the metal-semiconductor junction.

Regina PASZKIEWICZ received her M.Sc. degree in Electrical Engineering from St. Petersburg Electrotechnical University, St. Petersburg, Russia in 1982 and Ph.D. degree from the Wroclaw University of Technology in 1997. Now she is full professor at WrUT. Her research is focused on the technology of (Ga, Al, In)N semiconductors, microwave and optoelectronic devices technological processes development.

27

INTERFACIAL ROUGHNESS AND TEMPERATURE DEPENDENCE OF NARROW BAND THIN FILM FILTERS FOR THE DWDM PASSIVE OPTICAL NETWORKS

Lubomir SCHOLTZ, Libor LADANYI, Jarmila MULLEROVA

Institute of Aurel Stodola, Faculty of Electrical Engineering, University of Zilina,
Kpt. Nalepku 1390, 031 01 Liptovsky Mikulas, Slovak Republic

scholtz@lm.uniza.sk, ladanyi@lm.uniza.sk, mullerova@lm.uniza.sk

Abstract. *In the design of new components for passive optical networks (PONs), the non-ideal properties are worth considering. In this paper the influence of interface roughness and temperature changes on final transmittance of downstream channels blocking filters for next generation dense wavelength division multiplexing passive optical networks(DWDM-PONs) is shown. The transmittance as the filter transfer characteristicswas calculated with the transfer matrix method. The roughness was expressed by root mean square deviations from an ideally smooth surface and was taken into account in the modified Fresnel coefficients. It is demonstrated how the interfacial roughness may increase the insertion loss and decrease the channel bandwidth which results in reduction of transmitted light energy through the filter.*

Keywords

DWDM, passive optical networks, PON, transfer matrix method, wavelength blocking filters.

1. Introduction

Currently considerable interest is given to future developments of passive optical networks (PONs) satisfying the demands for increasing traffic, higher bandwidth and extended reach [1], [2], [3]. The specific wavelength bands for present and future PON technologies are or should be allocated by recommendations of International Telecommunication Union (ITU). However, it is necessary to protect present and future PON signals in optical network units (ONUs) at the end of a subscriber from interference in case of coexisting PON technologies. To guarantee this, a precise scheme of the wavelength allocation together with the so-called guard bands and the implementation of specific wavelength blocking filters are generally accepted and recommended. Thin-film interference filters (TFFs) are suitable, low-cost, coexisting (i.e. ONU-independent) candidates [4], [5].

It is expected that the current gigabit-capable (G-PON) and 10 Gb·s^{-1} PON (XG-PON), the so-called next generation PON networks stage 1 (NG-PON1) will be followed by the NG-PON stage 2. As a primary solution for NG-PON2 TWDM (time wavelength division multiplexing) technology is planned. More candidates for NG-PON2 networks e.g. WDM (wavelength division multiplexing), OFDM (orthogonal frequency-division multiplexing) are possible, but TWDM is the most preferred. This also follows from the meeting of FSAN (Full Service Access Network) community in the April 2012 [6]. Since the TWDM-PON will be extended after several years of use it is necessary to study new possible options for improvement of these types of PON networks [7].

Just the use of TWDM technology is an evolutionary step to the future PONs. Because of the increasing demands for network capacity it is assumed that others wavelength pairs (resp. channels) will be added to existing PON infrastructure. Gradual increase of number of wavelength pairs should lead to the replacement of TWDM technology with pure WDM with dense division multiplexing. In WDM each subscriber has its own assigned wavelength pair. And thus signals are distributed through the optical distribution network (ODN) from the optical line terminal (OLT) to ONUs of all particular users separately. Dense wavelength division multiplexing passive optical networks (DWDM-PON) are considered as one of very promising technologies after NG-PON1 and NG-PON2 mass deployment. DWDM-PON may use the same ODN as GPON and NG-PON using colorless ONUs. However to secure coexistence between the various technologies it is necessary that subscriber's ONU will detect only the signals intended for this specific ONU. Therefore the unnecessary signals should be blocked. This can be ensured by using optical band filters in ONUs. This is a strong advantage for any new emerging technology from the cost saving point of view. Nowadays various groups deal with the concept of DWDM-PON searching for improvements [9], [10], [11]. For example in [9], [10] the modulation formats in DWDM-PON were examined since the use of modulation formats in access optical networks is the additional step to increase their capacities.

In our previous paper [8] narrow TFFs used at different angles of incidence are proposed to be used in future DWDM-PON to coexist with current PONs. The multilayer structures of filter were numerically designed with amorphous silicon (a-Si) as high refractive index material (H), SiO_2 as low refractive index material (L) and ZnO as middle refractive index material (M). The ZnO was used for the polarization independence of the filter. The polarization independence of the filter is important for our proposal of using the same filter in various ONUs in DWDM-PON where the filter is tuned by the angle of light incidence for the specific wavelength in a particular ONU. The setting of the specific angle of incidence is achieved by the filter rotation towards the normal. If in one subscriber's ONU the filter is set to the angle ϕ_x the filter is tuned to the channel x. In another subscriber's ONU the filter is set to the angle ϕ_y towards the normal what means that the filter is tuned to the channel y. Thus in DWDM-PON each ONU can be equipped with the same filter but in each of them at a different angle of incidence. This could be a strong advantage and cost-saving solution due to the fact that in ONUs for all subscribers only TFFs secure the selection of the right channel. This type of blocking filters can provide the coexistence with older PON technologies without replacing ONUs with new models.

The design presented in [8] presumes ideal smooth interfaces of the multilayer structure of filter. This paper deals with the influence of temperature and rough interfaces on the transmittance of narrow band pass TFF filters proposed in [8] for the following DWDM-PON downstream channels: channel A at the central wavelength 1540.5 nm ($\phi_A = 12.2$ °), channel B at the central wavelength 1521.7 nm ($\phi_B = 24.4$ °), channel C at the central wavelength 1502 nm ($\phi_C = 33.2$ °). These central wavelengths correspond to the Rec. ITU-T G.694.1 [12], with 100 GHz channel spacing. If the interface between two layers is not ideally smooth or the ambient temperature changes, the transfer charac-

teristics of TFF filter may be negatively affected, what is the consequence of various undesirable effects.

2. Experimental

Transfer characteristics of multilayer filters can be calculated using the well-known transfer matrix method (TMM). The method expresses the relation between amplitudes of the forward E^+ and backward E^- electric field at the interface of the k-th and $(k-1)th$ layer of the structure depending on the parameters of layers and interfaces between layers. The phase change $\delta_{(k-1)}$ of light in the $(k-1)th$ layer depends on the thickness $d(k-1)$, on the complex refractive index $N(k-1)$ of $(k-1)th$ layer, on the wavelength of light λ and on the angle $\phi_{(k-1)}$ of incidence at $(k-1)th$ interface. It is well known that the complex refractive index N depends on λ and also on temperature T. This may be expressed as:

$$N(\lambda, T) = n(\lambda, T) + ik(\lambda, T), \qquad (1)$$

where is the extinction coefficient expressing absorption of light in optical medium. Then the light transmission and reflection at the interface between k-th and $(k-1)th$ layer can be expressed in the matrix form as follows:

$$\begin{pmatrix} E^+_{(k-1)} \\ E^-_{(k-1)} \end{pmatrix} = \frac{1}{t_k} \begin{pmatrix} e^{i\delta_{(k-1)}} & r_k e^{i\delta_{(k-1)}} \\ r_k e^{-i\delta_{(k-1)}} & e^{-i\delta_{(k-1)}} \end{pmatrix} \begin{pmatrix} E^+_{(k)} \\ E^-_{(k)} \end{pmatrix}, \quad (2)$$

where r_k and t_k are the Fresnel coefficients representing the amplitude reflectance and transmittance of the k-th layer. Each interface is expressed with its own individual matrix. With these individual matrices, the final matrix and subsequently the total transmittance T and reflectance R for the whole structure can be computed. The total T and R are then given by:

$$R = \frac{E^-_{(0)} E^{-*}_{(0)}}{E^+_{(0)} E^{+*}_{(0)}}, \qquad (3)$$

$$T = \frac{N_{(m+1)}}{N_0} \frac{E^+_{(m+1)} E^{+*}_{(m+1)}}{E^+_{(0)} E^{+*}_{(0)}}, \qquad (4)$$

where m is the total layer number, N_0 is the refractive index of the substrate, N_{m+1} is the refractive index of the ambient medium, usually air. $E^+_0 (E^+_{m+1})$ is the electric field amplitude of incident (outgoing) light, $E^-_{(0)}$ is the electric field amplitude of reflected light. The quantities denoted by a raised asterisk are complex conjugates of amplitudes of electric fields [13].

As the complex refractive index N is directly changed with the material density and the density normally varies inversely with temperature, it is not surprising that the refractive index varies inversely with temperature. Hence the change of temperature has influence on optical properties of a material. The change of the refractive index with temperature is defined by the thermo-optical coefficient $\frac{dN}{dT}$. The change of the refractive index resp. the extinction coefficient is different for each material.

If an interface which is not perfectly smooth occurs in the multilayer structure the final characteristics are obviously modified. Due to the scattering the wave propagations through rough and smooth interfaces differ. The resulting beam propagation or reflection is influenced by the random interface texture. Then the reflected and transmitted light power can be divided into two parts. The direct incident light is scattered into diffused components (diff) in reflection and in transmission, whereas the rest of light does not scatter, and is assigned to the specular components (spec). If the direct incident light is coherent, the specular component in reflection and in transmission has to preserve the coherence.

Intentionally created or modified roughness between two layers can by described by the modified amplitude Fresnel coefficients [13], [14], [15] as:

$$
\begin{aligned}
r_{k-1,k}^{(spec)} &= r_{k-1,k}^{(0)} e^{\left(-\left(\frac{2\pi Z N_{k-1}}{\lambda}\right)^2\right)}, \\
r_{k,k-1}^{(spec)} &= r_{k,k-1}^{(0)} e^{\left(-\left(\frac{2\pi Z N_k}{\lambda}\right)^2\right)},
\end{aligned}
\tag{5}
$$

where the superscripts (0) denote the Fresnel coefficients of smooth interfaces, Z is the root mean square (rms) roughness and N_k is the complex refractive index of k-th layer. The modified coefficients represent the phase differences in the reflected and transmitted beams and are based on the Gaussian distributions of the height irregularities. These coefficients are implemented into the specular part of directed light. The diffused part of light is easily calculable from equations [13], [14], [15]:

$$
\begin{aligned}
R^{(diff)} &= R^{(0)} - R^{(spec)}, \\
T^{(diff)} &= T^{(0)} - T^{(spec)},
\end{aligned}
\tag{6}
$$

where $R^{(0)}(T^{(0)})$ is total reflectance (transmittance) at a smooth interface, $R^{(spec)}(T^{(spec)})$ is specular reflectance (transmittance) at a rough interface.

Rms roughness Z is usually obtained from the AFM measurements and is given e.g. by the standard deviation of the values of surface heights of a measured sample area. Then Z is expressed by:

$$
Z = \sqrt{\frac{\sum_{n=1}^{P}(z_n - \bar{z})^2}{P - 1}},
\tag{7}
$$

where $\bar{z}$ is the average of the surface heights within the given area, z_n is the current height value, and P is the number of data points within the given area.

More definitions characterizing the surface roughness, for example the mean roughness or the peak-to-valley distances are commonly used, too [16].

These coefficients are included in the individual matrices. The relations for modified Fresnel coefficients are valid for the following conditions:

- The transmission and reflection are coherent.

- The planes (thin film layers) in structure are mutually parallel.

- The rms roughness of the roughness features is much smaller than wavelength of incidence light.

- The dimensions of illuminated parts of the boundaries are much larger than the wavelength of incidence light.

- The materials used in thin film multilayer structure are homogenous and isotropic from optical point of view.

As reported surface roughness of a thin film depends on deposition technology, surface roughness of the substrate, the film thickness, cleaning process, etc. and may reach several nanometers in case of the PECVD deposited a-Si layers [17]. For our simulations typical values of interfacial roughness of a-Si were used from [17], [18]. Therein a possible value of rms roughness of a Si layer of the thickness of 800 nm deposited on a thick substrate equals 5 nm.

3.   Results and Discussion

The influence of rough interfaces and temperature changes was demonstrated upon TFF filter composed from 201 layers. The designed TFF structure is described by Eq. (8):

$$
\begin{aligned}
&(HMLM)^{12}4.375H(MLMH)^{13} \\
&MLM(HMLM)^{12}4.375H(MLMH)^{12}.
\end{aligned}
\tag{8}
$$

The geometric thickness of the layers was set as follows: 145 nm for L-layer, 204 nm for M-layer and 160 nm for H-layer. The spacer H-layer is 4.375-times thicker than other H-layers. The values of all necessary refractive indices of the materials were taken from [19] and at the close vicinity of 1550 nm are for H-layer (a-Si) $N = 3.48$, for L-layer (SiO$_2$) $N = 1.44$, for M-layer (ZnO) $N = 1.92$. For substrate (fused silica) $N = 1.5$. We suppose that absorption in all materials involved is negligible, i.e. k is expected to be zero.

The filter structure was optimized to achieve as steep filter characteristics as possible and to keep the insertion loss of the filter less than -5 dB within the pass band. Maximum insertion loss -5 dB for downstream channels is required according to ITU-T Recommendation G.984.2 and also insertion loss outside the required pass-band width should be at least -32 dB [4]. These requirements we can observe in Fig. 1. In the figure ideal filterwhich has minimum insertion loss between cut-on and cut-off wavelengths and maximum possible insertion loss out of these wavelengths is shown (magenta). Required pass-band width at -32 dB as $\Delta\lambda_{32}$ and required pass-band width at -5 dB as $\Delta\lambda_5$ were set according to ITU-T Recommendation G.984 series corresponding to the power budget -32 dB for C+ grade of ODN [20]. The ratio of these parameters is the so-called abruptness coefficient. For an ideal band pass filter the abruptness coefficient is equal to one. On the other hand the filter contrast expresses how steep the edges of the filter are near the cut-off and cut-on wavelengths.

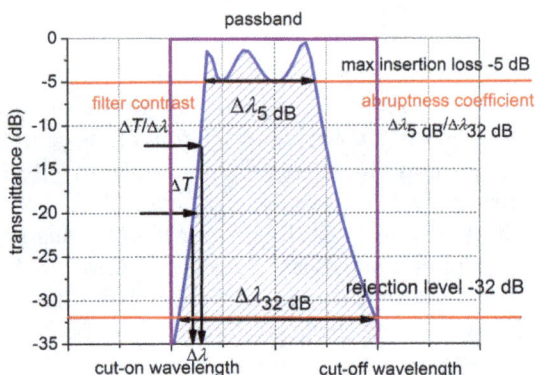

Fig. 1: Requirements for filter properties.

3.1. The Influence of Interface Roughness

It is known that the performance of a multilayer structure depends on optical properties of individual layers and their thicknesses. An ideal case predicts a homogeneous layer with ideally smooth and parallel interfaces. It is obvious to expect that any changes in surface and interface roughness of individual layers influence the performance of the whole structure.

If the roughness of substrate is not equal to zero in the production process, it is then distributed through the whole filter. This is due to subsequently depositing the filter layer by layer. In Fig. 2 the effect of interface roughness of the filter for channel A at the wavelength 1540.5 nm is depicted. If the interface roughness at each interface is equal to zero the filter has the minimal insertion loss which is approximately equal to -0.5 dB. The filter contrast is 96 dB·nm^{-1}. If the interface roughness at each interface is not equal to zero

the insertion lossesare increased and the filter contrast is decreased. If rms is equal to 1 nm the filter contrast is decreased to 65 dB·nm^{-1} and insertion losses are increased to -2.5 dB. In case of rms roughnessof 2 nm the insertion losses are increased up to the value of -8 dB and the filter contrast is decreased to 43 dB·nm^{-1}. In this instance the insertion losses are higher as required by the ITU-T and its use is not appropriate. We conclude that filters with the values of rms roughness > 1 nm at each interface are not applicable.

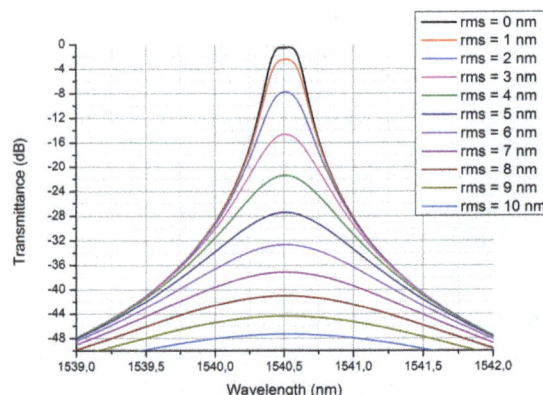

Fig. 2: Influence of interface roughness on TFF if roughness is applied at each interface. The case of channel at the wavelength 1540.5 nm.

Figure 3 shows the dependence of transmittance versus rms roughness. This figure was obtained from the simulations at the central wavelength of 1540.5 nm. From the figure it is clear that the transmittance decreases with increasing rms roughness in a nonlinear manner. The largest change in transmittance is seen for rms roughness from 1 nm to 6 nm. The filter with rms roughness of 6 nm has the transmittance smaller than -32 dB and thus the filter exceeds the power budget what means that does not meet the previously defined requirements.

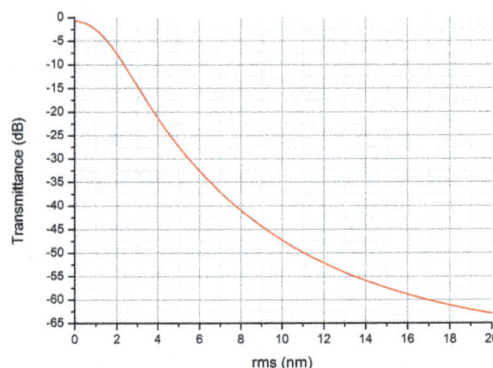

Fig. 3: Dependence of transmittance of filter on rms roughnessof the filter designed at the wavelength 1540.5 nm.

The objective of our further simulationsis to show that the sensitivity of particular interfacesto interface roughness differs. To prove this we change the value of roughness through the filter so that only at one interface the roughness is expected. The results are depicted in Fig. 4. From the figure it can be observed that the filter is minimum sensitive tothe interface roughness if it is applied at the filter boundary and at middle layers. On the contrary the highest sensitivityto the interface roughnesswas found at interfaces 48 and 152. These interfaces corespond to the interfaces between layers M and 4.375H. The layer 4.375H is the thickest in the structure of the filter what means that minimizing the interface roughness at this specific interface the decrease of the total transmittance of the filter can be minimized. Further we can see that at rms roughness of 5nm applied only at the 48th interface the total transmittance is decreased to value −4 dB however with rms equal to 10 nm at the same interface the total transmittance at 1540.5 nm is smaller than ∼10 dB. Due to this the higher roughness at the only one interface may also cause incompetence of the filter.

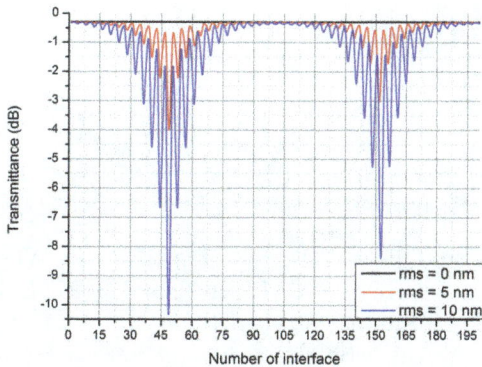

Fig. 4: Dependence of transmittance of filter to rms roughness for the wavelength 1540.5 nm in case if the rms roughness is gradually applied at only one interface.

Since the filter was designed for the application in DWDM-PON and the selection of the right channel is allowed by the rotation of the filter we must expect different impact of the roughness on the total transmittance for different rotations of the filter and in different wavelength regions. In the case the same rms roughness of 1 nm applied at each interface for the channel A at the central wavelength 1540.5 nm the increase of the insertion loss from −0.5 dB to approximately −2.5 dB can be seen (Fig. 5). However the same roughness at the same filter causes the insertion loss −2 dB for the channel C at the central wavelength 1502 nm. An even greater difference can be seen in the case of higher values of rms roughness. If rms roughness at each interface is equal to 3 nm the insertion losses are equal to −11.8 dB for channel C, −13.3 dB for channel Band −14.6 dB for channel A. Thus we conclude that for this

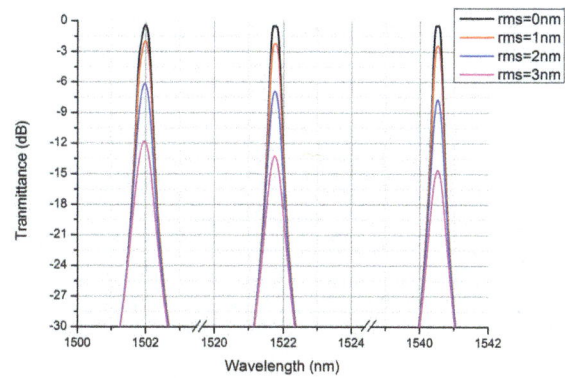

Fig. 5: Dependence of transmittance of filter on the rms roughness for the channels A at the central wavelength 1540.5 nm, B at the central wavelength 1521.7 nm and C at the central wavelength 1502.0 nm.

type of filter the roughness has the major impact on the filter designed for channels at higher wavelengths.

3.2. The Influence of Temperature Changes

Since ONUs may not be strictly set at the PON premises where the stable temperature is secured it is necessary to study the impact of temperature changes of ambient on the filter performance.

In this section we present the results of the simulations of this effect. The thermo-optical coefficients of the materials used for the simulations were taken from the references and are as follows:

- fused silica substrate $\frac{dN}{dT} = 0.0005 \cdot 10^{-4}$ K^{-1},

- a-Si $\frac{dN}{dT} = 2.3000 \cdot 10^{-4}$ K^{-1},

- SiO$_2$ $\frac{dN}{dT} = 0.1000 \cdot 10^{-4}$ K^{-1} [21],

- ZnO $\frac{dN}{dT} = 0.3200 \cdot 10^{-4}$ K^{-1} [22].

The results are illustrated in Fig. 6 where the transmittance of the channel A at 1540.5 nm is depicted. It can be seen that even if the temperature influence on the filter contrast factor and insertion losses are negligible the significant changes of the TFF central wavelength position are important. With the variations of temperature the central wavelength of the channel is changed. If the temperature drops from the room temperature 293.15 K up to 283.15 K the central wavelength is shifted so much that even the switching to the neighboring channel with the central wavelength of 1539.7 nm occurs. The switching to the completely different channel is highly undesirable. The same situation occurs at the temperature of 273.15 K. The temperature change of about 10 K causes the wavelength shift of 0.6 nm for this type of filter.

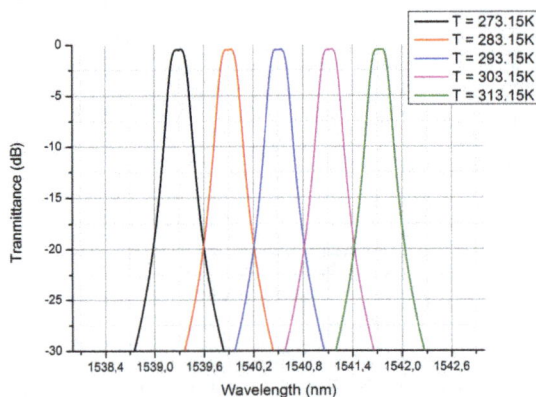

Fig. 6: Dependence of the transmittance of the filter on temperature changes for the channel A at the central wavelength 1540.5 nm.

The similar behaviour exhibits also the transmittance of the filter at the channel C as can be seen in Fig. 7. As in the previously mentioned case the temperature influences on the filter contrast factor and insertion losses are negligible but the wavelength shift of the filter transmittance spectrum at this channel is also equal to 0.6 nm. These numerical studies emphasize that the temperature protection is highly necessary for TFFs used in ONUs in this type of PONs.

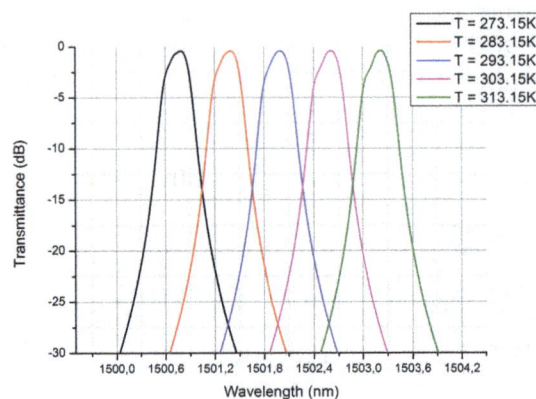

Fig. 7: Dependence of the transmittance of the filter ontemperature changes for the channel C at the central wavelength 1502.0 nm.

4. Conclusions

The numerical investigation of TFF filters designed for DWDM-PON definitely proved the negative effect of interfacial roughness on filter characteristics, especially the insertion loss and the filter bandwidth. It was shown that the impact of the interface roughness

may differ according to the interface position and the spectral region and therefore must be studied very carefully. Moreover the influence of temperature of ambient on the transfer characteristics of filters was presented. It was shown that temperature changes result in wavelength shifts. These effects could be a drawback when using these filters as wavelength blocking filters in DWDM PONs. These effects must be anticipated via thorough simulations and suppressed during the deposition as much as possible. However, it should be added that the impact of interface roughness and temperature changes may be different for different structures of filter sand different materials used in the filter design.

Acknowledgment

This work was partly supported by the Slovak Research and Development Agency under the project APVV-0025-12 and partly by the Slovak Grant Agency under the project No. 2/0076/15.

References

[1] KANI, J. and K. SUZUKI. Standardization Trends of Next-generation 10 Gigabit-class Passive Optical Network Systems. *NTT Technical Review.* 2009, vol. 7, pp. 1–6. ISSN 1348-3447.

[2] ANDRADE, M. D., G. KRAMER, L. WOSIN-SKA, J. CHEN, S. SALLENT. and B. MUKHERJEE. Evaluating strategies for evolution of passive optical networks. *Communications Magazine* 2011, vol. 49, no. 7, pp. 176–184. ISSN 0163-6804. DOI: 10.1109/MCOM.2011.5936171.

[3] KORCEK, D. and J. MULLEROVA. Wavelength Protection within Coexistence of Current and Next-Generation PON Networks. In: *15th International Conference on Transparent Optical Networks (ICTON)*. Cartagena: IEEE, 2013, pp. 1–4. ISBN 978-1-4799-0682-6. DOI: 10.1109/IC-TON.2013.6602980.

[4] UEHARA, A. N., R. OTOWA and R. OKUDA. Advanced Band Separation Thin-Film Filters for Coexistence-Type Colorless WDM-PON. In: *Optical Fiber Communication/National Fiber Optic Engineers Conference*. San Diego: IEEE, 2008, pp. 1–3. ISBN 978-1-55752-856-8. DOI: 10.1109/OFC.2008.4528711.

[5] MULLEROVA, J. and D. KORCEK. Super-separation thin film filtering for coexistence-type colorless WDM-PON networks. In: *13th International Conference on Transparent Optical*

Networks (ICTON). Graz: IEEE, 2011, pp. 1–4. ISBN 978-1-4577-0881-7. DOI: 10.1109/ICTON.2011.5970978.

[6] LUO, Y., X. ZHOU, F. EFFENBERGER, X. YAN, G. PENG, Y. QIAN and Y. MA. Time- and Wavelength-Division Multiplexed Passive Optical Network (TWDM-PON) for Next-Generation PON Stage 2 (NG-PON2). Journal of Lightwave Technology. 2013, vol. 31, iss. 4, pp. 587–593. ISSN 0733-8724. DOI: 10.1109/JLT.2012.2215841.

[7] SCHOLTZ, L., D. KORCEK, L. LADANYI and J. MULLEROVA. Tunable thin film filters for the next generation PON stage 2 (NG-PON2). In: ELEKTRO. Rajecke Teplice: IEEE, 2014, pp. 98–102. ISBN 978-1-4799-3720-2. DOI: 10.1109/ELEKTRO.2014.6847880.

[8] SCHOLTZ, L., D. KORCEK and J. MULLEROVA. Design of a Novel Wavelength Scheme for DWDM-PON Coexisting with Current PON Technologies and Protected Against Signal Interference. In: 16th International Conference on Transparent Optical Networks (ICTON). Graz: IEEE, 2014, pp. 1–4. ISBN 978-1-4799-5600-5. DOI: 10.1109/ICTON.2014.6876418.

[9] LATAL, J., J. VITASEK, P. KOUDELKA, P. SISKA, R. POBORIL, L. HAJEK, A. VANDERKA and V. VASINEK. Simulation of modulation formats for optical access network based on WDM-PON. In: 16th International Conference on Transparent Optical Networks (ICTON). Graz: IEEE, 2014, pp. 1–7. ISBN 978-1-4799-5600-5. DOI: 10.1109/ICTON.2014.6876473.

[10] CHOW, C. W. and C. H. YEH. Using Downstream DPSK and Upstream Wavelength-Shifted ASK for Rayleigh Backscattering Mitigation in TDM-PON to WDM-PON Migration Scheme. Photonics Journal. 2013, vol. 5, iss. 2, pp. 1–8. ISSN 1943-0655. DOI: 10.1109/JPHOT.2013.2247588.

[11] CHOVAN, J., F. UHEREK, R. KURINEC, A. SATKA, J. PAVLOV and D. SEYRINGER. Temperature characterization of passive optical components for WDM-PON FTTx. Advances in Electrical and Electronic Engineering. 2011, vol. 9, no. 3, pp. 143–149. ISSN 1336-1376. DOI: 10.15598/aeee.v9i3.512.

[12] ITU-T Recommendation G.694.1. Spectral grids for WDM applications: DWDM frequency grid. Geneva: ITU-T, 2012. Available at: https://www.itu.int/rec/TREC-G.694.1/en.

[13] SCHOLTZ, L., L. LADANYI and J. MULLEROVA. Influence of Surface Roughness on Optical Characteristics of Multilayer Solar Cells. Advances in Electrical and Electronic Engineering. 2014, vol. 12, no. 6, pp. 631–638. ISSN 1336-1376. DOI: 10.15598/aeee.v12i6.1078.

[14] DOMINE, D., F.-J. HAUG, C. BATTAGLIA and C. BALLIF. Modeling of light scattering from micro- and nanotextured surfaces. Journal of Applied Physics. 2010, vol. 107, iss. 4, pp. 044504-044504-8. ISSN 0021-8979. DOI: 10.1063/1.3295902.

[15] ZEMAN, M., R. A. C. M. M. VAN SWAAIJ, J. W. METSELAAR and R. E. I. SCHROPP. Optical modeling of a-Si: H solar cells with rough interfaces. Journal of Applied Physics. 2000, vol. 88, iss. 11, pp. 6436–6443. ISSN 0021-8979. DOI: 10.1063/1.1324690.

[16] FRIEDMAN, D. J. Progress and challenges for next-generation high-efficiency multi-junction solar cells. Current Opinion in Solid State and Materials Science. 2010, vol. 14, iss. 6, pp. 131–138. ISSN 1359-0286. DOI: 10.1016/j.cossms.2010.07.001.

[17] PODRAZA, N. J., C. R. WRONSKI and R. W. COLLINS. Model for the amorphous roughening transition in amorphous semiconductor deposition. Journal of Non-Crystalline Solids. 2006, vol. 352, iss. 9–20, pp. 950–954. ISSN 0022-3093. DOI: 10.1016/j.jnoncrysol.2005.12.013.

[18] NASRULLAH, J., G. L. TYLER and Y. NISHI. An Atomic Force Microscope Study of Surface Roughness of Thin Silicon Films Deposited on SiO_2. IEEE Transactions on Nanotechnology. 2005, vol. 4, iss. 3, pp. 303–311. ISSN 1536-125X DOI: 10.1109/TNANO.2005.847007.

[19] PALIK, E. Handbook of Optical Constants of Solids. 1st ed. Cambridge: Academic Press, 1998. ISBN: 978-0-12-544415-6.

[20] ITU-T Recommendation G.984.2. Gigabitcapable passive optical networks (GPON): Physical Media Dependent (PMD) layer specification. Geneva: ITU-T, 2003. Available at: https://www.itu.int/rec/TREC-G.984.2/en.

[21] SCHOLTZ, L., D. KORCEK, L. LADANYI and J. MULLEROVA. Temperature dependence of TWDM narrow band thin film filters for the Next generation PON stage 2 (NG-PON2). In: Applied physics of condensed matter. Strbske Pleso: APCOM, 2014, pp. 232–235. ISBN 978-80-227-4179-8.

[22] JOHNSON, J. C., H. YAN, P. YANG and R. J. SAYKALLY. Optical Cavity Effects in ZnO Nanowire Lasers and Waveguides. Journal of Physical Chemistry B. 2003, vol. 107,

iss. 34, pp. 8816–8828. ISSN 1520-6106. DOI: 10.1021/jp034482n.

About Authors

Lubomir SCHOLTZ received his M.Sc. from the Department of Telecommunications and Multimedia, the Faculty of Electrical Engineering, the University of Zilina in 2013. His research interests include optical communication networks and systems and thin film filters for passive optical networks.

Libor LADANYI received his M.Sc. in 2011 and his Ph.D. in telecommunications in 2014 both from the Department of Telecommunications and Multimedia, the Faculty of Electrical Engineering, the University of Zilina. His research interests include optical communication networks and systems.

Jarmila MULLEROVA received her M.Sc. degree in experimental physics in 1980 and her Ph.D. in quantum electronics in 1988, both from the Comenius University, Bratislava. Since 2003 she has been with the University of Zilina. Since 2014 she has been a professor in the field of electro-technology and materials. Her ongoing research interests comprise optical properties of solids and thin films optics.

Permissions

All chapters in this book were first published in AEEE, by VSB-Technical University of Ostrava; hereby published with permission under the Creative Commons Attribution License or equivalent. Every chapter published in this book has been scrutinized by our experts. Their significance has been extensively debated. The topics covered herein carry significant findings which will fuel the growth of the discipline. They may even be implemented as practical applications or may be referred to as a beginning point for another development.

The contributors of this book come from diverse backgrounds, making this book a truly international effort. This book will bring forth new frontiers with its revolutionizing research information and detailed analysis of the nascent developments around the world.

We would like to thank all the contributing authors for lending their expertise to make the book truly unique. They have played a crucial role in the development of this book. Without their invaluable contributions this book wouldn't have been possible. They have made vital efforts to compile up to date information on the varied aspects of this subject to make this book a valuable addition to the collection of many professionals and students.

This book was conceptualized with the vision of imparting up-to-date information and advanced data in this field. To ensure the same, a matchless editorial board was set up. Every individual on the board went through rigorous rounds of assessment to prove their worth. After which they invested a large part of their time researching and compiling the most relevant data for our readers.

The editorial board has been involved in producing this book since its inception. They have spent rigorous hours researching and exploring the diverse topics which have resulted in the successful publishing of this book. They have passed on their knowledge of decades through this book. To expedite this challenging task, the publisher supported the team at every step. A small team of assistant editors was also appointed to further simplify the editing procedure and attain best results for the readers.

Apart from the editorial board, the designing team has also invested a significant amount of their time in understanding the subject and creating the most relevant covers. They scrutinized every image to scout for the most suitable representation of the subject and create an appropriate cover for the book.

The publishing team has been an ardent support to the editorial, designing and production team. Their endless efforts to recruit the best for this project, has resulted in the accomplishment of this book. They are a veteran in the field of academics and their pool of knowledge is as vast as their experience in printing. Their expertise and guidance has proved useful at every step. Their uncompromising quality standards have made this book an exceptional effort. Their encouragement from time to time has been an inspiration for everyone.

The publisher and the editorial board hope that this book will prove to be a valuable piece of knowledge for researchers, students, practitioners and scholars across the globe.

List of Contributors

Michaela Solanska, Miroslav Markovic and Milan Dado
Department of Telecommunications and Multimedia, Faculty of Electrical Engineering, University of Zilina, Univerzitna 8215/1, 010 26 Zilina, Slovakia

Jacek Gryglewicz, Wojciech Macherzy-Nski, Andrzej Stafiniak, Bogdan Paszki-Ewicz and Regina Paszkiewicz
Department of Microelectronics and Nanotechnology, Faculty of Microsystem Electronics and Photonics, Wroclaw University of Technology, Janiszewskiego 11/17, 50-372 Wroclaw, Poland

Tran Hoang Quang Minh, Nguyen Huu Khanh Nhan, Thoai Phu Vo and Nguyen Doan Quoc Anh
Department of Electronics and Telecommunications, Faculty of Electrical and Electronics Engineering, Ton Duc Thang University, 19 Nguyen Huu Tho Street, Ho Chi Minh City, Vietnam

Saptadip Saha, Priyanath Das, Ajay Kumar Chakraborty, Ruchira Debbarma and Sharmistha Sarkar
Department of Electrical Engineering, National Institute of Technology Agartala, Jirania, West Tripura 799046, India

Milos Kozak and Leos Bohac
Department of Telecommunication Engineering, Faculty of Electrical Engineering, Czech Technical University in Prague, Technicka 2, 16000 Prague, Czech Republic

Brigitte Jaumard
Department of Computer Science & Software Engineering, Faculty of Engineering and Computer Science, Concordia University, 1515 Rue Sainte-Catherine, H3G 2W1 Montreal, Canada

Jan Nedoma, Marcel Fajkus, Petr Siska, Martin Novak, Lukas Bednarek, Jaroslav Frnda, Jan Zavadil, Stanislav Zabka, Frantisek Perecar, Stanislav Kepak, Jakub Cubik and Vladimir Vasinek
Department of Telecommunications, Faculty of Electrical Engineering and Computer Science, VSB–Technical University of Ostrava, 17. listopadu 15/2172, 708 33 Ostrava-Poruba, Czech Republic

Nguyen Doan Quoc Anh, Tran Hoang Quang Minh and Nguyen Huu Khanh Nhan
Faculty of Electrical and Electronics Engineering, Ton Duc Thang University, 19 Nguyen Huu Tho Street, Tan Phong Ward, District 7, Ho Chi Minh City, Vietnam

Jens Kobelke, Kay Schuster, Joerg Bierlich, Sonja Unger, Anka Schwuchow, Tino Elsman, Jan Dellith, Claudia Aichele, Ron Fatobene Ando and Hartmut Bartelt
Leibniz Institute of Photonic Technology, Albert-Einstein-Strasse 9, 07745 Jena, Germany

Dana Seyringer
Research Centre for Microtechnology, Vorarlberg University of Applied Sciences, Hochschulstrasse 1, 6820 Dornbirn, Austria

Michal Lucki
Department of Telecommunication Engineering, Faculty of Electrical Engineering, Czech Technical University in Prague, Technicka 2, 16627 Prague 6, Czech Republic

Catalina Butscher
Research Centre for Microtechnology, Vorarlberg University of Applied Sciences, Hochschulstrasse 1, 6820 Dornbirn, Austria
Department of Telecommunication Engineering, Faculty of Electrical Engineering, Czech Technical University in Prague, Technicka 2, 16627 Prague 6, Czech Republic

Martin Valica
Department of Ecochemistry and Radioecology, Faculty of Natural Sciences, University of SS. Cyril and Methodius, Namesti J. Herdu 2, 917 01 Trnava, Slovak Republic

Dusan Chorvat Jr.
Department of Biophotonics, International Laser Center, Ilkovicova 3, 841 01 Bratislava, Slovak Republic

Tibor Teplicky, Miroslava Danisova and Alzbeta Marcek Chorvatova
Department of Biophotonics, International Laser Center, Ilkovicova 3, 841 01 Bratislava, Slovak Republic
Department of Ecochemistry and Radioecology, Faculty of Natural Sciences, University of SS. Cyril and Methodius, Namesti J. Herdu 2, 917 01 Trnava, Slovak Republic

Arpad Kosa, Lubica Stuchlikova, Ladislav Harmatha and Jaroslav Kovac
Institute of Electronics and Photonics, Faculty of Electrical Engineering and Information Technology, Slovak University of Technology, Ilkovicova 3, 812 19 Bratislava, Slovakia

Beata Sciana, Wojciech Dawidowski and Marek Tlaczala
Division of Microelectronics and Nanotechnology, Faculty of Microsystem Electronics and Photonics, Wroclaw University of Science and Technology, Janiszewskiego 11/17, 50-372 Wroclaw, Poland

Joanna Prazmowska, Kornelia Indyki-Ewicz, Bogdan Paszkiewicz and Regina Paszkiewicz
Department of Microelectronics and Nanotec-hnology, Faculty of Microsystem Electronics and Photonics, Wroclaw University of Science and Technology, Wybrzeze Wyspianskiego 27, 503 70 Wroclaw, Poland

Johann Zehetner
Research Centre for Microtechnology, University of Applied Sciences, Hochschulstrasse 1, 6850 Dornbirn, Austria

Gabriel Vanko, Jaroslav Dzuba and Tibor Lalinsky
Institute of Electrical Engineering, Slovak Academy of Sciences, Dubravska cesta 9, 841 04 Bratislava, Slovakia

David Solus, Lubos Ovsenik, Jan Turan, Tomas Ivaniga, Jakub Oravec and Michal Marton
Department of Electronics and Multimedia Communications, Faculty of Electrical Engineering and Informatics, Technical University of Kosice, Park Komenskeho 13, 042 01 Kosice, Slovak Republic

Nela Strbikova, Jan Vanus, Radana Kahankova and Radek Martinek
Department of Cybernetics and Biomedical Engin-eering, Faculty of Electrical Engineering and Computer Science, VSB–Technical University of Ostrava, 17. listopadu 15/2172, 708 33 Ostrava, Czech Republic

Homer Nazeran
Department of Electrical and Computer Engineering, College of Engineering, University of Texas El Paso, 500 W University Ave, El Paso, TX 79968, United States of America

Petr Janku
Department of Gynecology and Obstetrics, Faculty of Medicine, Masaryk University and University Hospital Brno, Jihlavska 20, 625 00 Brno, Czech Republic

Homer Nazeran
Department of Electrical and Computer Engineering, College of Engineering, University of Texas El Paso, 500 W University Ave, El Paso, TX 79968, United States of America

Tomas Stratil and Petr Koudelka
Department of Telecommunications, Faculty of Electrical Engineering and Computer Science, VSB–Technical University of Ostrava, 17. listopadu 15, 708 33 Ostrava, Czech Republic

Tomas Novak
Department of General Electrical Engineering, Faculty of Electrical Engineering and Computer Science, VSB–Technical University of Ostrava, 17. listopadu 15, 708 33 Ostrava, Czech Republic

Dinara Sobola, Petr Sadovsky, Nikola Papez and Lubomir Grmela
Physics Department, Faculty of Electrical Engineering and Communication, Brno University of Technology, Technicka 8, 616 00 Brno, Czech Republic

Stefan Talu
Department of Automotive Engineering and Transports, Discipline of Descriptive Geometry and Engineering Graphics, Faculty of Mechanical Engineering, Technical University of Cluj-Napoca, 103 105 B-dul Muncii Street, 400641 Cluj-Napoca, Romania

Wojciech Macherzynski, Jacek Grygl-Ewicz, Andrzej Stafiniak, Joanna Prazmowska and Regina Paszkiewicz
Department of Microelectronics and Nanotech-nology, Faculty of Microsystem Electronics and Photonics, Wroclaw University of Technology, Janiszewskiego 11/17, 50-370 Wroclaw, Poland

Joanna Prazmowska, Wojciech Macher-Zynski and Regina Paszkiewcz
Department of Microelectronics and Nanotec-hnology, Faculty of Microsystem Electronics and Photonics, Wroclaw University of Technology, Janiszewskiego 11/17, 50-370 Wroclaw, Poland

Lubomir Scholtz, Libor Ladanyi and Jarmila Mullerova
Institute of Aurel Stodola, Faculty of Electrical Engineering, University of Zilina, Kpt. Nalepku 1390, 031 01 Liptovsky Mikulas, Slovak Republic

Index

www.ingramcontent.com/pod-product-compliance
Lightning Source LLC
Chambersburg PA
CBHW050443200326
41458CB00014B/5043